Ionic Processes in the Gas Phase

NATO ASI Series

Advanced Science Institutes Series

A series presenting the results of activities sponsored by the NATO Science Committee, which aims at the dissemination of advanced scientific and technological knowledge, with a view to strengthening links between scientific communities.

The series is published by an international board of publishers in conjunction with the NATO Scientific Affairs Division

A	Life Sciences	Plenum Publishing Corporation
B	Physics	London and New York
C	Mathematical and Physical Sciences	D. Reidel Publishing Company Dordrecht, Boston and Lancaster
D	Behavioural and Social Sciences	Martinus Nijhoff Publishers
E	Engineering and Materials Sciences	The Hague, Boston and Lancaster
F	Computer and Systems Sciences	Springer-Verlag
G	Ecological Sciences	Berlin, Heidelberg, New York and Tokyo

Series C: Mathematical and Physical Sciences Vol. 118

Ionic Processes in the Gas Phase

edited by

M. A. Almoster Ferreira
Faculty of Sciences, University of Lisbon, Portugal

D. Reidel Publishing Company

Dordrecht / Boston / Lancaster

Published in cooperation with NATO Scientific Affairs Division

Proceedings of the NATO Advanced Study Institute on
Chemistry of Ions in the Gas Phase
Vimeiro, Portugal
September 6-17, 1982

Library of Congress Cataloging in Publication Data

NATO advanced study institute on Chemistry of ions in the gas phase (1982 : Vimeiro,
    Lisbon, Portugal)
    Ionic processes in the gas phase.

    (NATO advanced science institutes series. Series C, Mathematical and physical
sciences : v. 118)
    "Proceedings of the NATO advanced study institute on Chemistry of ions in the
gas phase, Vimeiro, Portugal, sept. 6—17, 1982"—verso t. p.
    "Published in cooperation with NATO scientific affairs division."
    Includes index.
    1. Ions—Congresses.   2.   Gas dynamics—Congresses.   I.   Ferreira, M. A.
Almoster.   II.   North Atlantic Treaty Organization. Scientific affairs division.
III.   Title.   IV.   Series.
QD561.N36   1982   541'.0424   83-22980
ISBN-13:978-94-009-7250-6   e-ISBN-13:978-94-009-7248-3
DOI: 10.1007/978-94-009-7248-3

Published by D. Reidel Publishing Company
P.O. Box 17, 3300 AA Dordrecht, Holland

Sold and distributed in the U.S.A. and Canada
by Kluwer Academic Publishers,
190 Old Derby Street, Hingham, MA 02043, U.S.A.

In all other countries, sold and distributed
by Kluwer Academic Publishers Group,
P.O. Box 322, 3300 AH Dordrecht, Holland

D. Reidel Publishing Company is a member of the Kluwer Academic Publishers Group

CONTENTS

DEDICATION

To J. L. Franklin and Henry M. Rosenstock

This volume is jointly dedicated to the memories of two of the preeminent pioneers of research in ion chemistry, J. L. Franklin and Henry M. Rosenstock.

Shortly before the beginning of the proceedings which are the subject of this book, the participants received word of the death of Joe Franklin. Many of those present for the N.A.T.O. ASI in Portugal had heard Professor Franklin deliver the introductory lecture at a N.A.T.O. ASI on "Interactions between Ions and Molecules," held eight years earlier in Biarritz, France, and it seemed particularly appropriate that the group of his colleagues and friends assembled in Portugal honor him by dedicating this volume to his memory. Joe Franklin received his Ph. D. in physical chemistry from the University of Texas in 1934, and went to work for the Humble Oil and Refining Company at Baytown, Texas, where he remained until joining the faculty of the Chemistry Department at Rice University in 1964. In the early 1950's, he became active in the then-emerging field of mass spectrometry, and, in collaboration with Drs. Frank Field, Fred Lampe, and Burnaby Munson, carried out much of the significant pioneering work on ionization potential and appearance potential measurements and on ion-molecule reactions. In 1957, there appeared "Electron Impact Phenomena" by F. H. Field and J. L. Franklin, the first really comprehensive treatise on the subject of ion chemistry. In more recent years, he has continued to make so many important contributions to every area of the field that it would be difficult or impossible to name them all here. It is likely that most readers of this volume will have some personal familiarity with the work of Joe Franklin since his name is, and will continue to be, synonomous with ion chemistry.

One of the participants at the N.A.T.O. ASI on "The Chemistry of Ions in the Gas Phase" in Portugal was Henry M. Rosenstock, who had helped to organize the ASI. In the early 1950's when Joe

Franklin was beginning his work in mass spectrometry, Henry
Rosenstock was a graduate student at the University of Utah,
working on developing a theoretical picture of the dissociation
of ions under collision-free conditions. The resulting Quasi-
Equilibrium Theory, or so-called QET, is still an important tool
for ion chemists, referred to many times in the proceedings of
the ASI. After receiving his Ph. D. in 1952, Henry Rosenstock
worked for several years at the Oak Ridge National Laboratory and
in private industry before joining the National Bureau of Standards in 1960, where he directed a research team working in the
areas of photoionization mass spectrometry, and, later, photoion-
photoelectron coincidence measurements. From 1966 to 1968 he
served as chairman of the American Society for Testing Materials
Committee E-14 on Mass Spectrometry. He was succeeded as chairman by Joe Franklin, and together, they shared a large part of
the responsibility for the establishment and early leadership of
the American Society for Mass Spectrometry, now a major profess-
ional organization for scientists in the field of ion chemistry.
In 1963, Henry Rosenstock initiated a program of compiling and
evaluating data on ion thermochemistry, building initially on a
small compilation which had appeared as an Appendix in Field and
Franklin's "Electron Impact Phenomena." The initial publication
resulting from this effort in 1969, "Ionization Potentials, Appearance Potentials, and Heats of Formation of Gaseous Positive
Ions" by J. L. Franklin, J. G. Dillard, H. M. Rosenstock, J. T.
Herron, K. Draxl, and F. H. Field, is still the most widely cited
single publication in ion chemistry. Henry Rosenstock participated actively and enthusiastically in the proceedings in Portugal, sitting in the front row at every lecture, asking questions
and making comments, and obviously enjoying himself very much.
On the morning of Tuesday, September 14, 1982, he suffered a heart
attack and died that afternoon shortly after being taken to a
hospital in Lisbon. His death, no less than that of Joe
Franklin, will be felt by the ion chemistry community of the
world as a great loss.

M.A. Almoster Ferreira
P. Ausloos
R. Botter
A. Ferrer Correia
K. Jennings
S. Lias

PREFACE

The field of gas phase ion chemistry has undergone spectac-
ular growth and development in the last decade. New experimental
techniques have developed and theoretical advancements in quantum
chemistry and chemical dynamics have begun to provide quantitative
information, deeper insights and important guideposts for experiments. The growing interaction between theoreticians and experi-
mentalists is yielding scientific information of both basic and
applied significance.

The NATO Advanced Study Institute on Chemistry of Ions in
the Gas Phase held at Vimeiro, Portugal, from 6 to 17 September
1982 took place four years after the 1978 Advanced Study Institute on Kinetics of Ion-Molecule Reactions held at la Baule,
France. New theoretical approaches and experimental techniques
developed since then are included in this book. Fundamental
aspects concerning the kinetics and dynamics of formation and
decomposition of gaseous positive and negative ions, their
structural characterisation, energetics and thermochemistry are
also discussed. In addition, some recent innovative developments
in the areas of photodissociation, multiphoton ionisation, photoelectron-photoion coincidence spectroscopy, selected ion flow
tube technique, and other techniques will be found.

The success of the NATO Advanced Institute on Chemistry of
Ions in the Gas Phase was mostly due to the high quality of the
lectures, panel sessions , research review contributions and the
ensuing important discussions which would informally continue
after the scheduled sessions. All the participants were ultimately responsible for the attainment of the aims of the
Institute.

I would like to express my appreciation for the great assistance and support I always received from the members of the Organising Committee P. Ausloos, R. Botter, A. Ferrer-Correia, K. Jennings and S. Lias. I am very grateful for the efforts and enthusiastic help of my research group in the organisation of the meeting and for the efficient secretarial work of Mrs. Isabel Freire who also helped with the typing of parts of this book.

On behalft of all the participants I want to thank the Scientific Affairs Division of the North Atlantic Treaty Organisation whose grant made it possible to organise the meeting. Financial support from Instituto Nacional de Investigação Cientifica, Portugal, and student travel grants from the National Science Foundation, USA, are also gratefully acknowledged.

Finally I should like to mention the understanding and valuable guidance I always found in Dr. Mario di Lullo and in Dr. Craig Sinclair of the Scientific Affairs Division of the North Atlantic Treaty Organisation.

Lisbon, May 1983

M.A. Almoster Ferreira

POTENTIAL ENERGY SURFACES AND THEORY OF UNIMOLECULAR

DISSOCIATION

J.C. Lorquet and B. Leyh

(Département de Chimie, Université de Liège,
Sart-Tilman, B-4000 Liège 1, Belgium).

Reaction mechanisms can be determined from ab initio
calculations of potential energy surfaces. Their complicated
nature explains the frequent success of statistical theories,
but nonstatistical behaviour is also accounted for. A certain
correlation between structure and reactivity can be established,
subject to many qualifications. Information on the nuclear motion on the potential energy surfaces of an ionized molecule can
be extracted from a photoelectron spectrum by a Fourier transform operation.

REACTION MECHANISMS

        Is the rate constant of a unimolecular reaction really
determined by a single quantity, viz. the total intramolecular
energy E, as assumed by QET, or does it depend on additional factors, related to the specificity of the initial excitation ?
This may well be the cardinal question of unimolecular reaction
dynamics. An answer to that question is provided by the photoion-
photoelectron coincidence spectroscopy. Most of the time, the
evidence is in favour of the statistical model. PIPECO experimen-
tation has revealed that, for most molecules, the ion yield curves are smooth functions of energy alone. This constitutes very
convincing support of the validity of the assumption according
to which the molecular ion has reached a state of microcanonical
equilibrium. However, a few counterexamples are known, which are
rather loosely referred to as "isolated state decay". The exact

1

M. A. Almoster Ferreira (ed.), Ionic Processes in the Gas Phase, 1–6.
© 1984 by D. Reidel Publishing Company.

meaning of this expression has become clear only recently.

A comparison between the cases of $C_2H_4^+$ and CH_3OH^+ has been attempted. The reaction paths of $C_2H_4^+$ have been determined by ab initio calculations. It has been shown that they are so complicated (i.e., the system has to fulfill such a large number of steps in a prescribed order) that a tendency towards energy redistribution follows naturally (1-3). Since the dissociation mechanism involves several nonadiabatic steps, an extension of QET had to be devised to calculate the ion yield curves (4). Agreement with experiment is very satisfactory (4).

CH_3OH^+ is known to offer a counterexample and the reasons of its behaviour have been understood only recently (5). The dissociation mechanism is here distinctly bimodal. Most of the ions are produced by a statistical RRKM/QET mechanism, except the CH_3^+ ions. This results from a branching in the reaction path brought about by an avoided crossing between two potential energy surfaces. One of the reactive paths gives rise to the RRKM/QET component, whereas the other leads directly to the CH_3^+ + OH dissociation asymptote (Fig. 1). Production of CH_3^+ ions thus takes place by a fast, diatomic-like, non-statistical mechanism which escapes energy redistribution. "Isolated state decay" thus corresponds to dissociation of the non-randomized fraction of the population, and not to isolation, i.e., to lack of radiationless

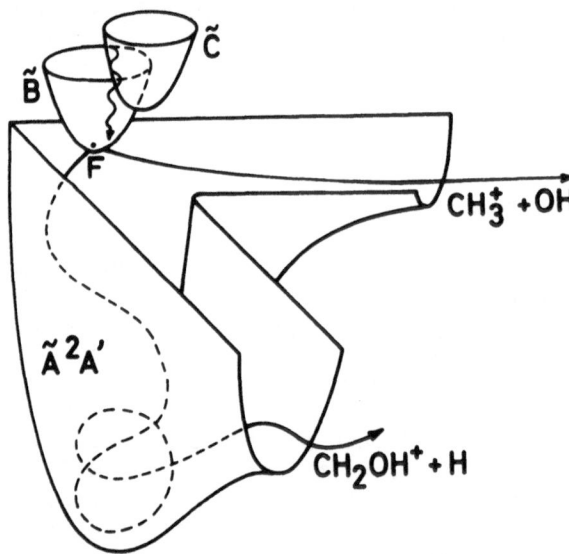

Fig. 1.   Isolated state decay corresponds to a branching in the reactive flux between the statistical and non-statistical paths.

transitions to the lower electronic states.

STRUCTURE AND DYNAMICS

Do structural studies provide a shortcut to potential energy surface calculations and do they help to understand the reactivity of a particular ion ? In principle (6), chemical reactivity is an inherently dynamic process which cannot be expected to reduce to a simple corollary of structural chemistry. A structure is an information concerning a particular point of the potential energy surface, whereas a reaction can be visualized as a trajectory leading from reactants to products, i.e., as a succession of structures. A structure can thus at most provide an initial tendency which may be altered and even reversed by the complicated nature of a reaction mechanism. Contrarily to what is often asserted in the literature, there is as a rule no correlation between the bonding or antibonding nature of a particular MO and the fragmentation of the ionic state which results from electron removal from this MO (6). The reason is that the bonding or antibonding nature is calculated at the equilibrium position of each state, whereas the nature of the wave function at the dissociation asymptote may be entirely different. In such a case, no correlation is to be expected between electronic structure and reactivity.

However, there exist cases in which, to a first approximation, the outcome of a chemical process is controlled by, or at least depends on the properties of a particular point of the potential energy surface or surfaces. This happens in two important circumstances : (i) when the transition state model is valid ; (ii) in nonadiabatic processes (i.e., in which two electronic states are involved) which are governed by localized surface crossings or interactions. In such cases, chemical reactivity turns out to be controlled by specific points or regions of space of the potential energy surfaces. Photochemists, who have to deal with much of the same problems as mass spectrometrists have coined a new word and designate these particular points as "funnels" (6, 7). In brief, a funnel is a curve crossing or a localized region of strong nonadiabatic interaction between two potential energy surfaces.

Thus, a distinction has to be established between, on the one hand, equilibrium structures which provide a basis for the understanding of static properties, and, on the other hand, transition states and funnels which are the key features which control the dynamic properties and thus the chemical reactivity.

INTRAMOLECULAR DYNAMICS

        Recently, a new experimental method has been developed
(8, 9) to obtain information on the unimolecular processes in
ionized molecules. It has been shown by Heller (10) that the
Fourier transform of an electronic spectrum leads to an autocor-
relation function C(t) which describes the evolution in time of
the wave packet created by the Franck-Condon transition, as it
propagates on the potential energy surface of the electronic
upper state. This correlation function is equal to the modulus
of the overlap integral between the initial position of the wave
packet and its instantaneous position at time t. When applied
to a photoelectron spectrum, the method provides information
about the nature of the nuclear motions on the potential energy
surface of an ionized molecule. The original data resulting from
an experimentally determined spectral profile must be corrected
for finite energy resolution, rotational, and spin-orbit effects
(8). The behaviour of the system can then be followed up to a
time of the order of 10^{-13} s, i.e., during the first few vibrations which follow immediately the electronic transition. An
oscillatory pattern of the correlation function indicates that
the nuclear motion is taking place in a bound potential (Fig. 2).

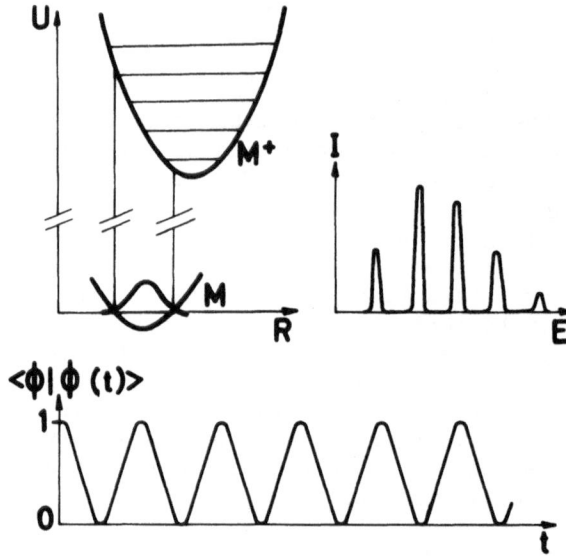

Fig. 2. A Franck-Condon transition to a stable electronic state
         leads to a well-resolved photoelectron spectrum whose
         Fourier transform gives a time-dependent correlation
         function with oscillatory character.

On the contrary, a time decrease of the correlation function is associated with an irreversible evolution. For a polyatomic molecule, the method provides a certain number of informations related to the multidimensional motion on a potential energy surface. The following problems can be studied :
(1) Coupling among vibrational normal modes and rate of energy redistribution ;
(2) Nonadiabatic interactions such as predissociations and internal conversions.

The correlation function is separable into a product of partial functions pertaining each to a single vibrational mode if the normal coordinates are identical for the initial and final states, except for a difference in the equilibrium positions ; in other words, if there is no Duschinsky effect. This is found to be the case for the $\tilde{X}\ ^2B_1$ ground state of the H_2O^+ ion. The case of the $\tilde{X}\ ^2B_{3u}$ ground state of the $C_2H_4^+$ ion is more complex : although the separable approximation is acceptable, a slight but noticeable vibrational coupling takes place.

In the case of HCN^+ (8), the vibrational normal modes of the $\tilde{B}\ ^2\Sigma^+$ state are entirely different from those of the ground state of the neutral molecule. One of them is an in phase stretching motion of the CH and CN bonds and the other is in fact a reaction coordinate. A measure of the rate of flux into the latter degree of freedom can be derived from the correlation function (8).

The method can be extended to study nonadiabatic processes (9). The predissociation of the $\tilde{A}\ ^2\Sigma^+$ state of HBr^+ by a repulsive quartet can be monitored. Part of the wave packet has flowed into the continuum after half a vibrational period while the remainder remains trapped in the bound potential. A somewhat analogous wave packet splitting is brought about by a surface crossing in the case of the $\tilde{B}\ ^2B_2$ state of H_2O^+. A similar process, also brought about by a conical intersection takes place in the first excited state $\tilde{A}\ ^2B_{3g}$ of the $C_2H_4^+$ ion. From the correlation function, it can be deduced that the lifetime of state $\tilde{A}$ with respect to internal conversion to the ground state $\tilde{X}\ ^2B_{3u}$ is of the order of $0.8\ 10^{-14}$ s.

ACKNOWLEDGEMENT

This work has been supported by the Belgian Government (Action de Recherche Concertée) and by the Fonds de la Recherche Fondamentale Collective.

REFERENCES

1. J.C. Lorquet, C. Sannen, and G. Raşeev, J. Am. Chem. Soc.
    102, 7976 (1980).
2. C. Sannen, G. Raşeev, C. Galloy, G. Fauville, and J.C.
    Lorquet, J. Chem. Phys. 74, 2402 (1981).
3. J.C. Lorquet, Org. Mass Spectrom. 16, 469 (1981).
4. M. Desouter-Lecomte, C. Sannen, and J.C. Lorquet, to be pu-
    blished.
5. C. Galloy, C. Lecomte, and J.C. Lorquet, J. Chem. Phys., in
    press.
6. J.C. Lorquet, to appear in "Mass Spectrometry", a Specialist
    Periodical Report, ed. by R.A.W. Johnstone, The Royal Society
    of Chemistry, London, Vol. 7.
7. J. Michl, Top. Curr. Chem., 1974, 46, 1.
8. A.J. Lorquet, J.C. Lorquet, J. Delwiche, and M.J. Hubin-
    Franskin, J. Chem. Phys. 76, 4692 (1982).
9. B. Leyh, Mémoire de Licence, Université de Liège (1982).
10. E.J. Heller, J. Chem. Phys. 68, 2066, 3891 (1978).

SCANNING METHODS FOR DOUBLE-FOCUSSING MASS SPECTROMETERS

K.R. Jennings

Department of Chemistry and Molecular Sciences,
University of Warwick, Coventry, CV4 7AL, England.

INTRODUCTION

Twenty-five years ago, one could purchase magnetic deflection mass spectrometers fitted with metastable suppressors which ensured that the mass spectrum consisted only of peaks due to ions formed in the source which reached the detector without further decomposition. Over the last decade, however, increasing use has been made of scans in which the only ions collected are those formed in decompositions occurring in one of the field-free regions of the flight tube. Such scans are proving particularly useful in the analysis of mixtures which have received the minimum of treatment prior to their being introduced into the mass spectrometer. Although they may be used to study the products of unimolecular decompositions which give rise to metastable transitions, their main use is in the study of the products of collision-induced decompositions which occur in a collision cell located close to the source slit or the intermediate focal point. This article aims to summarise the principles underlying the various scans currently in use, indicating briefly their strengths and weaknesses.

Double-focussing mass spectrometers are either forward geometry instruments, in which the electric sector precedes the magnetic sector, or reverse geometry instruments in which the magnetic sector precedes the electric sector. In each case, the three important parameters which control the type of mass spectrum recorded are V, the accelerating voltage, E, the electric sector field strength and B, the magnetic sector field strength. A normal mass spectrum is obtained by keeping V and E constant, the ratio V/E being determined by the dimensions and geometry of the instrument, and scanning B so as to bring ions of different mass-to-charge ratio (m/z) to focus

7

M. A. Almoster Ferreira (ed.), Ionic Processes in the Gas Phase, 7–21.
© *1984 by D. Reidel Publishing Company.*

at the collector slit.   In a forward geometry instrument, if m_1^+
ions fragment to give m_2^+ ions in the field-free region between the
two sectors, diffuse peaks of low intensity ("metastable peaks")[1]
are observed in the mass spectrum at an apparent mass $m*$ where
$m* = m_2^2/m_1$.   Such peaks are absent from the mass spectrum given
by a reverse geometry instrument since the m_2^+ ions are not transmitted by the electric sector.   In order to understand this and
to prepare the ground for a description of other scans, it is
convenient to consider briefly the operation of magnetic and electric
sectors.   Unless otherwise stated, the discussion will be limited
to singly-charged ions of charge e, where e is the electronic charge.

The Magnetic Sector

If an accelerating voltage, V, is applied to ions of mass m_1,
they acquire a translational energy given by

$$V_1 e = m_1 v_1^2/2 = (m_1 v_1)^2/2m_1 \tag{1}$$

where v_1 is the velocity of the ion of mass m_1.   The energy $V_1 e$ is
independent of the mass of the ion but the momentum of the ion is
given by

$$m_1 v_1 = (2V_1 e m_1)^{\frac{1}{2}} \tag{2}$$

so that for a fixed value of V_1, the relative momenta of ions of
different masses, m_1^+, are proportional to $m_1^{\frac{1}{2}}$.   This forms the
basis of the separation of ions of different mass-to-charge ratio
by a magnetic sector: when the accelerated ions enter a magnetic
field of strength B_1, they follow a circular path of radius R where
R is given by

$$R = m_1 v_1/B_1 e \quad or \quad m_1 v_1 = RB_1 e \tag{3}$$

For a fixed value of R, a scan of B therefore brings ions of different momenta to focus at the collector slit and from equation
(2), the separation is proportional to $m^{\frac{1}{2}}$.   By combining the above
equations, one obtains

$$m_1/e = R^2 B_1^2/2V_1 \tag{4}$$

At constant accelerating voltage V_1,

$$m_1/e = (R^2/2V_1) \cdot B_1^2 = KB_1^2 \tag{4a}$$

so that a scan linear in m is proportional to B^2, or alternatively,
a scan linear in B is proportional to $m^{\frac{1}{2}}$.   At a constant magnetic
field strength, B,

$$m_1/e = (R^2 B_1^2/2)(1/V_1) = K^1/V_1 \tag{4b}$$

so that a scan linear in V_1 is proportional to $1/m_1$ or mV = constant.

If m_1^+ ions fragment to give m_2^+ ions in the field-free region before the magnetic sector, the velocity of the m_2^+ ions equals v_1, that of the m_1^+ ions from which they were formed, neglecting effects due to the release of internal energy as translational energy of the fragments. Because the momentum m_2v_1 of these ions is less than that of the m_1^+ ions from which they were formed, m_1v_1, they are transmitted by the magnetic sector only if the field strength is reduced to B* which would transmit hypothetical ions of mass m* formed in the source. From the above equations,

$$(m_2v_1)^2 = (RB^*e)^2 = 2m^*V_1e = 2m^*. \ m_1v_1^2/2 = m^*m_1v_1^2 \qquad (5)$$

from which $m^* = m_2^2/m_1$.

The Electric Sector

The ability of a magnetic sector to separate ions of different mass is reduced by the spread of translational energy in the ion beam. The mass resolving power can be improved if the energy spread is reduced by passing the beam through an electric sector either before or after it passes through the magnetic sector. For ions to pass through a cylindrical electric sector along a path of radius r across which an electric field E is applied, it is necessary that

$$Ee = m_1v_1^2/r = 2Ve/r = 2 \times (\text{Ion Energy})/r \qquad (6)$$

from which

$$r = 2V/E \qquad (7)$$

Ions having energies greater or lower than Ve follow paths of radii greater or less than r so that an electric sector acts as an energy filter. A more detailed analysis of the paths of ions through an electric sector shows that it also has a focussing action. It acts as an energy-focussing device in that a mono-energetic, divergent beam of ions emanating from the source is brought to focus. A magnetic sector acts as a direction-focussing device and certain combinations of electric and magnetic sectors can focus an ion beam which is both divergent and inhomogeneous in energy, achieving both direction and velocity focussing, thereby giving rise to the term "double-focussing" mass spectrometer.

If m_2^+ ions are formed from m_1^+ ions in the field-free region before an electric sector, their translational energy is given by $(m_2/m_1)V_1e$ and if the ratio V/E is set to transmit m_1^+ ions, these m_2^+ ions are not transmitted. There are various methods of modi-

fying the instrumental parameters to allow the collection of m_2^+
ions as described in the following sections.   The earlier, single
paramater scans of V and E are described first, followed by a discussion of scans in which two of the three parameters V, E and B
are varied simultaneously.

Scans of a Single-Parameter

1.   The V-scan[2-4]If V_1 and E_1 are the accelerating voltage and
electric sector field strength respectively required for the transmission of the main beam, the energy of the transmitted ions is V_1e.
This energy controls the magnetic field strength B required to
transmit ions of a chosen mass-to-charge ratio so that if the mass
scale is not to alter, all ions which pass through the electric
sector must have an energy of V_1e.   If $m_1^+ \rightarrow m_2^+$ in the region between the source and electric sector of a   forward geometry instrument, the energy of the m_2^+ ions is now $(m_2/m_1)V_1e$ so that they
are not transmitted by the electric sector.   If the accelerating
voltage is now raised to a new value V_2 such that $V_2 = (m_1/m_2)V_1$,
then the energy of the m_2^+ ions formed in this way is given by

$$\text{Energy of } m_2^+ = (m_2/m_1)V_2e = (m_2/m_1)(m_1/m_2)V_1e = V_1e \qquad (8)$$

thereby allowing them to be transmitted by the electric sector.

This is the basis of the V-scan in which the instrument is
adjusted so that m_2^+ ions formed in the ion source are collected
at a reduced accelerating voltage V_1 and a scan is initiated in
which V increases, the maximum value usually being the maximum of
which the instrument is capable of delivering.   For a chosen
daughter ion m_2^+, m_2^+ ions are collected whenever equation (8) is
satisfied by different parent ions, m_1^+.   All signals are given
by m_2^+ ions of the same energy, V_1e, and the spectrum is essentially
an energy spectrum of the various m_1^+ ions which fragment to give
the chosen daughter ion, m_2^+.   This energy spectrum is converted
into a mass spectrum by using the above relationships.

The practical upper limit of V imposed by instrumental performance usually means that V cannot be varied by a factor of more
than 2-4; even over this range, the tuning of the source is often
markedly dependent on V so that the relative intensities of peaks
in this type of spectrum will vary with the source tuning conditions.
The peaks are broadened by release of internal energy during fragmentation and this is discussed in more detail in a later section.

2.   The E-Scan[5,6]   This is used on instruments of reverse geometry
in which ions are mass analysed prior to their being passed through
the electric sector.   The instrument is adjusted so that parent ions
m_1^+ are collected under normal operating conditions (V_1,E_1,B_1).
If $m_1^+ \rightarrow m_2^+$ in the field-free region between the two sectors, the

resulting m_2^+ ions are transmitted by the electric sector only if E is reduced from E_1 to E_2 such that

$$E_1/E_2 = m_1/m_2 \text{ or } E_2 = (m_2/m_1)E_1 \qquad (9)$$

If E is scanned downwards from E_1, different daughter ions m_2^+ formed from m_1^+ ions are transmitted by the electric sector whenever eqn. (9) is satisfied. The E-scan therefore produces a spectrum of m_2^+ ions formed from m_1^+ ions; the spectrum is in fact an ion kinetic energy spectrum of m_2^+ ions which can be converted to a mass spectrum by the use of eqn. (9). Because the accelerating voltage V is constant throughout this scan, there are no variations of source tuning and no discrimination effects of the type associated with a V-scan. In principle, E_1/E_2 may be very large but a practical limitation is that the sensitivity of detection falls as E is reduced because of the fall in energy of the ions striking the electron multiplier.

For the detection of the products of metastable transitions, the E-scan is likely to be less sensitive than a V-scan because observations are made on decompositions which occur later in the flight path. On the other hand, it has some advantages in the study of collision-induced decompositions because a collision cell placed at the focal point between the two sectors may be floated at a potential above earth potential in order to separate the peaks arising from metastable transitions from those arising from collision-induced decompositions. In a forward geometry instrument in which the collision cell precedes the magnetic sector, floating of the cell is impracticable since it would lead to a change in the mass scale. As is the case in other single parameter scans, the peaks are broadened by the release of internal energy during fragmentation.

Scans in which Two Parameters are Scanned Simultaneously

Five scans of this type have been described in four of which the accelerating voltage V remains constant while B and E are scanned in such a manner as to maintain a constant relationship between them throughout the scan. The oldest of these scans, however, is that in which B is held constant and V and E are scanned simultaneously, and this will be described first.

1. The V/E^2 Scan[7-9] This is used only with instruments of forward geometry. The primary ion m_1^+ is focussed under normal operating conditions (B_1, E_1, V_1) but on a high mass scale so that both V_1 and E_1 are, for example, only one quarter of their maximum value. A linked scan of V and E is then initiated from essentially zero values of V and E such that throughout the scan the ratio V/E^2 remains constant. The scan yields a spectrum of the daughter ions m_2^+ formed by the decomposition of the chosen parent ion m_1^+, in

the field-free region between the source and the electric sector. The choice of $m_1{}^+$ fixes the initial value of B_1, and V_1 and E_1 are the values of accelerating voltage and electric sector field strength at which the main beam is transmitted.

Because B is fixed by the choice of $m_1{}^+$ and remains constant throughout the scan, the product mV is a constant for ions formed in the ion source and which are transmitted by the magnetic sector at magnetic field strength B_1. In order that the main beam should pass through the electric sector, eqn. (7) must be satisfied so that by combining these two requirements, one obtains

$$m_1E_1 = m_2E_2 \quad \text{so that} \quad E_2 = (m_2/m_1)E_1 \tag{10}$$

Although this equation was derived for ions formed within the ion source, it must also be valid for $m_2{}^+$ ions formed from $m_1{}^+$ ions in the field-free region between the source and the electric sector, since the passage of ions through the two sectors must be independent of the mode of formation of the ions.

Suppose that E_2 is the electric sector field-strength which allows the transmission of $m_2{}^+$ ions formed from $m_1{}^+$ ions in the field-free region when V_2 was the accelerating voltage. The energy of the $m_2{}^+$ ions equals $(m_2/m_1)V_2e$ and using equation (6) :

$$(r/2)E_2e = (m_2/m_1)V_2e$$

From eqn. (10), $m_2/m_1 = E_1/E_2$ so that

$$(m_2/m_1)V_2 = (E_1/E_2)V_2 = (r/2)E_2 \tag{11}$$

If the main beam is transmitted when $E = E_1$, eqn. (11) can be re-written in the form

$$V_2/E_2{}^2 = r/2E_1 = \text{constant} \tag{12}$$

the value of the constant being determined by the choice of E_1.

The scan is carried out by allowing E to increase linearly with time and since E is inversely proportional to the mass of the ion transmitted (from eqn. (10)) :

$$m_1t_1 = m_2t_2 \quad \text{so that}$$

$$m_2 = m_1(t_1/t_2) = C/t_2 \tag{13}$$

where $C = m_1t_1$ and t_1 and t_2 are the times after the initiation of the scan at which $m_1{}^+$ and $m_2{}^+$ ions are transmitted by the electric sector.

The major disadvantages of the scan are that the upper instrumental limit of V usually requires that $m_1/m_2 \geq 2$ and the need to vary V over a factor of four to accomplish this may lead to problems associated with source tuning.   It gives relatively narrow peaks, essentially independent of release of internal energy during fragmentation, but since many of the advantages of this scan without the attendant disadvantages can be achieved by other means, it has not been widely used.

2.   The B/E Linked Scan[10-12]   This is the first of four scans to be described which have the advantage that throughout the scan, V and hence the source tuning conditions remain constant.   The first three scans may be used with instruments of either forward or reverse geometry and in the B/E linked scan, B and E are scanned simultaneously such that the ratio B/E is constant throughout the scan.   This gives a spectrum of all daughter ions m_2^+ formed from m_1^+ ions in the field-free region between the source and the first sector and under typical operating conditions, the resolving power for m_1^+ is 300-400 and that for m_2^+ is 500-1000.

The instrument is adjusted so that m_1^+ ions are collected under conditions defined by V_1, E_1 and B_1.   Since the choice of m_1^+ fixes the value of B_1, this also fixes the constant value $B/E = B_1/E_1$. A scan of B and E is then initiated, either under computer control or by using the output of a Hall probe to generate the reference voltage for E.   Because the electric sector acts as an energy analyser, m_1^+ and m_2^+ ions are transmitted when $E_1/E_2 = m_1/m_2$, from eqn. (9).   Similarly, the magnetic sector acts as a momentum separator so that

$$B_1/B_2 = m_1 v_1 / m_2 v_1 = m_1/m_2 \tag{14}$$

since after fragmentation, m_2^+ ions will have the velocity v_1 of the m_1^+ ions from which they were formed, neglecting the release of internal energy during fragmentation.   Combining equations (9) and (14), therefore,

$$B_1/B_2 = E_1/E_2 \quad \text{or} \quad B/E = \text{constant} \tag{15}$$

care must be taken to ensure that the instrument is adjusted to transmit the centre of the peak given by m_1^+ ions, thereby accurately defining B/E, since if this ratio is slightly incorrect, the relative intensities of different peaks given by the scan may be irreproducible.

3. B^2/E Scan[10,13] This scan complements the previous scan in that it yields a mass spectrum of all parent ions of a chosen daughter ion and therefore gives the same information as the V-scan but without the problems which arise when V is varied over a large range.   The instrument is adjusted so that m_2^+ ions are collected

under normal operating conditions specified by V_2, E_2 and B_2.   B
and E are then scanned, either under computer control or by using
the output of a Hall probe to generate a suitable reference voltage
for E, such that the ratio B^2/E remains constant throughout the
scan.   As before, care must be taken in setting up the scan to
avoid irreproducibility of the relative intensities of peaks in the
scan; by tuning to collect the middle of the peak given by the m_2^+
ions, the constant B^2/E is accurately defined.   In this scan, the
peaks are broadened by the release of internal energy as translational energy of the fragments, for reasons which are discussed
in a later section.

If at the fixed accelerating voltage V_2, m_2^+ ions formed by
the fragmentation of m_1^+ ions are transmitted by the electric sector
when $E = E_1$ and by the magnetic sector when $B = B_1$.   Each ion has
a velocity v_1 characteristic of m_1^+ ions formed in the source so
that if m_2^+ ions are to be transmitted by the two sectors,

$$RB_1e = m_2v_1$$
$$E_1e = m_2v_1^2/r$$

which, when combined, lead to

$$m_2/e = R^2(B_1^2/E_1)/r \quad \text{or} \quad B_1^2/E_1 = m_2r/R^2e \qquad (16)$$

Thus, for a chosen value of m_2, B_1^2/E_1 is a constant and all m_1^+
ions which fragment to give m_2^+ ions give signals.   Although the
resolving power with which m_2^+ can be defined is very high, the
broadening of the peaks means that the resolving power with which
m_1^+ may be identified is typically 100-200.

4.   The Constant Neutral Loss Scan[14] The above two scans allowed
one to select either m_1^+ ions and observe all m_2^+ ions formed from
them or to select m_2^+ ions and observe all the parent ions m_1^+ from
which they are formed.   In the constant neutral loss scan, as its
name implies, attention is focussed on the mass of the neutral
fragment lost and the scan allows one to collect all daughter ions
m_2^+ formed by processes in which the neutral species lost is of a
chosen constant mass.

The instrument is adjusted so that m_1^+ ions formed in the
source are collected under conditions specified by V_1, E_1 and B_1.
If $m_1^+ \to m_2^+ + m_n$ in the field-free region between the source and
the first sector, the m_2^+ ions will be transmitted by the electric
sector only if eqn. (9) is satisfied so that one may write

$$m_2/m_1 = E_2/E_1 = E' = 1 - m_n/m_1 \qquad (17)$$

On rearranging, this give

$$m_1 = m_n/(1-E^-) \tag{18}$$

If the $m_2{}^+$ ions are to be transmitted by the magnetic sector, eqn. (5) must be satisfied so that they are collected when the magnetic field strength is set equal to B_2 for the collection of hypothetical ions of mass-to-charge ratio m* $(=m_2{}^2/m_1)$.

Combining this with the above two equations gives

$$m^* = m_2{}^2/m_1 = m_2 E' = (m_1-m_n)E'$$
$$= (m_n/(1-E') - m_n)E' = m_n E'^2/(1-E') \tag{19}$$

But from equation (5)

$$m^*/e = m_n E'2/(1-E')e = R^2 B_2{}^2/2V_1$$

which may be rearranged to give

$$B_2{}^2 (1-E')/E'^2 = (2V_1 e/R^2)m_n \tag{20}$$

which is constant for a fixed, chosen value of m_n. If B and E are scanned, therefore, under computer control such that $(B/E')^2/(1-E')$ is constant throughout the scan, $m_2{}^+$ ions are collected only if the chosen neutral mass m_n is lost in the fragmentation.

5.   The B^2E Scan [15] This is the most recently described of the linked scans and is for use with reverse geometry instruments only. It is also the only linked scan in which the products of decomposition occurring in the field-free region between the sectors are collected. The instrument is set up under normal operating conditions specified by V_1, E_1 and B_1, so that the chosen daughter ion, $m_2{}^+$. is collected. Scans of B and E are initiated such that B is increased and E is decreased in such a manner that the product B^2E is constant at all times during the scan. This allows the collection of $m_2{}^+$ ions formed from a variety of $m_1{}^+$ ions which decompose in the field-free region between the two sectors.

If a magnetic sector field strength B_2 is required for the transmission of $m_1{}^+$ ions, this can be related to the magnetic field strength B_1 required to transmit $m_2{}^+$ ions, when both are formed in the source, by means of the equation

$$m_1/m_2 = B_2{}^2/B_1{}^2 \tag{21}$$

If, after transmission by the magnetic sector, $m_1{}^+$ ions fragment to give $m_2{}^+$ ions, these are transmitted by the electric sector only if the field strength is reduced from E_1 (at which both $m_1{}^+$

and m_2^+ ions formed in the source are transmitted) to E_2 where

$$E_1/E_2 = m_1/m_2 \qquad (22)$$

Remembering that B_1 and E_1 are conditions under which m_2^+ ions formed in the source are transmitted, and B_2 and E_2 are conditions under which m_2^+ ions formed from m_1^+ ions in the field-free region between the sectors are transmitted, (21) and (22) may be combined to give

$$B_2^2/B_1^2 = E_1/E_2 \quad \text{or} \quad B^2E = \text{constant} \qquad (23)$$

The spectrum of m_2^+ ions formed from different m_1^+ ions is therefore akinetic energy spectrum which may be converted to a mass spectrum of m_1^+ ions by means of (22).

The output of a Hall probe is used to provide a voltage proportional to B^2 and this, together with a reference voltage is used to generate a reference voltage for the electric sector field. This is made equal to the internal reference voltage when the electric sector transmits the main beam; throughout the scan, the voltage generated is such as to ensure the constancy of the products B^2E.

Effects arising from Translational Energy Release during Fragmentation

For simplicity, in deriving all the above relationships, it has been assumed that the velocity of m_2^+ ions is the same as the m_1^+ ions from which they are formed, i.e. release of energy as translational energy of the fragments has been neglected. It is probable that this is never strictly true and in the fragmentation of singly-charged ions, translational energy release from near zero up to about 1.5eV is found. Momentum is conserved during fragmentation and the relative velocities imparted to m_2^+ and m_n as a result of the energy release are in inverse proportion to the masses of the two fragments. [2]

It is convenient to define the x-axis as the direction of the ion beam, the y-axis as the direction of the electric sector field and the z-direction as the direction of the magnetic sector field. If all the energy were to be released in the y and z directions, the velocity of m_2^+ ions in the x-direction would be unchanged and the above relationships would hold. In general, of course, there is a range of total energies released and a range of incremental velocities to be added to or subtracted from the velocity in the x-direction. A full treatment of the effects this energy release has on the peak shapes observed in each scan is too lengthy for this review, [16, 17] but a qualitative picture may be gained from the following considerations.

In most instruments, there is no focussing in the z-direction so that ions formed with appreciable components of velocity in the z-direction are collected only if long slits are used. If the release is primarily in the y-direction, the effect on peak shape depends on slitwidth and the type of scan in use. In single - parameter scans, the beam is swept across a slit so that ions with velocity increments in the y-direction pass through the slit over a small range of values of the parameter being scanned. Similarly, for velocity increments in the x-direction, the velocity of m_2^+ ions can be represented as $v_1 \pm \delta v$ and as before, m_2^+ ions are collected over a small range of values of the parameter being scanned. Thus all single-parameter scans give peaks broadened by the release of translational energy and in none of them is it possible to specify the ratio m_1/m_2 very precisely.

When two parameters are scanned simultaneously, the effect of translational energy release on peak shape depends on the scan in use. For ions of a given mass, the relationship between the parameter and the velocity of ions transmitted is

$$V \propto v^2 \; ; \quad E \propto v^2 \; ; \quad B \propto v \qquad\qquad (24)$$

If the relationship between the two parameters in the linked scan is such that the transmission of the two sectors has the same dependence on ion velocity, ions with the full distribution of velocities will be collected and the peaks will be broad. Since $B^2 \propto v^2$ and $E \propto v^2$, a scan in which the ratio B^2/E remains constant will transmit ions with the full range of velocities, giving broad peaks (B^2/E independent of v). In the other linked scans in which E is scanned, $V^{\frac{1}{2}}/E$ and B/E are proportional to $1/v$ and for these scans, transmission will be highest when the relationships derived earlier are satisfied but falls sharply as the velocity increment or decrement rises. Thus, whereas in a V scan and E scan, the resolution with which m_1^+ and m_2^+ respectively can be observed is usually within the range 40 – 80, the resolution with which the two linked scans allow one to identify m_2^+ formed from a chosen m_1^+ ion is usually 300 – 500 and may exceed 1000 in favourable cases. [18, 19] The situation is more complicated for the constant neutral loss scan but the performance is similar to that of the B/E linked scan.

The situation is different in the case of the B^2E scan[15] in which the sectors are scanned in opposite senses and m_1^+ ions are transmitted by the magnetic sector and m_2^+ ions by the electric sector. In this scan, very narrow peaks are produced since for values of E which would allow the collection of ions with significant velocity increments, the values of B are such that m_1^+ ions are not transmitted by the intermediate slit. The resolving power with which m_1^+ may be observed is typically 3000 but the resolution with which m_2^+ ions can be resolved is very much less since this is determined only by the scan of the electric sector.

Artefact Peaks

All linked scans give spectra which may contain artefact peaks, i.e. peaks which do not arise from the collection of m_2^+ ions formed from m_1^+ in the field-free region of interest. The type of artefact peak varies with the type of scan [20-22] but although they are often held to detract from the reliability of data obtained from linked scans, they can usually be recognised quite readily. The origins of some of the artefact peaks in linked scans of forward geometry instruments can be seen from the accompanying figure.

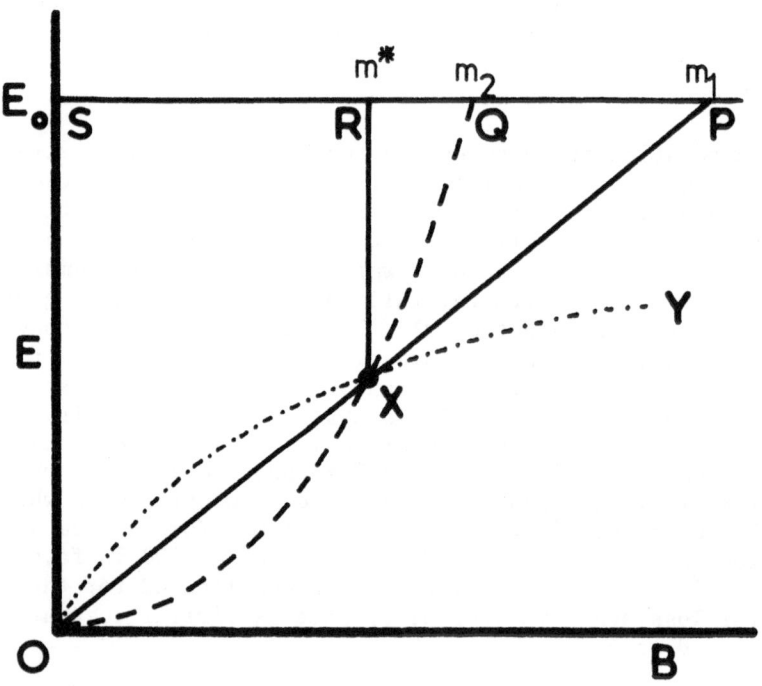

The B, E Plane of a Forward Geometry Instrument

illustrating various scanning modes.

The figure illustrates the B,E plane in which a normal mass spectrum is represented by the scan PQRS. A B/E linked scan is represented byPXO, a B^2/E scan is represented by QXO, and a constant neutral loss scan by YXO. Points which represent conditions under which m_2^+ ions can be collected are Q (formed in the source), X (formed in the first field-free region), R (formed in the second field-free region) and P (formed in the third field-freeregion). Any m_2^+ ions formed during acceleration will be collected under conditions the locus of which forms the line QX. Similarly, m_2^+ ions formed within the electric sector may be collected under conditions which form the line XR, although if the neutral species lost formed an appreciable fraction of the mass of m_1^+, the m_2^+ ions will not be transmitted by the electric sector. The line RQP represents the locus of points giving conditions under which m_2^+ ions formed within the magnetic sector can be collected; again, if the mass of the neutral species lost is an appreciable fraction of the mass of m_1^+, the m_2^+ ions will not be transmitted by the magnetic sector. For every fragmentation, therefore, there will be low intensity regions of the type QX, possibly accompanied by further low intensity regions of the type XR. If other scans cut through these regions of low intensity, artefact peaks will result (e.g. a B/E scan for an ion of mass between m_1 and m_2 would cut both QX and XR). Furthermore, although X is represented as a point, release of translational energy causes the intensity to be smeared out slightly along the line QXO so that, for example, a B/E scan of $(m_1 -1)^+$ would cut QXO sufficiently close to X to give an artefact peak.

A similar diagram can be constructed for a reverse geometry instrument, and the origins of the various types of artefact peaks identified. In either case, problems arising from peaks of this type can readily be sorted out by obtaining spectra given by two or more linked scans.

Metastable Ion Mapping

From the figure, it is clear that the various linked scans are simply different methods of investigating the intensity distribution within the B,E plane. Earlier methods of investigating this intensity distribution, either manually or under computer control, were made before the usefulness of linked scans was appreciated.[23-25] The method developed at Warwick[26] is a development of that described by Kiser[23] and makes use of an MS50 instrument and a DS55 data system based on a NOVA II minicomputer. A MINC-11 minicomputer was used to control reference voltages for V and E but other equally valid methods may also be used.[27]

The mass spectrometer scans B in the conventional manner over the range of interest and is first calibrated at low resolution for time to mass conversion. The magnetic field B is then scanned

repetitively and after each scan, the electric sector field E is
decremented by the MINC-11 by a chosen fraction of the normal value
E_o. The intensity versus time file generated in the scan is converted in the usual way to an intensity versus mass file, stored
on disc and numbered. In this way, the whole or a chosen part of
the B,E plane can be rapidly scanned. The scans for which $E=E_o$
give a normal mass spectrum; when E is less than E_o, products of
metastable transitions or collision-induced decompositions may be
collected. The data may then be manipulated off-line and displayed in various ways, the generally most useful being that in
which masses of parent and daughter ions may be read off the axes
directly.

The advantage of this procedure is that once the intensity
distribution within the B,E plane has been recorded and stored in
the data system, further programs may be used[26] which simulate the
B/E, B^2/E and constant neutral loss linked scans for any mass of
parent or daughter ion so that a full investigation of all the
fragmentations may be carried out subsequently off-line. If a
particular region of the B,E plane exhibits some interesting
features, it can be investigated at higher resolution by reducing
the size of the decrements of E, the electric sector field strength.
Although a comparatively large amount of sample is needed for an
investigation of the whole plane, it is at least as sensitive and
considerably more rapid than the carrying out of a large number of
linked scans. Furthermore, the origin of any artefact peaks is
readily established.

REFERENCES

(1) Hipple, J.A., Fox, R.E., and Condon, E.U.: 1946, Phys Rev,
69, p.347.
(2) Cooks, R.G., Beynon, J.H., Caprioli, R.M., and Lester, G.R.:
1973, Metastable Ions, Elsevier Scientific Pub Co.
(3) Jennings, K.R.: 1965 , J Chem Phys. 43, p.4176.
(4) Barber, M., and Elliott,R.M.: 1964, ASMS 12th Annual
Conference.
(5) Futrell, J.H., Ryan, K.R., and Sieck, L.W.: 1965, J Chem
Phys. 43, p. 1832.
(6) Beynon, J.R., Caprioli, R.H., and Ast, T.: 1971, Org Mass
Spectrom. 5, p. 229.
(7) Evans, S., and Graham, R.: 1974, Adv Mass Spectrom. 6, p.429.
(8) Lacey, M.J., and MacDonald, C.G.: 1975, J Chem Soc, Chem
Comm. p. 421.
(9) Weston, A.F., Jennings, K.R., Evans, S., and Elliott, R.M.:
1976, Int J Mass Spectrom Ion Phys. 20, p. 317.
(10) Boyd, R.K., and Beynon, J.H.: 1977, Org Mass Spectrom. 12
P. 163.
(11) Millington, D.S., and Smith, J.A.: 1977, Org Mass Spectrom.
12, p. 264.

(12) Bruins, A.P., Jennings, K.R., Stradling, R.S., and Evans, S.: 1978, Int J Mass Spectrom and Ion Phys. 26, p. 395.

(13) Evers, E.A.I.M., Noest, A.J., and Akkermann, O.S.: 1977, Org Mass Spectrom. 12, p.419.

(14) Haddon, W.F., 1980, Org Mass Spectrom, 15, p. 539.

(15) Boyd, R.K., Porter, C.J. and Beynon, J.H.: 1981, Org Mass Spectrom. 16, p.490.

(16) Holmes, J.L. Osborne, A.D., and Weese, G.M.: 1976, Int J Mass Spectrom Ion Phys. 19, p.207.

(17) Beynon, J.H., Fontaine, A.E., and Lester, G.R.: 1972, Int J Mass Spectrom Ion Phys. 8, p. 341.

(18) Porter, C.J., Brenton, A.G., and Beynon, J.H.: 1980, Int J Mass Spectrom Ion Phys. 36, p. 69.

(19) Stradling, R.S., Jennings, K.R., and Evans, S.: 1978, Org Mass Spectrom. 13, p.429.

(20) Jennings, K.R. 1978, A C S Symposium Ser 70, High Performance Mass Spectrometry: Chemical Applications, Ed M L Gross. p.1.

(21) Morgan, R.P., Porter, C.J., and Beynon, J.H.: 1977, Org Mass Spectrom. 12, p. 735.

(22) Haddon, W.F.: 1979, Anal Chem. 51, p. 987.

(23) Kiser, R.W., Sullivan, R.E., and Lupin, M.S.: 1969, Anal Chem. 41, p. 1958.

(24) Coutant, J.C., and McLafferty, F.W.: 1972, Int J Mass Spectrom Ion Phys. 8, p.323.

(25) Shannon, T.W., Mead, T.E., Warner, C.G., and McLafferty, F.W.: 1967, Anal Chem. 39, p.1748.

(26) Farncombe, M.J., Mason, R.S., Jennings, K.R. and Scrivens, J.: 1982, Int J Mass Spectrom Ion Phys. 44, p.91.

(27) Warburton, G.R., Stradling, R.S. Mason, R.S., and Farncombe, M.J.: 1981, Org Mass Spectrom. 16, p. 507.

CHARGE EXCHANGE MASS SPECTROMETRY

Alex. G. Harrison

Department of Chemistry
University of Toronto
Toronto, Canada

The techniques used and the problems encountered in the use of charge exchange mass spectrometry to determine breakdown graphs for polyatomic molecules are reviewed. A survey of the results obtained using both tandem mass spectrometers and chemical ionization techniques shows that breakdown graphs can be derived which adequately represent the energy dependence of the fragmentation of polyatomic molecular ions. The analytical uses of charge exchange ionization also are reviewed.

Charge exchange reactions

$$R^+ + M \rightarrow M^+ + R \qquad\qquad (1)$$

have been investigated extensively with respect to the mechanism of reaction, the role of Franck-Condon factors and the dependence of the cross-section on the energy mismatch. These aspects have been reviewed recently [1]. The present review of charge exchange mass spectrometry will be confined to two aspects, first, the use of charge exchange experiments in examining the energy dependence of the fragmentation of polyatomic molecular ions and, second, the use of charge exchange as an analytical technique in chemical ionization mass spectrometry.

Energy Dependence of the Fragmentation of Polyatomic Ions

When a gaseous sample of a polyatomic molecule, M, of ionization energy IE, is bombarded with a beam of electrons of energy E, the molecular ions produced possess internal energies ranging

23

M. A. Almoster Ferreira (ed.), Ionic Processes in the Gas Phase, 23–40.
© *1984 by D. Reidel Publishing Company.*

from 0 to E-IE, with a distribution which is unknown but depends
on the energy of the electrons and on the identity of the polyatomic molecule. The mass spectrum observed following electron
ionization is the result of a series of competing and consecutive
unimolecular fragmentation reactions starting from the molecular
ion, each reaction having a rate which depends on the internal
energy (2). The net results of this kinetic fragmentation scheme
are summed over the unknown internal energy distribution. In effect, in this summation process much of the primary kinetic data,
particularily the effect of internal energy on reaction competition, are lost. Although qualitative information on reaction com-
etition can be obtained by varying the energy of the impacting
electrons, more quantitative data can be realized only by obtaining the breakdown graph.

The breakdown graph, ie, the fractional ion intensities as
a function of the internal energy of the initial M^+ ions produced,
provides detailed information on the fragmentation reactions occurring and on the relative rates of these reactions as a function
of the internal energy of M^+. Such breakdown graphs, can be obtained, with difficulty, by determining either the normalized
second derivatives of the ion signals as a function of electron
energy for electron impact experiments (3) or the normalized first derivatives of the ion signals as a function of the photon energy for photon impact experiments (4). In general such breakdown graphs are reliable only for 3-4 volts internal energy above
the ground state of the molecular ion; in addition, they are ser-
iously distorted if autoionization processes are significant.

While breakdown graphs can be obtained only indirectly from
electron and photon impact experiments, both charge exchange and
photoelectron photoion coincidence (PEPICO) experiments give direct information on breakdown graphs. In the latter method the
translational energy of the electron ejected during the ionization process

$$h\nu + M \rightarrow M^+ + e \qquad (2)$$

is measured, allowing the energy of the fragmenting ion to be
determined through the relationship

$$E(M^+) = h\nu - IE(M) - KE(e) \qquad (3)$$

If the masses of the ionized species produced in coincidence with
the energy-analyzed electrons are measured, construction of the
breakdown graph directly is possible (5,6). In spite of considerable experimental difficulties the PEPICO technique appears to
be the best method for determining breakdown graphs since internal energies are well-defined and can be varied continuously.

In principle, in charge exchange experiments the internal energy of M^+ is given by

$$E(M^+) = RE(R^+) - IE(M) \qquad (4)$$

where $RE(R^+)$ is the recombination energy of the reactant ion, ie, the energy liberated in the process

$$R^+ + e \rightarrow R \qquad (5)$$

By choosing reactant ions of different recombination energies the internal energy of M^+ can be varied and the breakdown graph may be derived directly from the charge exchange mass spectra normalized to the total ion current derived from the species M. The charge exchange method of obtaining breakdown graphs predates the PEPICO method and, although it gives less precise results, it is simpler experimentally and still serves a useful role in the armoury of techniques designed to study the energy dependence of the fragmentation of gaseous ions. The present paper reviews some of the techniques used, difficulties encountered, and results obtained in such charge exchange studies.

Breakdown graphs from charge exchange mass spectra first were reported by Lindholm and co-workers (7,8) using a tandem mass spectrometer such as is illustrated schematically in Figure 1. The primary reactant ions are produced by electron impact and are mass selected in the first stage of the tandem instrument. They then are retarded in a system of electrostatic lenses to the desired kinetic energy and are introduced into a collision cell containing the neutral M at $\sim 10^{-4}$ torr pressure. The secondary ions produced in the collision cell are extracted by a weak eletrical field perpendicular to the direction of travel of the incident ions and are mass analyzed in the second stage of the tandem mass spectrometer. The perpendicular arrangement discriminates against observation of the products of reaction in which kinetic energy transfer from the reactant ions to be products has occurred (see below).

Charge exchange also has been observed in conventional mass spectrometers using either modified sources (9) or modified source potentials (10). Finally, with the advent of high pressure chemical ionization mass spectrometry a number of groups have used charge exchange ionization in analytical applications (11-16). However, with the exception of two brief reports by Einolf and Munson (17,18) no attempts were made to develop reagent ion systems covering a range of recombination energies suitable for the derivation of breakdown graphs. These attempts have been made in our laboratory over the past few years and some of the results obtained will be discussed later.

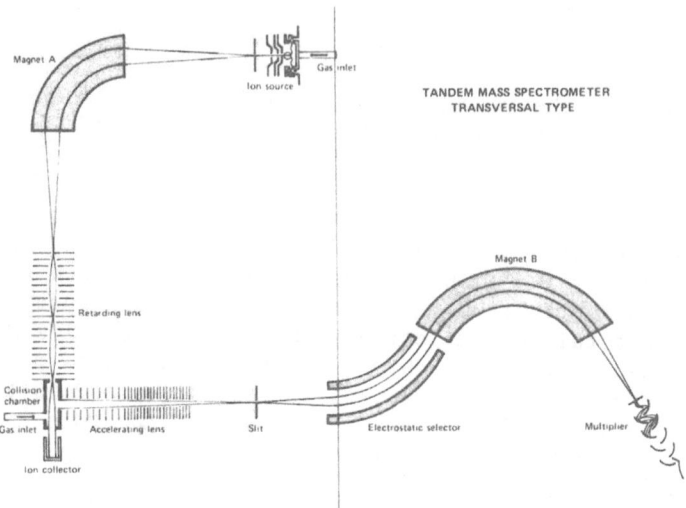

Figure 1. Tandem mass spectrometer for charge exchange experiments (from ref. 8).

There are a number of problems which arise in determining breakdown graphs from charge exchange mass spectra. First, only a limited number of primary ions are available with the consequence that the results are discontinuous; the extent of resolution of fine structure in the graphs depends directly on the number of charge exchange experiments carried out. In addition, in certain cases, other ion/molecule reaction, such as hydride ion abstraction or proton transfer (perhaps followed by fragmentation) also may occur and may distort the breakdown graph. However, by far the biggest problem is the uncertainty in the internal energy of M^+ produced in the initial charge exchange reaction. This uncertainty arises from uncertainties in the recombination energies of the reactant ions and from the possibility of kinetic-internal energy interconversion.

The recombination energy of R^+ is given by the exothermicity of reaction (5) and clearly depends on the initial state of R^+ and the final state of R. Lindholm (7) has summarized the available knowledge concerning recombination energies and perusal of this summary indicates that in some cases multiple recombination energies are involved. For example, the following recombination reactions are possible for C^+ ions produced by electron impact

$$
\begin{array}{lll}
 & & \text{RE} & \\
C^+(^2P) + e \rightarrow C(^3P) & 11.26eV & (6) \\
\rightarrow C(^1D) & 10.00eV & (7) \\
\rightarrow C(^1S) & 8.58eV & (8) \\
C^+(^4P) + e \rightarrow C(^3P) & 16.58eV & (9) \\
\rightarrow C(^5S) & 12.40eV & (10)
\end{array}
$$

Lindholm has estimated the ratio $C^+(^2P): C^+(^4P) = 0.60:0.40$. Clearly charge exchange mass spectra from such a reactant ion cannot be used to establish breakdown graphs; rather reactant ions with well-defined single REs or a small range of REs must be used; fortunately there are numerous ions which meet this criterion.

A further problem arises if there is conversion of kinetic energy of the reactant ion into internal energy of the M^+ ion. Table 1 compares spectra measured for the reaction of Ar^+ with C_2H_2 obtained by Maier (19) using an in-line tandem instrument and by Lindholm et al (2) using a perpendicular tandem instrument. Formation of C_2H^+ is 1.4 eV endothermic and formation of CH^+ is 6 eV endothermic for ground state reactant Ar^+ ions. From the observation of these products it is clear that considerable con-version of the kinetic energy of Ar^+ into internal energy of $C_2H_2^+$ has occurred particularly in the in-line arrangement which does not discriminate against reactions involving kinetic energy transfer. This problem is much less severe in single source chem-ical ionization experiments where the reactant ions have very low kinetic energies.

Table 1

PRODUCT DISTRIBUTION FROM Ar^+ + C_2H_2 REACTION

	In-Line		Perpendicular	
KE(Ar^+)	30eV	4eV	30eV	4eV
$C_2H_2^+$	51	96	74	93
C_2H^+	31	4	24	7
CH^+	13	--	1	--

A further possible problem affecting the internal energy of M^+ is the conversion of internal energy of R^+ into kinetic energy of the products. For example, Mauclaire et al have shown (21) that in the reaction

$$Ar^+ + O_2 \rightarrow Ar + O_2^+ \quad \Delta H = -3.7eV \quad (11)$$

almost 2eV (or 50%) of the reaction exothermicity shows up as kinetic energy of the product species rather than as

internal energy of O_2^+. Clearly, if this is a frequent occurence in charge exchange reliable breakdown graphs will be difficult to establish. The majority of experiments carried out to
date have involved charge transfer to relatively simple molecules which have few excited ionic states and hence little chance
for accidentally resonant charge exchange. For polyatomic species
where the density of ionic states should be much higher the probability of accidental resonance where all of the reaction exothermicity is retained as internal energy is likely to be much higher[*]

The majority of the early tandem charge exchange experiments
carried out by Lindholm and colleagues involved detailed study of
relatively simple molecules. This early work has been reviewed
by Lindholm (7,8). Despite the problems discussed above breakdown graphs were obtained which gave fragment ion onset energies
in agreement with other measurements and which, when combined
with assumed, but reasonable, internal energy distributions gave
calculated mass spectra in agreement with experimental mass spe-
tra. In particular there was no evidence that effective recombination energies were influenced significantly by internal-kinet-
ic energy interconversion processes.

A particularily interesting breakdown graph from this earlier work is that of ethanol (22) which showed double maxima in
the $m/_z$ 45 and $m/_z$ 31 intensities indicating the possible participation of an isolated electronic state of the molecular ion.
Recent PEPICO experiments (23) have not observed these double
maxima and it is probable that their observation in the charge
exchange experiments may reflect the occurrence of hydride ion
abstraction reactions with some reactant ions. Some of the discrepancies between the experimental and calculated breakdown
graphs for propane (24) also may be the result of the occurence
of hydride ion transfer reactions in the charge exchange experiments. A more rigorous assessment of the experimental breakdown
graphs is provided by the recent studies of Sunner and Szabo
which have involved a careful comparison of breakdown graphs for
i-butane (25) and n-propanol and i-propanol (26) with QET calculations (27). Good agreement between theory and experiment was
found at low internal energies but substantial deviations were
observed at energies >6eV, which were attributed to over-simpli-
fied decomposition schemes used in the calculations. Figure 2
shows the experimental and calculated breakdown graphs for i-propanol.

The success of the tandem mass spectrometeric experiments
in establishing breakdown graphs which elucidate clearly the reaction competition involved in the fragmentation of polyatomic
[*] For a discussion of the more recent studies on energy disposal
in charge exchange reactions see the chapter by R. Marx in this
volume.

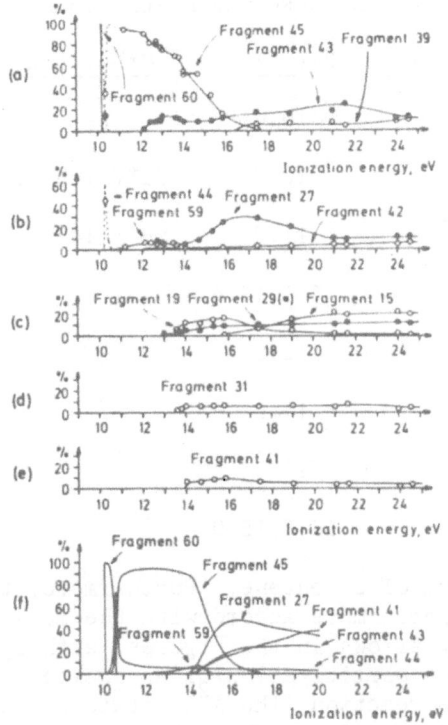

Figure 2. Experimental (a) - (e), and calculated (f) breakdown graphs for i-propanol. From ref. 26.

ions prompted us to undertake similar charge exchange studies using a single-source mass spectrometer under chemical ionization conditions. Without mass selection of the primary reactant ion one is more limited in the number of reactant ions that can be used. Table 2 summarizes the chemical ionization reagent systems developed, and the major reactant ions observed under high pressure (0.2-0.3 torr) conditions, along with the recombination energy of the major ion.

In our initial study (28) using the chemical ionization method we examined the charge exchange mass spectra of isomeric C_5H_{10} molecules, primarily because the fragmentation behaviour and the energetics of this system are relatively well understood. Breakdown graphs for 1-pentene and 2-pentene are shown in Figure 3 and are in agreement with the observation that $C_3H_6^+$ is the base peak in the EI mass spectrum of 1-pentene while $C_4H_7^+$ is the base peak

Table 2

CHEMICAL IONIZATION CHARGE EXCHANGE REAGENT IONS

Reagent Gas	Reagent Ion	RE(eV)	Other Ions
$CO_2/10\%\ C_6F_6$	$C_6F_6^+$	~10.0	--
$N_2/10\%CS_2$	CS_2^+	~10.2	$S_2^+(<10\%)$, $(CS_2)+_2(<10\%)$
$CO/10\%\ COS$	COS^+	11.2	$S_2^+(\sim14\%)$, $(COS)_2^+(\sim4\%)$
Xe	Xe^+	12.1,13.4	--
$N_2/10\%N_2O$	N_2O^+	12.9	
CO	CO^+	14.0	
N_2	N_2^+	15.3	
Ar	Ar^+	15.8,15.9	

in the EI mass spectrum of 2-pentene. Furthermore, the onset of
fragmentation is in approximate accord with the measured onset
energies (indicated by arrows in the Figures) although it appears
that there may be minor participation of an excited state of CS_2^+
in the charge exchange reaction (The $A^2\Pi_u$ state of CS^+_2 (RE=12.7
eV) has a lifetime of ~4 s(29)). There is no evidence from these
breakdown graphs that a significant fraction of the exothermicity
of the charge exchange reaction is ending up as kinetic energy of
the products rather than as internal energy of the $C_5H_{10}^+$ molecular ions. For most systems significant yields of $C_5H_9^+$ were observed which are not included in the diagrams since $C_5H_9^+$ may
arise by H^- abstraction rather than charge exchange; the EI mass
spectra show only very low $C_5H_9^+$ intensities. Despite the fact
that the two breakdown graphs in Figure 3 are different the molecular ions of the two isomers show very similar metastable ion
characteristics (28,30) and collisional activation mass spectra
(31).

The next members of the olefinic series, the C_6H_{12} isomers,
are of interest in that they are the simplest olefins for which
the molecular ions show substantially different metastable ion
characteristics (30,32). The linear hexenes show almost equal
abundances for loss of CH_3 and C_2H_4 from M^+, the 2-methyl-1-pen-
tene and 4-methyl-1-pentene molecular ions show dominant loss of
C_2H_4, while the remaining isomers show dominant loss of CH_3 in
metastable fragmentation of the molecular ion. These differences,
as well as the differences in the EI mass spectra, are reflected
in the breakdown graphs reported (32). Again the onsets for frag-

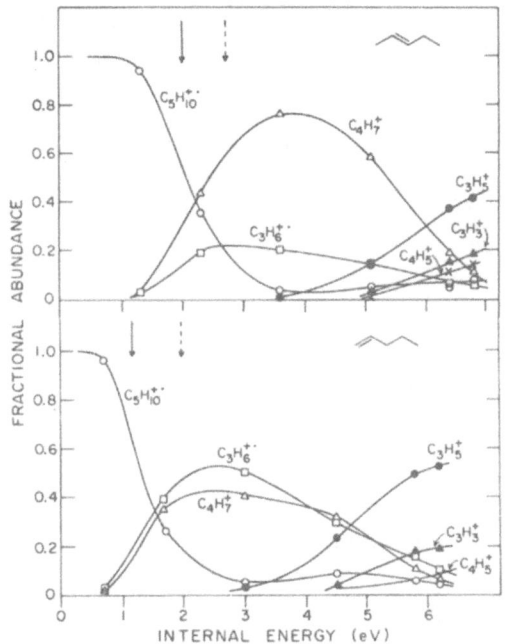

Figure 3. Breakdown graphs for 1-pentene and 2-pentene (ref.28)
Solid arrow gives thermochemical onset for $C_3H_6^+$ and $C_4H_7^+$. Dotted
arrow gives onset for $C_3H_5^+$.

mentation which were observed were in agreement with measured values and there was no evidence that the internal energies imparted
to the molecular ions were significantly different from the values
estimated from the reported recombination energies and ionization
energies.

The 4-methyl-1-pentene system is of particular interest in
that, although in the metastable ion timeframe $C_4H_8^+$ is the major
fragment ion, $C_3H_7^+$ forms the base peak in the 70 eV EI mass
spectrum (32). The breakdown graph, Figure 4, shows that formation of $C_4H_8^+$ is favoured at low internal energies but that formation of $C_3H_7^+$ is favoured between 2 to 4.2 eV internal energy.
Figure 4 also shows the EI internal energy distribution as approximated by the HeI photoelectron spectrum (33). Note that a large
portion of the energy region where $C_4H_8^+$ dominates the breakdown
graph corresponds to a gap in the internal energy distribution
while there is a high population in the energy region where $C_3H_7^+$
dominates the breakdown graph. Thus it is not surprising that
$C_3H_7^+$ is the base peak in the EI mass spectrum.

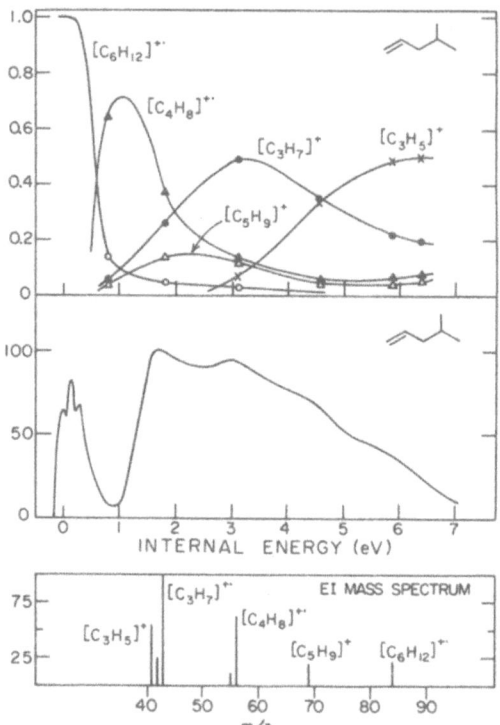

Figure 4.   Breakdown graph, internal energy distribution and
mass spectrum of 4-methyl-1-pentene.

    The C$_7$H$_{14}$ branched olefins with a terminal double bond show
(34) a strong preference for loss of ethylene from M$^+$ for molecular ions fragmenting as metastable ions.   For example, for the
2-methyl-1-hexene and 2,4-dimethyl-1-pentene molecular ions loss
of C$_2$H$_4$ accounts for >90% of the total metastable ion intensity.
Despite this intense metastable the resulting C$_5$H$_{10}$$^+$ fragment ion
is not prominent in the 70 eV EI mass spectra.   To examine the
reasons for this behaviour we have determined the breakdown graphs
for a number of C$_7$H$_{14}$ isomers (35).   Figure 5 shows the results
obtained for 2,4-dimethyl-1-pentene.   Formation of C$_5$H$_{10}$$^+$ obviously is the favoured fragmentation reaction energetically.   However,
the rapid drop-off in intensity of this product with increasing
internal energy indicates that the frequency factor for formation
of C$_5$H$_{10}$$^+$ must be very unfavourable.   Similar results were obtained for 2-methyl-1-hexene and 5-methyl-1-hexene which also show
an abundant metastable for formation of C$_5$H$_{10}$$^+$.   These results
point to a complex rearrangement process (or processes) in the

formation of this fragment; an isotopic labelling study (36) of
the fragmentation of 2,4-dimethyl-1-pentene indicates a complex
mechanism (with a highly structured transition state) which leads
to loss of C_3 and C_5 as ethylene.

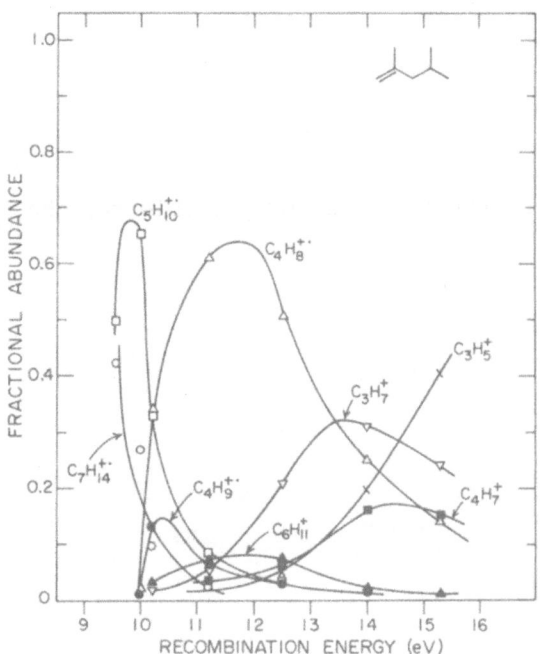

Figure 5. Breakdown graph for 2,4-dimethyl-1-pentene

Another system which we have studied comprises the isomeric
chloroanisoles (37). Figure 6 shows the breakdown graphs obtained for the three isomers. For the ortho and para isomers the
dominant fragmentation reaction of $\overline{M^+}$ is loss of CH_3, which is
followed by loss of CO at higher internal energies. By contrast,
for the meta isomer rearrangement elimination of CH_2O is the dominant low energy fragmentation route of M^+ and loss of CH_3 by simple bond cleavage only becomes competitive at higher internal
energies where further elimination of CO also occurs. These differences in behaviour are also reflected in metastable and mass
spectral ion abundances. Obviously the critical energy for CH_3
loss is lower for the ortho and para isomers than for the meta
isomer presumably as a result of a result of resonance stabilization of the resulting cation, viz

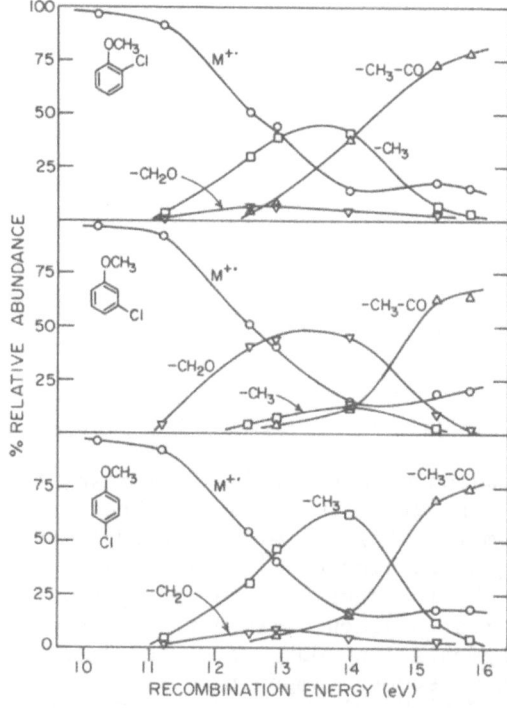

Such resonance stabilization is not possible for the meta-chloro-phenoxy cation.

Figure 6. Breakdown graphs for chloroanisole isomers.

Aktar et (38) have shown that in the photon-impact mass spectra of stereoisomeric 4-methyl-cyclohexanols water elimination is more facile from the molecular ion of the trans isomer than from

the molecular ion of the cis isomer, presumably because of a
favourable sterospecific 1,4-cis water elimination reaction for
the trans isomer.   Figure 7 compares the breakdown graphs obtained for the two isomers (39).   Beyond 11 eV recombination energy
the breakdown graphs are essentially identical; the only significant difference lies in the region around 10 eV recombination energy where for the trans isomer there is a sharp peak in the abundance of $C_7H_{12}^+$ (M^+-H_2O) which is absent for the cis isomer.
This peak presumably corresponds to the energetically favourable
cis-1,4-H_2O elimination reaction which rapidly drops off in importance with increasing internal energy because of an unfavourable frequency factor arising from the highly-ordered transition
state necessary.

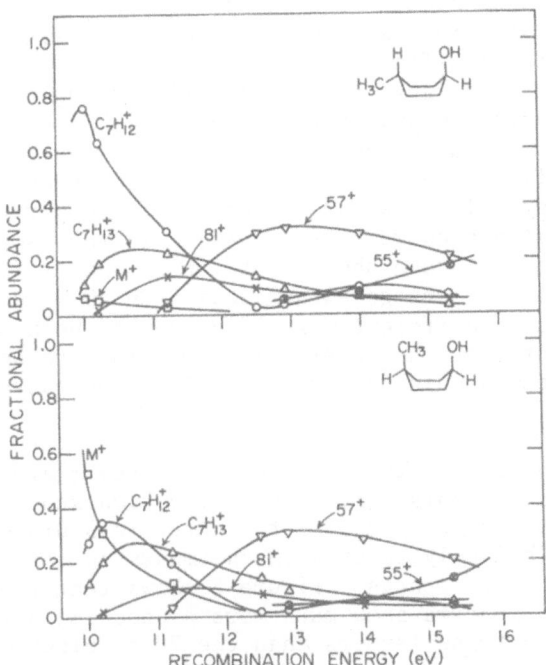

Figure 7.   Breakdown graphs for cis-and trans-4-methyclcyclohexan-
ol(major ions only are shown)

    Finally, if the breakdown graph is known from independent
measurements, the charge exchange mass spectrum can be used to
estimate the recombination energy of the reactant ion.   Using
this approach Bieri and Jonsson (40) have shown that the m/z 27
fragment ion from CH_3NH_2 and CH_3NC does not have the HCN^+ structure but rather has the HNC^+ structure.   A recombination energy of
12.5 ± 0.2 eV was derived for the latter structure.

Analytical Applications of Charge Exchange Mass Spectrometry

Charge exchange has not seen extensive use as a chemical
ionization method. Since the fragmentation products following
charge exchange ionization are the same as those following electron impact it appears, at first glance, that charge exchange
ionization is unlikely to enhance structural elucidation capabilities. However, this is not entirely true and there is considerable evidence that low energy charge exchange ionization enhances
the differences in the mass spectra of isomeric components. Thus,
Sunner and Szabo have shown that the low energy charge exchange
mass spectra of n-butane and i-butane (25) as well as n-propanol
and i-propanol (26) show greater differences than the 70 eV electron impact mass spectra. A particularly interesting case are
the cis and trans isomers of 2-butene which show indistinguishable
electron impact mass spectra and indistinguishable charge exchange mass spectra with reactant ions of single recombination energy
but show significant differences in the spectra obtained with
reagent ions having a range of recombination energies (41). This
result can be explained qualitatively by small differences in the
photo-electron spectra of the isomers if it is assumed that these
spectra also reflect the variation of charge exchange cross sections with energy. This latter assumption is supported by the
results of Tedder and co-workers (42-44). In the same vein
Sunner and Szabo (16) have shown that ionization of four isomeric
C_6H_{14} hydrocarbons by charge exchange with CH_3COH^+ (RE~10.5 eV)
produces distinctly different mass spectra with both molecular
ions and fragment ions characteristic of structure being observed
while giving spectra which are much simpler than conventional EI
mass spectra.

Figure 7 shows that the differences in the fragmentation
behaviour of the stereoisomeric 4-methyl-cyclohexanols is concen-
trated in the low energy peak for the water loss process for the
trans isomer. Obviously spectra which emphasize this region will
show the largest differences. Figure 8 shows that the $C_6F_6^+$ charge exchange mass spectra show much more pronounced differences between the cis and trans isomers than the 70 eV electron impact
mass spectra which integrate ion intensities over a wide internal
energy range. In the same vein Table 3 compares the 70 eV electron impact and COS^+ charge exchange mass spectra of C_5H_{10} isomers. The differences among isomers are much more pronounced in
the charge exchange mass spectra and all isomers can be readily
identified from their charge exchange mass spectra except for
the pair 2-methyl-1-butene and 2-pentene which show identical
charge exchange mass spectra.

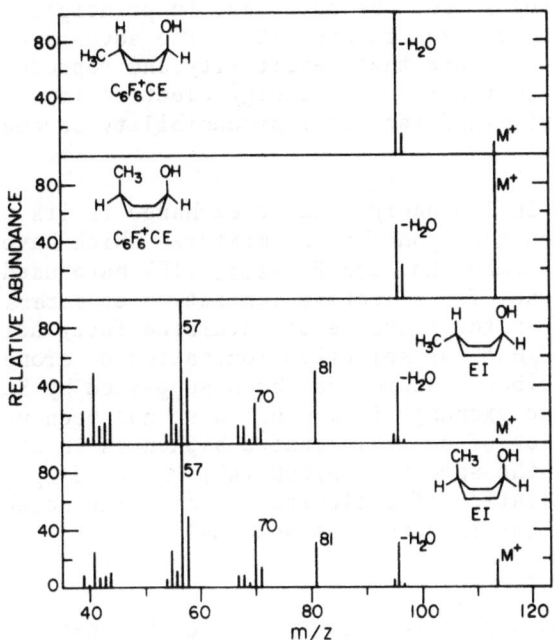

Figure 8. $C_6F_6^+$CE and EI mass spectra of cis and trans-4-methycy-clohexanol.

Table 3

EI & COS$^+$CE Mass Spectra of C_5H_{10} Isomers

	$C_5H_{10}^+$		$C_4H_7^+$		$C_3H_6^+$	
	EI	CE	EI	CE	EI	CE
Cyclopentane	30	100	29	1	100	7
1-pentene	32	73	58	94	100	100
3-Me-1-butene	26	46	100	100	28	36
2-Me-1-butene	26	75	100	100	27	39
2-pentene	34	84	100	100	46	44
2-Me-2-butene	35	100	100	77	31	29

In all of these examples one is making use of charge exchange ionization to produce ions of relatively low internal energy content. Similar results could be obtained, in principle, by low energy electron impact experiments (45). The advantages of charge exchange ionization are that sensitivity and reproducibility are maintained in contrast to low energy electron impact studies where low sensitivity and lack of reproducibility become serious problems.

The ultimate in low energy charge exchange is its use to ionize selectively those components of mixtures which have low ionization energies. Subba Rao and Fenselau (15) have used $C_6H_6^+$ (produced in benzene) for selective ionization of esters of unsaturated fatty acid in the presence of saturated fatty acid esters. A similar use of $C_6H_6^+$ for selective ionization of aromatic hydrocarbons in hydrocarbon mixtures has been suggested by Sunner and Szabo (16). Charge exchange from $C_6H_6^+$ also has been used for selective ionization of polychlorinated biphenyls in atmospheric samples (46). In the same vein Sieck (47) has used $C_6H_{12}^+$ (produced by photoionization of cyclohexane) to ionize selectively aromatic hydrocarbons in gasolines and fuel oils.

Summary

The breakdown graph expresses directly the energy dependence of the fragmentation of polyatomic ions. The results available show that these breakdown graphs can be obtained directly by charge exchange experiments using either tandem mass spectrometers or chemical ionization techniques, although the latter method provides less detail. By careful selection of the reactant ion(s) the problem of multiple recombination energies can be avoided while studies at low kinetic energies of the incident ion avoid problems of conversion of kinetic energy to internal energy of the fragmenting species. In addition, the results of charge exchange studies show that conversion of internal energy to kinetic energy during the charge exchange process is not significant in these polyatomic species. Consequently the internal energy of the fragmenting species can be estimated from the recombination energy of the reactant ion and the ionization energy of the neutral molecule.

In addition to its use in deriving breakdown graphs charge exchange ionization has useful analytical applications. The major application is in low-energy ionization where the mass spectral difference among isomeric compounds frequently are enhanced. As the ultimate, charge exchange can be used to ionize selectively those components of complex mixtures which have the lowest ionization energies.

Acknowledgements
The author is indebted to the Natural Sciences and Engineer-

ing Research Council of Canada for financial support and to his co-workers, listed in the references, for their contribution to this work.

References

1. Baer, T, in Mass Spectrometry, Specialist Periodical Reports, Vol. 6, Royal Society Chemistry, London, 1981.
2. Rosenstock, H.M., Wallenstein, M.B., Wahrhaftig, A.L., and Eyring, H. Proc. Natl. Acad. Sci. U.S., 38, 667 (1952).
3. Chupka, W.A., J. Chem. Phys., 30 191 (1959).
4. Murad, E. and Inghram, M.G., J. Chem. Phys., 40, 3263 (1964).
5. Brehm, B. and Puttkamer, E.V., Z. Naturforsch, 22a, 8 (1967).
6. Brehm, B and Puttkamer, E.W., Adv. Mass Spectrom., 4, 591 (1968).
7. Lindholm, E. in Ion-Molecule Reactions in the Gas Phase, P. Ausloos, Ed., Am. Che. Soc. Washington, 1966.
8. Lindholm, E. in Ion-Molecule Reactions, Vol. 2, J.L. Franklin, Ed., Plenum Press, N.Y., 1972, Chap. 10.
9. Andlauer, B., and Ottinger, Ch., Z. Naturforsch, 27a, 293 (1972).
10. Cermak, V. and Herman, Z. Nucleonics, 19, 106 (1961).
11. Hunt, D.F. and Ryan, J.F., Anal. Chem., 44, 1306 (1972).
12. Jelus, B.L., Munson, B. and Fenselau, C., Biomed. Mass Spectrom. 1, 96 (1974).
13. Lee, M.L. and Hites, R.A., J. Am. Chem. Soc., 99, 2008 (1977).
14. Munson, B., Anal. Chem., 49 772A, (1977).
15. Subba Rao, S.C. and Fenselau, C., Anal. Chem., 50, 511 (1978).
16. Sunner, J. and Szabo, I, Adv. Mass Spectrum, 7, 1383 (1978).
17. Einolf, N. and Munson, B, Org. Mass Spectrom, 5, 397 (1971).
18. Einolf, N. and Munson, B, Org. Mass Spectrom, 7, 155 (1973).
19. Maier, W.B., J. Chem. Phys., 42, 1790 (1965).
20. Lindholm, E., Szabo, I., and Wilmenius, P. Arkiv. Fysik, 25, 417 (1963).
21. Mauclaire, G., Derai, R., Fenistein, S., and Marx, R.J. Chem. Phys. 70, 4017 (1979).
22. von Koch, H. and Lindholm, E., Arkiv. Fysik. 19, 132 (1961).
23. Niwa, Y., Nishimura, T., and Tsuchiya, T., Int. J. Mass Spectrom. Ion Phys., 42, 91 (1982).
24. Vestal, M.L., J. Chem. Phys., 43, 1356 (1965).
25. Sunner, J., and Szabo, I., Int. J. Mass Spectrom. Ion. Phys., 25, 241 (1977).
26. Sunner, J. and Szabo, I., Int. J. Mass Spectrom, Ion Phys., 25, 263 (1977).
27. Vestal, M. and Lerner, G., Fundamental Studies Relating to the Radiation Chemistry of Small Organic Molecules, ARL 67-0114, 1967.
28. Li, Y.-H., Herman, J.A., and Harrison, A.G., Can. J. Chem., 59, 1753 (1981).
29. Eland, J.H.D., Devoret, M., and Leach, S., Chem. Phys. Lett.,

43, 97 (1976).
30. Bowen, R.D. and Williams, D.H., Org. Mass Spectrom, 12, 453 (1977).
31. Nishishita, T. and McLafferty, F.W., Org. Mass Spectrom, 12, 75 (1977).
32. Herman, J.A., Li, Y.-H., and Harrison, A.G., Org. Mass Spectrom, 17, 143 (1982).
33. Meisels, G.G., Chen, C.T., Giessner, B.G., and Emmel, R.H., J. Chem. Phys., 56, 793 (1972).
34. Falick, A.M., Tecon, P., and Gaumann, T., Org. Mass Spectrom, 11, 409 (1976).
35. Lin, M.S., Li, Y.-H., Herman, J.A., and Harrison, A.G., to be published.
36. Stefani, A., Org. Mass Spectrom., 7, 17 (1973).
37. Reiner, E.J. and Harrison, A.G., unpublished results.
38. Akhtar, Z.M., Brion, C.E., and Hall, L.D., Org. Mass Spectrom, 7, 647 (1973).
39. Lin, M.S. and Harrison, A.G., unpublished results.
40. Bieri, G., and Jonsson, G.-O., Chem. Phys. Lett., 56, 446 (1978).
41. Sunner, J., Int. J. Mass Spectrom. Ion Phys., 32, 285 (1980).
42. Tedder, J.M. and Vidaud, P.H., Chem. Phys. Lett., 64, 81 (1979).
43. Tedder, J.M. and Vidaud, P.H., J.C.S. Faraday II, 75, 1648 (1979).
44. Jalonen, J., Tedder, J.M. and Vidaud, P.H., J.C.S. Faraday, II, 76, 1450 (1980).
45. Maccoll, A., Org. Mass Spectrom. 17, 1 (1982).
46. Thomson, B.A., Sakuma, T., Fulford, J., Lane, D.A., Reid, N.M., and French, J.B., Adv. Mass Spectrom., 8, 1422 (1980).
47. Sieck, L.W., Anal. Chem., 51, 128 (1979).

ISOTOPE EXCHANGE IN ION-MOLECULE REACTIONS

David Smith and Nigel G. Adams

Department of Space Research,
University of Birmingham,
Birmingham B15 2TT, England

1. INTRODUCTION

The body of kinetic data relating to appreciably exoergic
ion-molecule reactions at thermal and near-thermal energies has
grown rapidly during the last two decades largely due to the
exploitation of fast flow tube techniques (notably the flowing
afterglow, FA, and the selected ion flow tube, SIFT, techniques)
and of ion trap techniques (notably ion cyclotron resonance,
ICR, techniques). As this body of data increased and as more
complex reactions were studied, the obvious desire arose to
understand the mechanisms of the reactions. To this end,
isotopic labelling of reactants was often used to establish the
origin of the atoms in the product ions and neutrals in order to
distinguish which process or processes were occurring in the
reactions (see for example, Huntress and Elleman, 1970;
Fehsenfeld et al, 1973, 1974a,b; Kim and Huntress, 1975; Smith
and Futrell, 1977). In such studies it was tacitly assumed that
isotopic substitution in the reactants does not significantly
influence the speed or the mechanism of the reactions (i.e. there
are no kinetic isotope effects). We will return to this subject
in Section 2 but only briefly since in this paper we are
concerned largely with a discussion of isotope exchange in near-
thermoneutral ion-molecule reactions, that is in reactions in
which the nature of the reactants is not changed except that the
isotopes within them have been <u>effectively</u> interchanged. Such a
process can be represented by the generalised reaction:

$$^1A^+ + {}^2AB \; \underset{k_r}{\overset{k_f}{\rightleftharpoons}} \; {}^2A^+ + {}^1AB - \Delta H^o \tag{1}$$

M. A. Almoster Ferreira (ed.), Ionic Processes in the Gas Phase, 41–66.
© *1984 by D. Reidel Publishing Company.*

where ^1A and ^2A are isotopically labelled species. The magnitudes of the forward (exoergic) rate coefficients, k_f, and the reverse (endoergic) rate coefficients, k_r, cannot be predicted with any certainty (except perhaps at low temperatures, see Section 3). However k_f and k_r are related via the equilibrium constant, $K(= k_f/k_r)$, and thus to the enthalpy change, ΔH^o, and the entropy change, ΔS^o, for the reaction according to the standard thermodynamic relation:

$$\ln K = \ln \frac{k_f}{k_r} = - \frac{\Delta H^o}{RT} + \frac{\Delta S^o}{R} \qquad (2)$$

(R is the Universal Gas Constant and T is the absolute temperature).

In reactions such as (1) ΔH^o is largely (but not solely) equal to the zero-point-energy (zpe) differences between the reactants and products. In some reactions, especially those involving H/D exchange, ΔH^o can be sufficiently large that, even at room temperature, isotope exchange is strongly inhibited in the endoergic direction. The likely effect of an enthalpy change on a reaction at any particular temperature is more readily appreciated if $\Delta H^o/R$ (in degrees K) is given and so in this paper we often quote $\Delta H^o/R$ in preference to ΔH^o. Kinetic isotope effects become apparent when $\Delta H^o/R$ is an appreciable fraction of the kinetic temperature of the reactants. Obviously, when $\Delta H^o/R$ is small it is necessary to study the reaction at low temperatures to quantitatively investigate kinetic isotope effects and equilibria. The entropy change, ΔS^o, is also an important factor in these reactions and some of the data presented in this paper demonstrates the relationship between ΔS^o and the statistical factors involved in isotope exchange.

The study of the kinetics of isotope exchange reactions can provide insight into the fundamental processes occurring during ion-molecule interactions and can provide information on the factors which determine the course of the reactions (state-to-state correlations, potential surfaces, etc.). Valuable critical data such as the zpe in the species, relative bond strengths, etc. can also be obtained. We will also show how sufficient understanding of these reactions has now been obtained to allow the rate coefficients of many isotope exchange reactions to be predicted under specific conditions. The importance of this advance to the understanding of some aspects of interstellar ion chemistry will be mentioned.

Most of the data referred to in this paper has been obtained using the variable-temperature SIFT technique, although some of the data obtained from FA, ICR and ion-beam experiments will be discussed as appropriate. The SIFT technique has been described in detail in the literature (Adams and Smith, 1976a,b; Smith and

Adams, 1979) and has been applied to the study of a wide range of
ion-molecule reactions at thermal energies in several laboratories
(see the reviews by Smith and Adams, 1979 and Adams and Smith,
1983). One of its advantages over the otherwise similar FA
technique (in which the reactants are also truly thermalised) is
that the ion source gas is excluded from the flow tube in the
SIFT. This allows k_f and k_r to be accurately determined for
reactions such as (1) without complication. Many ICR experiments
suffer from the disadvantage of the simultaneous presence of the
ion source gas and the reactant gas in the reaction cell. As
well as this, they have the additional complication that the ion
energies are not sufficiently well defined to permit accurate
studies of thermoneutral isotope exchange reactions.

In this paper we first discuss briefly some exoergic ion-
molecule reactions in which isotopic labelling has been used and
then consider in some detail the various types of near-
thermoneutral isotope exchange reactions. Finally a summary of
the major features of the reactions is given and some general
conclusions are drawn concerning the nature of isotope exchange
reactions.

2. EXOERGIC REACTIONS OF ISOTOPICALLY LABELLED SPECIES

Isotope labelling has obvious value for the study of
exoergic reactions in which an atomic species is common to both
the reactant ion and the reactant molecule since it can be used
to distinguish which process or processes are occurring in the
reaction. For example, the reaction:

$$NH_3^+ + NH_3 \longrightarrow NH_4^+ + NH_2 \tag{3}$$

can be either proton transfer and/or atom abstraction; both
processes result in the production of NH_4^+. By using ND_3^+ as
the reactant ion in a SIFT experiment (Adams et al, 1980), it has
been shown that at 300 K the reaction proceeds thus:

$$ND_3^+ + NH_3 \longrightarrow NH_3D^+ + ND_2 \quad (85\%) \tag{4a}$$
$$\longrightarrow ND_3H^+ + NH_2 \quad (15\%) \tag{4b}$$

So in this exoergic reaction deuteron (proton) transfer is the
dominant mechanism. It must be stressed again that, in such
exoergic reactions, it is assumed that there are no kinetic
isotope effects, i.e. that reactions (3) and (4) proceed by
exactly similar paths and at the same rates. Certainly the rate
coefficients for both reactions are immeasurably different. This
was also observed to be the case for the reactions of NH^+ and

NH_2^+ with both H_2 and NH_3 and by deuterium labelling these
reactions have also been shown to proceed via parallel proton
transfer and atom abstraction (Adams et al, 1980). In none of
these exoergic reactions does isotopic scrambling occur at thermal
energies; for example in reaction (4) there is no $NH_2D_2^+$ product
ion. This is a common feature of most exoergic reactions implying
either that the intermediate ion-molecule complex does not have a
sufficiently long lifetime for isotopic scrambling to occur or
that there is no simple mechanism by which the bond re-arrangements
and motions of the atoms within the complex can take place. In
some of the near-thermoneutral isotope exchange reactions
discussed in Section 3.4, complete isotope scrambling is evident
in the products and we describe a mechanism by which this may
occur which requires amongst other things that the lifetime of
the intermediate complex is sufficiently long. In a recent SIFT
study (as yet unpublished) we have observed, somewhat surprisingly,
that isotope scrambling occurs in the much-studied normal
exoergic, but almost thermoneutral reaction:

$$CH_4^+ + CH_4 \longrightarrow CH_5^+ + CH_3 \tag{5}$$

This reaction has generally been assumed to proceed via direct
proton transfer and/or H-atom abstraction depending on the
interaction energy; however by observing the products of both the
$CH_4^+ + CD_4$ and $CD_4^+ + CH_4$ reactions, the reaction mechanism is seen
to be much more complex. For example, at 300 K the reaction:

$$CH_4^+ + CD_4 \longrightarrow CH_4D^+(10\%), CH_3D_2^+(22\%), CH_2D_3^+(43\%), CHD_4^+(25\%) \tag{6}$$

clearly proceeds via a long-lived $(C_2H_8^+)^*$ complex in which there
is a large amount of mixing before unimolecular decomposition
occurs to give the observed products. A similar result was
obtained for the reaction:

$$CH_3^+ + CH_4 \longrightarrow C_2H_5^+ + H_2 \tag{7}$$

A near statistical mixture of H and D atoms is observed in the
product $C_2H_5^+$ - like ions at 80 K when either the reactant ion
or reactant neutral are fully deuterated.

A complication which sometimes arises when studying
reactions of polyatomic species is that two or more of the
product ions have the same mass. Isotopic labelling can then
assist in correctly assigning the elemental composition of the
product ion and can again provide information on the reaction
mechanisms. Good examples of such are the exoergic reactions of

N^+, NH^+ and NH_2^+ with C_2H_4. The detailed SIFT study of these
reactions (Smith and Adams, 1980a) well illustrates what can be
achieved using deuterium labelling of reactants. In these
reactions multiple products were observed and to determine the
nature of these products it was necessary to use both the
hydrogenated and deuterated forms of the reactant ions and of the
molecules. Thus, for example, in the reaction of NH^+ with
C_2H_4 it was established that the product ions at mass 28 amu
were both $C_2H_4^+$ and H_2CN^+ and that one of the H atoms in the
H_2CN^+ product ion originated from the reactant ion and one
originated from the reactant neutral:

$$ND^+ + C_2H_4 \rightarrow HDCN^+(20\%), C_2H_3^+(25\%), C_2H_4^+(25\%), C_2H_2^+(10\%),$$
$$H_3CN^+(10\%), H_3C_2N^+(5\%), H_2DC_2N^+(5\%) \quad (8)$$

The complexity of this reaction is evident from the results and
from a discussion of a possible mechanism for it which are given
in the original paper. No evidence for isotopic scrambling was
obtained in any of these reactions.

3.   NEAR-THERMONEUTRAL REACTIONS OF ISOTOPICALLY LABELLED SPECIES

In this section, we are largely concerned with reactions in
which ΔH^o and ΔS^o result only from isotope exchange between the
reactant ions and neutrals. This is in contrast to the normal
exoergic reactions discussed in Section 2 in which new chemical
bonds are formed. For isotope exchange to occur a close ion-
molecule encounter is obviously necessary, that is in general an
intermediate complex ion must form within which bond rearrangement can take place. Also the lifetime of this complex against
unimolecular decomposition, τ_d, must be sufficiently long to
allow at least partial isotopic mixing to occur. Thus τ_d is an
important parameter in determining the efficiency of these
reactions. Estimates of the magnitude of τ_d for the intermediate
complexes formed in several reactions have been obtained from
SIFT studies of the corresponding ternary association reactions
(Adams and Smith, 1982). There are, however, two types of
reactions in which long-lived complexes do not necessarily need
to form but in which nevertheless isotope exchange might be
considered to have occurred. These are symmetrical electron
transfer and symmetrical proton transfer reactions between
isotopically labelled (but otherwise identical) molecules, the
former process possibly occurring at relatively large inter-
nuclear separations. Nevertheless, for such reactions in common
with those requiring complex formation, ΔH^o largely results
from the differences in the zero-point energies (zpe) and
recombination energies of the reactants and products and ΔS^o is
closely related to the statistics of isotopic partitioning

between the reactants and products. This is verified by the
bulk of the data presented here.

3.1 Charge transfer reactions

Symmetrical charge transfer reactions are exemplified by:

$$^{15}N_2^+ + {}^{14}N_2 \rightleftharpoons {}^{14}N_2^+ + {}^{15}N_2 \tag{9}$$

SIFT studies have shown that this reaction is rapid in both
directions and that k_f (9) = k_r (9) = 5.0 (-10) cm^3s^{-1}. Also
both k_f and k_r are significantly greater than half of k_c (the
hard sphere, collisional limiting value; Su and Bowers, 1979) and
are independent of temperature between 80 K and 300 K (Adams and
Smith, 1981a). This indicates that ΔH^0 for this reaction is
very small, as expected on the basis of zpe considerations, and
that a fraction if not all of the interactions proceed via long-
range electron transfer since for a hard (close) collision
$k \approx k_c/2$. No $^{14}N^{15}N^+$ product was observed in this reaction which
also suggests that a close interaction does not occur. Similar
results have been obtained for the CO^+ + CO reaction (Smith and
Adams, 1980b). McMahon et al (1976) have studied reaction (9)
in an ICR experiment and found that k_f (9) = k_r (9) = 6.6 (-10)
cm^3s^{-1}, i.e. somewhat greater than the SIFT values. They also
studied, using isotopic labelling, the symmetrical charge transfer
reactions CO^+ + CO and CO_2^+ + CO_2 and found that in both cases
k_f = k_r $\approx k_c/2$. Similar results were obtained by Dotan (1980)
for the O_2^+ + O_2 reaction.

The reaction of O^+ with O_2 has received considerable
attention because of its ionospheric importance (e.g. see
Ferguson et al, 1979; Smith and Adams, 1980c) and was originally
thought to proceed via exoergic ion-atom interchange rather
than exoergic charge transfer since it is a slow reaction at
300 K. However, in the FA experiments of Fehsenfeld et al (1974a)
only an exoergic charge transfer product was recognized:

$$^{18}O^+ + {}^{16}O_2 \rightarrow {}^{16}O_2^+ + {}^{18}O \tag{10}$$

Neither an ion-atom interchange channel (producing $^{16}O^{18}O^+$) nor
an isotope exchange channel (producing $^{16}O^+$) was observed even
though the charge transfer reaction occurred only on about 1%
of the collisions. However in a recent study of this reaction
using a selected ion flow drift tube (Dotan, 1980) in which
product identification is more straightforward, ion-atom interchange and isotope exchange channels have both been observed,
although over an appreciable energy range, charge transfer
remains the most important process, viz:

$$^{18}O^+ + {}^{16}O_2 \longrightarrow {}^{16}O_2^+ (\sim 75\%), \quad {}^{16}O^{18}O^+ (\sim 15\%), \quad {}^{16}O^+ (\sim 10\%) \quad (11)$$

This implies that at least a proportion of the interactions proceed via a long-lived intermediate complex.

The wide variety of ion-molecule interaction processes involving simple reactant species is further illustrated by the observation that O^+ ions do not charge transfer with NO at low energies even though this process is exoergic by several eV, and that the isotope exchange reaction:

$$^{18}O^+ + N^{16}O \longrightarrow {}^{16}O^+ + N^{18}O \qquad\qquad (12)$$

is also immeasurably slow. Poor Franck-Condon factors have been invoked for the absence of charge transfer in this reaction but if this were so then the reaction of H^+ with NO might also be expected to be slow; however the latter reaction is actually very fast. Fehsenfeld et al (1974a) have considered several reactions including (11) and (12) in terms of state-to-state correlations and conclude that when isotope exchange in these elementary reactions is slow (inefficient) then the potential energy surfaces involved in the reactions must be repulsive in character. A brief discussion of these aspects of isotope exchange reactions has been given by Talrose et al (1979).

3.2. Exchange reactions involving isotopes of carbon, nitrogen and oxygen

Some of the earliest measurements of these reactions were carried out by Fehsenfeld et al (1974a) in a FA. They used the isotopic species $^{15}N^+$ and $^{18}O^+$ to study isotope exchange processes at room temperature in reactions such as:

$$^{15}N^+ + {}^{14}N_2 \longrightarrow {}^{14}N^+ + {}^{14}N^{15}N \qquad\qquad (13)$$

No normal exoergic binary channels are possible in this reaction (charge transfer is endoergic by about 1 eV). The isotope exchange was seen to proceed with a rate coefficient of about 20% of the collisional value i.e. $k(13) \approx 0.2\ k_c$. The same result was obtained for the analogous reaction of $^{14}N^+$ with $^{15}N_2$ consistent with a very small ΔH^o for these reactions and a small τ_d for the N_3^+ intermediate complex. The reaction:

$$^{18}O^+ + C^{16}O \longrightarrow {}^{16}O^+ + C^{18}O \qquad\qquad (14)$$

was observed to proceed more rapidly with $k(14) \approx 0.5\ k_c$ at 300 K and this has also recently been found to be the case at 80 K (Smith and Adams, unpublished). This suggests as expected that ΔH^o is very small and that in this case reaction proceeds via a complex which has a τ_d sufficiently long even at 300 K

to allow for the statistical mixing of the products, and thus
allowing equally for $^{16}O^+$ and $^{18}O^+$ production.

In none of the isotope exchange reactions discussed so far
were both k_f and k_r measured and so ΔH^O and ΔS^O could not be
determined. However, ΔH^O can be calculated for these simple
reactions from a knowledge of the zpe and ionization potentials
of the species, and ΔS^O can be estimated from a consideration
of the statistics of isotopic mixing. Nevertheless it is
obviously desirable to experimentally measure ΔH^O and ΔS^Q and
such requires a study of equilibria in the reactions or
alternatively a determination of k_f and k_r as a function of
temperature. A start was made by Anicich et al (1976) who
measured k_f and k_r in an ICR experiment for the reaction:

$$^{13}C^+ + {}^{12}CO \underset{k_r}{\overset{k_f}{\rightleftharpoons}} {}^{12}C^+ + {}^{13}CO \tag{15}$$

They found under the conditions of their experiment that k_f (15)
= k_r (15) = 2 (-10) $cm^3 s^{-1}$. On the basis of the small ΔH^O for
this reaction, which has been calculated by Watson et al (1976)
to be -0.003 eV or -0.07 kcal mol^{-1} (equivalent to $\Delta H^O/R$ = -35 K),
it is to be expected for temperatures greater than a few hundred
degrees Kelvin that k_f (15) should indeed be equal to k_r (15).
More recently, Smith and Adams (1980b) using a variable-
temperature SIFT have determined k_f (15) and k_r (15) at 80, 200,
300 and 510 K. The data obtained are reproduced in Figure 1 since
they illustrate the trends which are common to the many isotope
exchange reactions which have been studied subsequently, some
of which are discussed below. As can be seen in Figure 1,
k_f (15) and k_r (15) both increase with decreasing temperature
and the sum ($k_f + k_r$) approaches k_c at the lower temperatures.
That ($k_f + k_r$) approaches k_c at the lowest temperatures accessible
to the experiment appears to be a general rule for these isotope
exchange reactions and is an important observation since it
facilitates the calculation of k_f and k_r at the even lower
temperatures of interstellar clouds (this feature of these
reactions and its implications to interstellar cloud chemistry
is discussed by Smith, (1981) and by Smith and Adams, (1981)).
An increasing k_f with decreasing temperature is also a common
feature of many isotope exchange reactions, but a decrease in k_r
must eventually set in at low temperatures for all such
reactions (i.e. when $|\Delta H^O|/R \gtrsim T$) as is indicated for reaction
(15) by the dotted line in Figure 1. At 300 K, k_f (15) and
k_r (15) are measurably different and both are somewhat larger
than the ICR values given above. This may be an indicator that
the ions in this particular ICR experiment were suprathermal.
A van't Hoff plot, i.e. $\ln k_f/k_r$ versus T^{-1}, in accordance with
equation (2), provides mean values for ΔH^O and ΔS^O over the
temperature range of measurement. Such a plot is given in

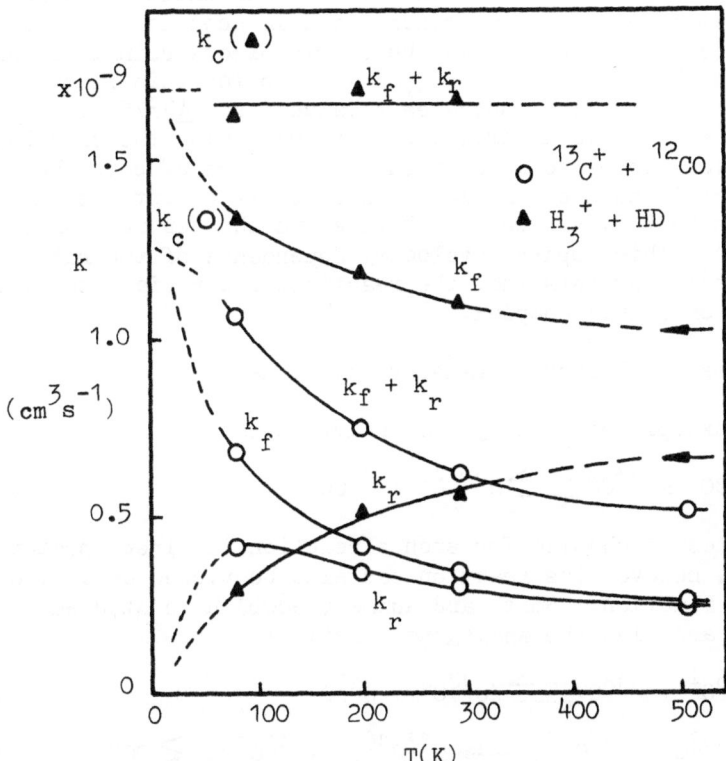

Figure 1. The forward and reverse rate coefficients, k_f and k_r, as a function of temperature for the reactions indicated. Also shown are the corresponding collisional rate coefficients, k_c, for the forward reactions. The dashed parts of the curves at low temperatures are calculated from the respective ΔH^o values assuming that $k_f + k_r = k_c$ in each case. Note that at the higher temperatures k_f ($\blacktriangle$) and k_r ($\blacktriangle$) approach the values predicted on the basis of simple statistics (arrowed) and also indicate that ($k_f(\blacktriangle) + k_r(\blacktriangle)$) remains equal to k_c at the higher temperatures. Actually, ($k_f + k_r$) must eventually become smaller than k_c at high temperatures when τ_d is too small to allow total mixing of H and D atoms in the intermediate $(H_4D^+)^*$ complex. As can be seen, a reduction in $k_f(O)$ and $k_r(O)$ does occur with increasing temperature.

Figure 2 (together with several other plots which are discussed
in the remainder of this section), and is seen to be linear with
an intercept passing through the origin of co-ordinates, which
indicates that within error ΔS° is zero for this reaction.
The slope of the line indicates a value for $\Delta H^{\circ}/R$ of
$-(40 \pm 6)$K which is in excellent agreement with the calculated
value given above and is a strong vindication of both the
experimental and the calculated values. It is interesting to
note that k_f (15) varies as $T^{-\frac{1}{2}}$ (i.e. as the centre-of-mass
velocity). This apparent velocity dependence of the rate
coefficients suggests that the reactions occur via a short-lived
(loosely orbiting) complex.

3.3. Symmetrical proton transfer reactions

An example of this type of reaction is

$$H^{12}CO^+ + {}^{13}CO \rightleftharpoons H^{13}CO^+ + {}^{12}CO \qquad (16)$$

The simplest mechanism for such a reaction is direct proton
transfer, however the reaction can also be viewed as an isotope
exchange reaction. Smith and Adams (1980b) have studied this
reaction and also the analogous reactions:

$$HC^{16}O^+ + C^{18}O \rightleftharpoons HC^{18}O^+ + C^{16}O \qquad (17)$$

$$H^{12}C^{18}O^+ + {}^{13}C^{16}O \longrightarrow H^{13}C^{16}O^+ + {}^{12}C^{18}O \,(\gtrsim 90\%) \qquad (18a)$$

$$\longrightarrow H^{13}C^{18}O^+ + {}^{12}C^{16}O \,(\lesssim 10\%) \qquad (18b)$$

The appearance of the minor channel (18b) indicates that some
isotopic mixing does occur but the overriding indication is that
these reactions do indeed proceed very largely via proton transfer.
The general features of the kinetic data obtained for reactions
(17) and (18) are very similar to those obtained for the $C^+ + CO$
reaction (see Figure 1) in that both k_f and k_r increase with
decreasing temperature over the experimentally available
temperature range, although the differences between the k_f and
k_r for all of these reactions are much smaller than those for
reaction (15) at all temperatures indicating much smaller ΔH°
values as is evident from the slope of the van't Hoff plot given
in Figure 2 ($\Delta H^{\circ} \sim -0.001$ eV or ~ -0.03 kcal mol^{-1}; see
Smith and Adams, 1980b). The ΔH_0° for reactions (16) and (17)
(and reaction (19) below) have been estimated by Henning et al
(1977) from their calculations of the zpe in the HCO$^+$ and N$_2$H$^+$
ions and they are close to the experimental values. The
analogous proton transfer reaction:

$${}^{14}N_2H^+ + {}^{15}N_2 \rightleftharpoons {}^{15}N_2H^+ + {}^{14}N_2 \qquad (19)$$

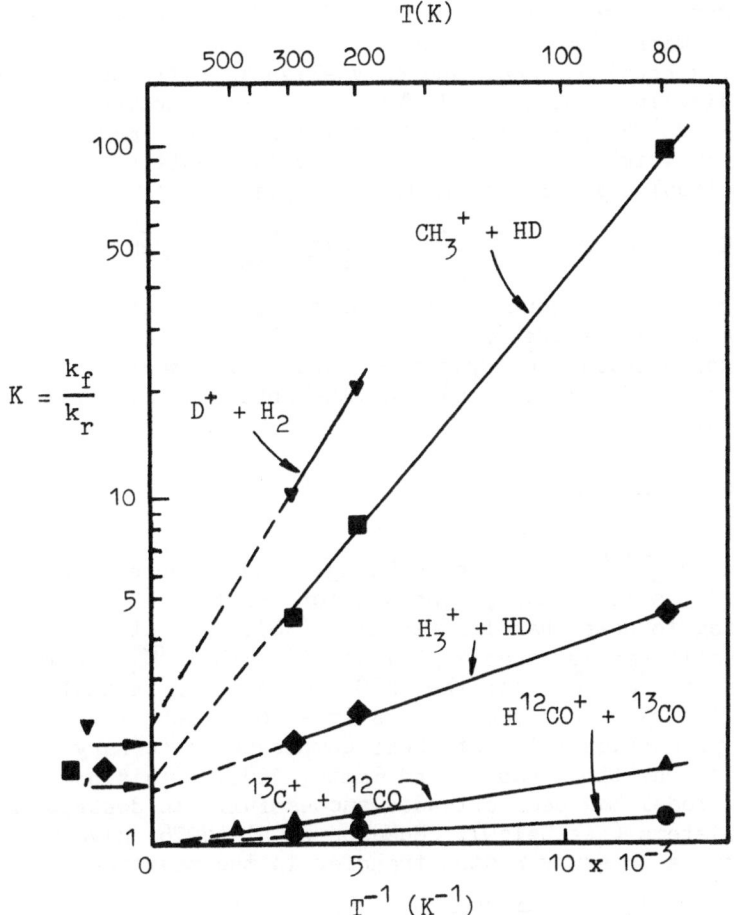

Figure 2. Van't Hoff plots for the reactions indicated. The $\Delta H^\circ/R$ values derived from the slopes of the lines range from about -10 K to -464 K for the reactions indicated by ● and ▼ respectively. Note that ΔS° is zero for the reactions indicated by ● and ▲ and is finite for the reactions indicated by ◆ , ■ and ▼ . The intercepts indicated by the arrows are those expected purely from the statistical decomposition of well-mixed intermediate complexes.

and the similar reactions involving $^{14}N^{15}N$ have been studied by
Adams and Smith (1981a). In these reactions, k_f and k_r are
essentially equal at 300 K and only slightly different at 80 K,
again indicating a very small ΔH^o for these reactions. Also k_f
(19) $\approx k_r$ (19) $\approx k_c/2$, which is consistent with a model of the
reaction in which a proton is envisaged to 'shuttle' between the
two N_2 molecules in an intermediate complex, viz:

$$^{14}N_2H^+ + {}^{15}N_2 \rightleftarrows ({}^{14}N_2 {-}{-}{-} \overset{+}{H} {-}{-}{-} {}^{15}N_2) \rightleftarrows {}^{14}N_2 + H^+{}^{15}N_2 \qquad (20)$$

If τ_s is the 'mean shuttle time' of the proton between the N_2
molecules and as before τ_d is the mean lifetime of the complex
against unimolecular decomposition then it is readily shown
(Pellerite and Brauman, 1980) that the observed rate coefficient,
k_{obs}, is given by:

$$k_{obs} = \frac{k_c}{2 + \tau_s/\tau_d} \qquad (21)$$

Thus when $\tau_d \ggg \tau_s$, i.e. for a "long-lived" complex, then
$k_{obs} \approx k_c/2$ as is actually observed for reaction (19). It is
interesting to note that for reaction (16), k_{obs} at 300 K is
significantly smaller than $k_c/2$ indicating that τ_d (or more
specifically τ_d/τ_s) for the (HCO$^+$.CO) complex is smaller than
that for the (N$_2$H$^+$.N$_2$) complex, as is also evident from the
relative production rates of these complexes in ternary
association reactions (Adams and Smith, 1982). This proton
shuttling model has been extended successfully to describe more
complex systems (see below). McMahon et al (1976) have used an
ICR apparatus to study proton transfer in the reaction:

$$^{14}NH_4^+ + {}^{15}NH_3 \rightleftarrows {}^{15}NH_4^+ + {}^{14}NH_3 \qquad (22)$$

and also find that $k_f = k_r \approx k_c/2$.

3.4 Hydrogen /deuterium exchange reactions

The change in the vibrational zpe of a molecule is relatively
large when an H/D exchange occurs because of the large mass change
and thus the correspondingly large change in the bond vibrational
frequencies. Therefore it is expected that enthalpic effects will
be more pronounced in H/D exchange reactions. In this section,
this is shown to be the case and several "reactive systems" are
discussed varying in complexity from the simple triatomic
$H^+ + H_2$ system to the eleven-atomic $CH_5^+ + CH_4$ system.

3.4.1. The $H^+ + H_2$ system. It is to be expected that
enthalpic effects will be very pronounced in the elementary
reactions:

$$D^+ + H_2 \rightleftharpoons H^+ + HD \qquad \Delta H_0^o/R = -462 \text{ K} \qquad (23)$$

$$D^+ + HD \rightleftharpoons H^+ + D_2 \qquad \Delta H_0^o/R = -541 \text{ K} \qquad (24)$$

The values given for $\Delta H_0^o/R$ were calculated from the accurately known bond energies of H_2, HD and D_2 and the recombination energies of H^+ and D^+ (Huber and Herzberg, 1979; Rosenstock et al, 1977). There have been several experimental and theoretical studies of these simplest-of-all chemical reactions, but most experimental studies have been carried out using ion beams at collision energies in excess of 1 eV. However, Fehsenfeld et al (1974a,b) using a FA have measured k_f (23) and k_r (24) at 80, 200 and 278 K and found, as expected, that k_f (23) $>$ k_r (24) and that k_f (23) was only about a factor of two smaller than k_c. Unfortunately k_r (23) and k_f (24) were not measured in the FA. Recently however, Henchman et al (1981) have measured k_f and k_r for both reactions (23) and (24) at 200 and 300 K in a SIFT. At 200 K, k_f (23) was within error equal to k_c whereas k_r (23) was some twenty times smaller, reflecting the large ΔH^o for the reaction. Equilibrium constants derived from the data agree well with theoretical van't Hoff plots calculated using statistical mechanics and the derived ΔH^o values were in excellent agreement with the calculated values given above. The magnitude of ΔS^o obtained for each reaction is a clear indication of the influence of statistical factors and symmetry numbers in the kinetics of these reactions. The effect of statistical factors is also clearly evident when k_f or k_r for reactions (23) and (24) are compared that is the rate coefficient obtained when HD is the reactant is a factor of about 2 smaller than with H_2 or D_2 as reactant as expected statistically. The van't Hoff plot for reaction (23) is given in Figure 2 and indicates the finite entropy change in the reaction ($\Delta S^o \approx R\ln 2$). The trend of increasing rate coefficient with decreasing temperature for the exoergic reactions (referred to in respect of the isotope exchange involving heavy elements which was discussed above) was also evident in these reactions and has been tentatively explained in terms of the partitioning of energy within the system according to the number of states available to the products and to the reactants (see Henchman et al, 1981). k_f (23) and k_r (24) have recently been measured at relative energies in the range 0.04 to 0.3 eV using a drift tube apparatus (Villinger et al, 1982; see also Lindinger and Smith, 1983).

3.4.2. The $H_3^+ + H_2$ system. This system has received considerable attention in several laboratories. It has been studied by partially or totally deuterating both the reactant ion and neutral in ICR experiments (McMahon et al, 1976; Smith and Futrell, 1975, 1976), in ion beam experiments (see the reviews by Gentry, 1979a,b, 1981) and in flow tube experiments (Adams and Smith, 1981b, Smith et al 1982a; Smith and Adams, 1983).

From the large amount of data obtained on these reactions, it is appropriate only to select a few examples to illustrate the essential features.

In the complete series of $H_3^+ + H_2$ reactions, ten different reactions can be studied. In <u>four</u> of these reactions only one binary product ion is possible whilst in the remaining <u>six</u> reactions two binary product ions are possible. Amongst the group of four, the reaction:

$$H_3^+ + HD \rightleftharpoons H_2D^+ + H_2 \qquad (25)$$

is of particular interest because of the rôle it is thought to play in the deuteration of interstellar molecules (see the reviews by Watson 1976, 1977, Dalgarno and Black, 1976, Winnewisser et al, 1979, Smith and Adams, 1981). An estimate of the rate coefficient for this reaction of 3 (-10) cm^3s^{-1} was given by Huntress and Anicich (1976) based on an ICR measurement of the rate coefficient for the deactivation of excited H_3^+ by H_2 (Kim et al, 1974). Recently however, direct measurements of both k_f (25) and k_r (25) have been made by Adams and Smith (1981b) in a SIFT at 80, 200 and 292 K and these give a value for k_f (25) which is several times larger than the ICR estimate. The rate coefficients obtained are shown graphically in Figure 1 and the familiar trends are again evident i.e. an increasing k_f and a decreasing k_r with decreasing temperature. Note that the forward (exoergic) direction is that resulting in an increase in the degree of deuteration of the ion (this is the case in all of these $H_3^+ + H_2$ reactions) and also note that k_f (25) at 80 K is a large fraction of k_c. The van't Hoff plot constructed from these data is shown in Figure 2 from which a mean value of $\Delta H^0/R$ of $-(90 \pm 10)$ K was obtained. This value must be compared with the value calculated from the known vibrational zpe in the reactants and products and which also takes into account the rotational energy content of each species, including the so-called 'rotational zpe' of H_3^+ (Porter, 1977; Carney and Porter, 1977). Thus at 300 K, $\Delta H^0/R$ is calculated to be -141 K. However, due to the relatively large spacings of the quantized rotational levels in these small molecules, $\Delta H^0/R$ varies appreciably with temperature below about 200 K having a minimum value of -117 K at a temperature of about 105 K (Smith et al, 1982a). Since the experimentally determined value of $\Delta H^0/R$ given above is strongly weighted by the 80 K data point, then the experimental $\Delta H^0/R$ should more closely approximate to the calculated value for 80 K which is -121 K. That the experimental value is significantly smaller than this calculated value has been attributed to the use of 'normal H_2' in the experiment which at 80 K has an excess of the ortho-H_2 above the equilibrium ortho-H_2 to para-H_2 ratio (for further discussion of this see Smith et al, 1982a). It is very important to be cognizant of such quantum effects when studying reactions of small molecules at low temperatures; such effects are usually of no consequence at temperatures above about 200 K in small molecules and at even lower temperatures in polyatomic species.

Since the calculations for reaction (25) show that ΔH^O varies somewhat over the temperature range for which the SIFT measurements were made, it is obviously not strictly valid to construct a linear van't Hoff plot using the experimental data for this reaction. However, since the variation in ΔH^O is not large and, in view of the reasonable agreement between the experimental and the calculated values of ΔH^O, then the procedure is not without value. Notwithstanding this unavoidable approximation, it is clearly evident from Figure 2 that there is a finite entropy change in reaction (25), the value obtained from the intercept of the line with the ln k_f/k_r axis indicating that $\Delta S^O \approx R \ln 1.4$. This is remarkably close to the value predicted simply on the basis of the statistical production of H_2D^+ and H_3^+ from the $(H_4D^+)^*$ intermediate complex (i.e. $H_2D^+ : H_3^+ :: 6:4$) which is R ln 6/4. This is again a clear indicator of the relationship between entropy change and statistical factors in these isotope exchange reactions (a detailed statistical mechanical treatment gives $\Delta S^O = R \ln 1.32$).

An example of a reaction for which two exoergic product channels are available is:

$$H_3^+ + D_2 \longrightarrow HD_2^+ + H_2 \qquad \Delta H^O/R = -241 \text{ K} \qquad (26a)$$

$$\longrightarrow H_2D^+ + HD \qquad \Delta H^O/R = -57 \text{ K} \qquad (26b)$$

The values of $\Delta H^O/R$ quoted here for this and for the following reactions in the $H_3^+ + H_2$ system which are discussed in this paper, have been calculated from the vibrational zpe only and do not include a contribution due to any differences in the rotational energy contents between the reactants and products. Statistics (entropy) favour H_2D^+ (67%) production which can be viewed as an atom exchange or isotopic scrambling process, whereas HD_2^+ (33%) production can be viewed as the simpler, more direct process of proton transfer. Douglass and Gentry (see Gentry 1979a, b, 1981) have studied this reaction in a merged beam experiment and observed only channel (26a) to occur with a high reaction probability (0.7) at low interaction energies. They suggest therefore that the direct mechanism is favoured because it can occur on a time scale which is short compared to that required for isotopic scrambling in a complex. In a recent detailed SIFT study of the reactions of H_3^+ and its partially and totally deuterated analogues with H_2, HD and D_2 at 80 and 295 K, Smith and Adams (1983) have shown that at both of these relatively low temperatures, channel (26a) is indeed

dominant and becomes increasingly so at higher temperatures, but
the channel (26b) is also significant ($\sim$10% at 295 K). The
overall rate coefficient for the reaction is large, essentially
equal to k_c at 80 K. These results also seem to support the
view that the direct mechanism is favoured,$_*$but the appearance
of channel (26b) indicates that the $(H_3D_2^+)$ complexes have
sufficiently large τ_d, particularly at the lowest temperatures,
to allow some H/D exchange to occur within a loosely-bound
proton-bound dimer (as discussed in Section 3.3). Under such
circumstances, at the lowest temperatures the most exoergic
channel is expected to be favoured whereas the direct mechanism
would be favoured at the higher temperatures where enthalpic
effects are unimportant. Coincidentally in this reaction both
circumstances favour the same product ion, HD_2^+. It is important
to note that the H_3^+ ions in the ion beam experiments were
vibrationally excited, which presumably reduces τ_d for any
intermediate complex formed and hence decreases the probability
of H/D exchange (recognised by Gentry (1979a,b)).

The reaction:

$$HD_2^+ + HD \longrightarrow D_3^+ + H_2 \qquad \Delta H^o/R = -230 \text{ K} \qquad\qquad (27a)$$

$$\longrightarrow H_2D^+ + D_2 \qquad \Delta H^o/R = + 105 \text{ K} \qquad\qquad (27b)$$

is interesting. Again there are two distinguishable product
channels, but in this case one is exoergic and one is endoergic.
Clearly D_3^+ formation involves an isotope exchange and presumably
the formation of an intermediate complex. Predictably, therefore,
D_3^+ is only a minor product in the ion beam experiment even at
low energies (Gentry 1979a,b, 1981) and is essentially absent
at higher energies. However this is not the case in the SIFT
experiments (Smith and Adams, 1983). The D_3^+ product channel
is the major channel (80%) at 80 K and is still a very
significant product channel (35%) at 295 K. Note also that
the total rate coefficient for the reaction is 8 (-10) cm^3s^{-1},
that is only about half of k_c at 80 K and is appreciably smaller
(5(-10) cm^3s^{-1}) at 295 K. These data can be explained in the
following way. The H_2D^+ ion is a small product relative to D_3^+
at 80 K in spite of the fact that it is strongly favoured
entropically and also that it can be formed by a direct proton
transfer. This is simply because H_2D^+ production in this reaction
is endoergic with $\Delta H^o/R$ greater than the temperature at which
the reaction is occurring, whereas D_3^+ production is very exoergic.
In effect at this temperature enthalpic effects swamp entropic
effects. At the higher temperature of 295 K, $\Delta H^o/R$ is less
than the reaction temperature and is therefore less of a
controlling factor. Hence, the H_2D^+ product channel becomes
relatively more important. That the total rate coefficient is
appreciably smaller than k_c even at 80 K reflects the high

probability for the regeneration of the reactant HD_2^+ ions.

Finally in this brief selection of $H_3^+ + H_2$ reactions, it is worth discussion the reaction:

$$D_3^+ + H_2 \longrightarrow H_2D^+ + D_2 \qquad \Delta H^\circ/R = 335 \text{ K} \qquad (28a)$$

$$\longrightarrow HD_2^+ + HD \qquad \Delta H^\circ/R = 230 \text{ K} \qquad (28b)$$

Both of the product channels are significantly endoergic and statistics favour the production of HD_2^+, yet the H_2D^+ can be formed by the simpler deuteron transfer mechanism. Smith and Futrell (1976) have studied this reaction in a <u>tandem</u> ICR experiment and were able to use a high pressure of D_2 in the ion source thus vibrationally relaxing the D_3^+ reactant ions (a procedure also adopted in the SIFT experiments). Under these conditions they observed both H_2D^+ (70%) and HD_2^+ (30%). Thus, at the ion energies of this ICR experiment, the direct deuteron transfer process is again favoured although, perhaps significantly for the vibrationally relaxed D_3^+ reactant ions, the atom exchange channel (which requires a longer interaction period) was also evident. The total rate coefficient for the reaction was determined to be $4(-10) \text{ cm}^3\text{s}^{-1}$ and this is about a factor of two smaller than the 295 K SIFT value ($8.8(-10) \text{ cm}^3\text{s}^{-1}$). This is probably a manifestation of kinetically-excited ions in the ICR experiment. Note also that in the SIFT experiment the H_2D^+ ion is the favoured product (60%) at 295 K (c.f. the ICR result above), whereas at 80 K the HD_2^+ product channel (65%) is favoured, presumably because it is the least endoergic channel and also because it is favoured entropically. These effects will be more evident when the temperature is sufficiently low such that an intermediate complex of appreciable τ_d can form. At the higher temperatures (and energies) the direct mechanism of proton or deuteron transfer will be the more likely mechanism.

3.4.3. <u>The $CH_3^+ + H_2$ system</u>. A detailed SIFT study has been made of isotope exchange in this system (Smith et al, 1982b). It is interesting for several reasons. The CH_5^+ ion is very stable even at high temperatures and readily forms in collisional association reactions of CH_3^+ with H_2 at relatively low pressures which indicates that the excited complex $(CH_5^+)^*$ has a relatively large τ_d. This suggests that isotope exchange in the complex might be very efficient since there should be no constraints due to complex lifetimes, yet, counter to this, a mechanism for isotope exchange is not obvious since in this case the intermediate complex cannot for example be seen as a symmetrical proton–bound dimer. The intermediate complex however apparently has a unique structure. Pople (1976) has indicated that the most stable structure of CH_5^+ contains a three–centred bond i.e. $[H_3C \longrightarrow\!\!\!< H_2]^+$ and this may facilitate isotope exchange.

Of the several reactions in the series, one of special interest is:

$$CH_3^+ + HD \rightleftharpoons CH_2D^+ + H_2 \tag{29}$$

since it is thought to be another important route to the deuteration of interstellar molecules (Watson, 1976; Smith and Adams, 1981). This reaction together with the other reactions of CH_3^+ and its partially and totally deuterated analogues with H_2, HD and D_2, was studied in a SIFT at 80, 205 and 295 K. k_f (29) was measured to be close to k_c at the lowest temperatures as is now expected. However for the reverse reaction, in addition to the isotope exchange channel, a parallel ternary association channel was observed at each of the three temperatures, i.e.:

$$CH_2D^+ + H_2 + He \longrightarrow CH_4D^+ + He \tag{30}$$

In this association reaction, the helium carrier gas atoms acted as the stabilizing third body for the long-lived, excited intermediate complex $(CH_4D^+)^*$. The association ion CH_4D^+ became an increasing fraction of the product ion distribution with decreasing temperature as the importance of the endoergic isotope exchange channel decreased. Parallel isotope exchange and association channels are a common feature in this series of reactions (specifically when the isotope exchange is endoergic). Fortunately it is apparent that under the SIFT experimental conditions, the parallel binary and ternary channels can be viewed as independent reactions and their separate rate coefficients can be obtained from the measured total rate coefficient and the product ion distribution (justification for this procedure is given by Smith et al, 1982b).

The forward and reverse rate coefficients for the isotope exchange channels of the reactions in this $CH_3^+ + H_2$ system have been used to construct a series of van't Hoff plots from which values of ΔH^o and ΔS^o have been determined. The plot corresponding to reaction (29) is given in Figure 2 and the slope indicates a mean $\Delta H^o/R$ of -335 K. This is somewhat greater than the value of -290 K derived by Blint et al (1976) from their calculations of the vibrational zpe in CH_3^+ and CH_2D^+ and the known zpe in HD and H_2. This large value of $\Delta H^o/R$ ensures that at low temperatures k_r is very small compared to k_f (and hence k_c). Actually, the difference in the zpe in the CH_3^+ and CH_2D^+ ions is itself of fundamental interest since it is a direct indicator of the difference in the strengths of the C–H and C–D bonds in CH_3^+ - like ions. Also van't Hoff plots constructed from the data for several other reactions in the series indicated a consistent value for the difference in the C–H and C–D bond strengths in CH_3^+-like ions independent of the degree of deuteration of the ions. The bond energy difference derived from these data is 0.068 eV (or 1.56 kcal mol^{-1}), only about 10%

lower than that calculated by Blint et al – remarkably good
agreement. It is interesting and important to note that this
energy difference is significantly greater than the difference
in the bond energies in the CH and CD radicals of 0.047 eV which
is obtained spectroscopically (Huber and Herzberg ,1979) and which
is often assumed to apply to C–H and C–D bonds in hydrocarbons
regardless of their complexity and structure.

The ΔS^o values indicated for these reactions from the
intercepts on the van't Hoff plots are again close to those
predicted on the basis of the statistical decomposition of the
intermediate CH_5^+-like ions. This and the many other interesting
features of the kinetics of these reactions have been discussed
by Smith et al (1982a,b).

3.4.4. <u>Systems involving polyatomic ions and neutrals.</u>
For H/D exchange reactions involving polyatomic neutrals, it is
to be expected that enthalpy changes will be much smaller than
for reactions involving H_2, HD or D_2 as the reactant neutral.
This is certainly the case for the H_3O^+ + H_2O system. The
reaction:

$$H_3O^+ + D_2O \longrightarrow H_2DO^+ + HDO \qquad \Delta H^o/R \approx + 50 \text{ K} \qquad (31a)$$

$$\longrightarrow HD_2O^+ + H_2O \qquad \Delta H^o/R \approx + 100 \text{ K} \qquad (31b)$$

and the mirror isotope exchange reaction D_3O^+ + H_2O have been
studied at 300 K in a SIFT (Smith et al, 1980). Both reactions
occur very rapidly, the rate coefficients being close to the
respective k_c values, i.e. no kinetic isotope effects were
detected at 300 K. This indicates that $|\Delta H^o|$ /R $<$ 300 K for
these reactions and this has recently been shown to be the case
by calculation (Henchman et al, 1982). The most interesting
and(at the time of the study) the most surprising result was the
appearance of two primary products in the ratio of H_2DO^+ :
HD_2O^+ :: 2:1, which is precisely in accordance with the
statistical decomposition of a well-mixed $(H_3D_2O_2^+)^*$ complex.
This has been explained in terms of the sequential 'shuttling'
within a proton bound dimer of protons and deuterons between the
two water molecules which are also able to rotate:

$$H_3O^+ + D_2O \rightarrow H_3O^+ . D_2O \xrightleftharpoons{H^+} H_2O.HD_2O^+ \xrightleftharpoons{D^+} H_2DO^+.HDO \qquad (32)$$

$$\downarrow \qquad\qquad\qquad \downarrow \qquad\qquad\qquad \downarrow$$

$$H_3O^+ \qquad\qquad HD_2O^+ \qquad\qquad H_2DO^+$$

Total scrambling can occur because of the relatively long lifetime,
τ_d, against unimolecular decomposition of the $(H_3D_2O_2^+)^*$ complexes.
The rate coefficients, k_{obs}, for the reactions have been parameterized by Adams et al (1982) in terms of τ_d and of the

shuttle times, τ_s, of the protons and deuterons in the complexes. In the absence of enthalpic effects (i.e. for $\Delta H^o/R \ll T$) then:

$$k_{obs} \approx \frac{k_c}{1 + \tau_s/\tau_d} \qquad (33)$$

The numerical constant in the denominator differs from that in equation (21) since the statistical probability for unimolecular decomposition of the intermediate complex back to the reactants is much less than unity for complex systems, whereas for the simple system represented by equation (20) it is actually 0.5. Thus when $\tau_d \gg \tau_s$, as is expected for the strongly-bonded $(H_3D_2O_2^+)^*$ complexes formed in reaction (32), then $k_{obs} \approx k_c$ as is actually observed for reaction (31) at 295 K. k_{obs} for this endoergic reaction will decrease below k_c at lower temperatures (as enthalpic effects become important) and also at higher temperatures where τ_d becomes small (or more correctly, where $\tau_d \lesssim \tau_s$). Also at higher temperatures τ_d will be too short for statistical mixing of the H and D atoms within the complex before it separates to the products. Under these circumstances, the product of the first stage of the shuttle chain, HD_2O^+, will become an increasing fraction of the product distribution.

A similar study has been made of the reactions $NH_4^+ + NH_3$ by Adams et al (1982). For the exoergic $NH_4^+ + ND_3$ reaction the three possible product ions (NHD_3^+, NH_3D^+, $NH_2D_2^+$) are approximately in their statistical distribution at reaction temperatures of 204, 295 and 475 K and $k_{obs} \approx k_c$ at 204 K decreasing to about 0.8 k_c at 475 K. This is again interpreted as being due to a reduction in τ_d/τ_s with increasing temperature. For the endoergic $ND_4^+ + NH_3$ reaction k_{obs} is somewhat smaller ($\approx 0.7 k_c$) at all three temperatures due to the appreciable ΔH^o for these reactions (calculated by Henchman et al, 1982) and changes in the product distribution occur at the lowest temperature in accordance with the different ΔH^o for each reaction channel.

The effects observed in the $NH_4^+ + NH_3$ system are much more pronounced in the $CH_5^+ + CH_4$ system. The exoergic reaction:

$$CD_5^+ + CH_4 \longrightarrow CH_4D^+,\ CHD_4^+,\ CH_3D_2^+,\ CH_2D_3^+ \qquad (34)$$

and the endoergic reaction:

$$CH_5^+ + CD_4 \longrightarrow CHD_4^+,\ CH_4D^+,\ CH_2D_3^+,\ CH_3D_2^+ \qquad (35)$$

have been studied at 80, 204, 295 and 475 K. At the highest temperature, $k(34) = 9.4(-11)$ cm^{3}s^{-1} and $k(35) = 1.6(-10)$ cm^{3}s^{-1} which are both much less than k_c (= 1.2(-9) cm^{3}s^{-1}). This is consistent with the very small τ_d (about 2 orders of magnitude smaller than for the $NH_4^+ + NH_3$ system) for the weakly-bonded intermediate complexes formed in these reactions (again τ_d was estimated from parallel studies of the $CH_5^+ + CH_4 + He \longrightarrow$ $CH_5^+.CH_4 + He$ association reaction). It is also significant that the major product ions for reactions (34) and (35) at the highest temperature were CH_4D^+ and CHD_4^+ respectively, which is consistent with a shuttling model analogous to reaction (32). Also the rate coefficients for these reactions at this high temperature are in the ratio $k(34): k(35)::1: \sqrt{2}$ ($\approx \sqrt{m_{H+}/m_{D+}}$) which is perhaps indicative of the different τ_s for protons and deuterons within the complexes. As the temperature decreases then the rate coefficients both increase towards k_c, reaching about 0.6 k_c at 80 K, although k(35) increases less rapidly as enthalpic effects become significant. Also major changes in the product distributions occur with temperature, again due to the different ΔH^o for each product channel. Detailed analysis of these data has resulted in an estimate of the zpe of CH_5^+ (Henchman et al, 1982).

The influence of τ_d on the rate of isotope exchange is also evident from a study of reactions in which no exchange is detected at thermal energies as, for example, in the reactions of HCO^+, H_2CN^+ and NH_4^+ with HD and D_2 (Huntress, 1977; Adams and Smith, 1981b). It is known from ternary association reaction studies that the bonding in the complexes formed in these reactions, e.g. $HCO^+.H_2$, is very weak and hence very small τ_d values are indicated for them which reduces the probability of isotope exchange. This correlation between τ_d and the efficiency of isotope exchange can be very useful when considering to what extent isotope exchange will occur in ion-molecule reactions. However other factors can exercise a controlling influence on isotope exchange reactions as is evident, for example, from the $N_2H^+ + N_2$ reaction (19) for which isotope (proton) exchange is rapid even though τ_d is small. Herein lies another clue to the nature of these reactions; the $N_2H^+ + N_2$ system is "symmetrical" i.e. the proton is able to shuttle between the two N_2 molecules without severe hindrance due to barriers on the potential surface (see Brauman, 1979). However in the reactions cited here for which isotope exchange does not occur, proton shuttling will be inhibited between the unlike molecules which have very different proton affinities and this can be viewed as due to the presence of large barriers. These and other factors which influence the efficiency of isotope exchange are discussed by Adams and Smith (1982).

3.5. Negative ion reactions

Although most of the studies of isotope exchange in ionic reactions have involved positive ions, a significant amount of data is becoming available relating to H/D exchange in <u>negative</u> <u>ion</u> reactions. Of special note is the work of DePuy and Bierbaum and their colleagues who are using FA and SIFT techniques to study the mechanisms of gas phase negative ion reactions by observing H/D exchange. Stewart et al (1977) have observed that a wide variety of organic negative ions participate in sequential deuterium exchange reactions with D_2O e.g.:

$$C_3H_5^- \xrightarrow{D_2O} C_3H_4D^- \xrightarrow{D_2O} C_3H_3D_2^- \xrightarrow{D_2O} C_3H_2D_3^- \xrightarrow{D_2O} C_3HD_4^- \xcancel{\xrightarrow{D_2O}} \quad (36)$$

Clear variations in the extent of deuteration were evident (note, for example, that the $C_3HD_4^-$ does not exchange the remaining H atom with D_2O) and such observations are leading to an understanding of the mechanisms of the reactions. These studies are also useful as probes of the acidity of the negative ions and this aspect is stressed in a subsequent paper (DePuy et al, 1978) in which the reactions of several organic negative ions with CH_3OD and CF_3CH_2OD are discussed. In a more recent paper, DePuy and Bierbaum (1981) discuss the general features of these types of reactions and indicate the extent to which these gas phase studies of organic reactions compliment and help to elucidate reactions of organic ions in the solution phase.

Other recent work on isotopic labelling of negative ions has been carried out by Fahey et al (1982) who used oxygen isotopes to elucidate the mechanisms of the exoergic reactions of $O_2^- \cdot (H_2O)_n$ with O_3. No oxygen scrambling occurred in these interactions, the mechanism being charge transfer with the simultaneous transfer of water molecules. Much more work could be carried out using isotope labelling to understand the mechanisms of negative ion-molecule reactions.

4.   SUMMARY AND CONCLUSIONS

In this paper we have been concerned with the kinetic and thermodynamic aspects of isotope exchange in ion-molecule interactions at truly thermal energies. The majority of the data discussed has been obtained by the exploitation of a variable-temperature SIFT apparatus. The data have led to a clearer understanding of the mechanisms of some of these reactions and show how the kinetic and thermodynamic aspects of these ionic reactions are related.

Perhaps the result most significant to the kinetics of isotope exchange revealed by the body of data is that without exception the forward (exoergic) rate coefficients, k_f, for these isotope exchange reactions increase towards their collisional-limiting values, k_c, at low temperatures i.e. when $T < |\Delta H^0|/R$.

Since the corresponding reverse (endoergic) rate coefficient, k_r, must obviously decrease with decreasing T in this temperature regime, then it follows that $k_f + k_r \rightarrow k_c$ at low temperatures. The recognition of this has greatly facilitated the predictions of rate coefficients for isotope exchange reactions at lower temperatures than are at present generally accessible in laboratory experiments. This has been particularly valuable in elucidating some aspects of interstellar ion chemistry.

The thermodynamic quantities ΔH^O and ΔS^O have been derived for many reactions from the kinetic data and, when these quantities are calculable, then the calculated and experimental values are in very good agreement. The ΔH^O values are dependent on differences in the vibrational zero-point energies and rotational energies of the reactant and product ions and neutrals. Quantum restrictions inevitably result in temperature dependent ΔH^O values at low temperatures. Amongst the reactions discussed in this paper, this phenomenon is particularly evident in the $H_3^+ + HD \rightleftharpoons H_2D^+ + H_2$ reaction (25), principally because of the large "rotational zpe" of H_3^+ but also because of nuclear spin restrictions (ortho/para forms) of H_2. For reactive systems involving polyatomic species, calculations of ΔH^O are either impossible or excessively difficult and time consuming, and for such reactions SIFT experiments have much to offer. For example, as referred to in this paper, the differences in the C–H and C–D bond strengths in CH_3^+-like ions have been derived from SIFT kinetic data and similarly the zpe of CH_5^+ has been estimated.

We have described many of these isotope exchange reactions in terms of the "shuttling" of protons or deuterons within loosely-bound complexes but which have appreciable lifetimes against unimolecular decomposition, τ_d. This mechanism is most readily envisaged in symmetrical proton-bound dimer intermediate complexes in which the proton shuttles with a characteristic time, τ_s, between two similar molecules which are also able to freely rotate. Estimates of τ_d can be obtained from studies of ternary association reactions and these estimates have been obtained for the majority of the reactions discussed in this paper (Adams and Smith, 1982). Thus it is generally observed that when τ_d is large, that is for relatively strongly-bonded complexes, then isotope exchange efficiently occurs (more correctly, the efficiency of isotope exchange, k_{obs}/k_c is parameterized in terms of τ_d and τ_s; see equations (21) and (33)). The distribution of the isotopes amongst the products of these reactions is determined by the temperature and ΔH^O and ΔS^O for each reactive channel. ΔS^O is clearly related to the statistics of mixing of the isotopes within the intermediate complex (or more precisely, to the partition functions of the reactants and products). Clearly this description of these reactions essentially involves the ideas of statistical theory

and transition state theory. Brauman (1979) has succinctly
stated these ideas in a somewhat more formal way in terms of
generalised potential surfaces (represented by reaction
coordinate diagrams). These reactions are seen to occur on
potential surfaces which in some cases can be very complex
containing several minima each relating to a possible
configuration of the transition state (intermediate complex).
The intervening potential barriers between these states exercise
an important and in some cases major influence on the course of
the reactions. This approach to isotope exchange reactions at
thermal energies has many appealing features and can explain
qualitatively the general features of these reactions. However
quantitative descriptions of the kinetics of these reactions
remain to be formulated.   Until this is achieved, further
exploitation of apparatuses such as the variable-temperature
SIFT apparatus will provide valuable data with which to advance
the understanding of this class of reactions. Recently a
variable-temperature SIFT-DRIFT apparatus has been constructed
in our laboratories and is expected to provide new information
on the rôle of internal and kinetic excitation of the reactant
ion on the rates and mechanisms of isotope exchange and indeed
other classes of ion-molecule reactions.

REFERENCES

Adams, N.G., and Smith, D.: 1976a, Int. J. Mass Spectrom. Ion.
    Phys. 21, pp. 349-359.
Adams, N.G., and Smith, D.: 1976b, J. Phys. B 9, pp. 1439-1451.
Adams, N.G., and Smith, D.: 1981a, Astrophys, J. (Letters). 247,
    pp. L123-L125.
Adams, N.G., and Smith, D.: 1981b, Astrophys, J. 248, pp. 373-379.
Adams, N.G., and Smith, D.: 1982, in preparation.
Adams, N.G., and Smith, D.: 1983, In "Reactions of Small
    Transient Species: Kinetics and Energetics", (A. Fontijn,
    ed.), in press, Academic, New York.
Adams, N.G., Smith, D., and Paulson, J.F.: 1980, J. Chem.
    Phys. 72, pp. 288-297.
Adams, N.G., Smith, D., and Henchman, M.J.: 1982, Int. J. Mass
    Spectrom. Ion Phys. 42, pp. 11-23.
Anicich, V.G., Huntress, W.T. Jr., and Futrell, J.H.: 1976,
    Chem. Phys. Letts. 40, pp. 233-236.
Blint, R.J., Marshall, R.F., and Watson, W.D.: 1976, Astrophys.
    J. 206, pp. 627-637.
Brauman, J.I.: 1979, In "Kinetics of Ion-Molecule Reactions"
    (P. Ausloos, ed.), pp. 153-164, Plenum, New York.
Carney, G.D., and Porter, R.N.: 1977, J. Chem. Phys. 66,
    pp. 2756-2758.
Dalgarno, A., and Black, J.H.: 1976, Rept. Prog. Phys. 39,
    pp. 573-612.

DePuy, C.H., and Bierbaum, V.M.: 1981, Accts. Chem. Res. 14, pp. 146-153.

DePuy, C.H., Bierbaum, V.M., King, G.K., and Shapiro, R.H.: 1978, J. Amer. Chem. Soc. 100, pp. 2921-2922.

Dotan, I.: 1980, Chem. Phys. Letts. 75, pp. 509-512.

Fahey, D.W., Böhringer, H., Fehsenfeld, F.C. and Ferguson, E.E.: 1982, J. Chem. Phys. 76, pp. 1799-1805.

Fehsenfeld, F.C., Dunkin, D.B., Ferguson, E.E., and Albritton, D.L.: 1973, Astrophys. J. (Letts.) 183, pp. L25-L26.

Fehsenfeld, F.C., Albritton, D.L., Bush, Y.A., Fournier, P.G., Govers, T.R., and Fournier, J.: 1974a, J. Chem. Phys. 61, pp. 2150-2155.

Fehsenfeld, F.C., Dunkin, D.B., and Ferguson, E.E.: 1974b, Astrophys. J. 188, pp. 43-44.

Ferguson, E.E., Fehsenfeld, F.C., and Albritton, D.L.: 1979, In "Gas Phase Ion Chemistry, Vol 1" (M.T. Bowers, ed.) pp. 45-82, Academic, New York.

Gentry, W.R.: 1979a, In "Gas Phase Ion Chemistry, Vol 2" (M.T. Bowers, ed.) pp. 221-297 Academic, New York.

Gentry, W.R.: 1979b, In "Kinetics of Ion Molecule Reactions" (P. Ausloos, ed.) pp. 81-102, Plenum, New York.

Gentry, W.R.: 1981, Nucl. Phys. A353, pp. 273c-286c.

Henchman, M.J., Adams, N.G., and Smith, D.: 1981, J. Chem, Phys. 75, pp. 1201-1206.

Henchman, M.J., Smith, D., and Adams, N.G.: 1982, Int. J. Mass Spectrom. Ion Phys. 42, pp. 25-32.

Henning, P., Kraemer, W.P., and Diercksen, G.H.F.: 1977, Max-Planck-Institut. Int. Rep. No. MPI/PAE Astro 135.

Huber, K.P. and Herzberg, G.: 1979, "Molecular Spectra and Molecular Structure. IV Constants of Diatomic Molecules", Van Nostrand-Reinhold, New York.

Huntress, W.T. Jr.: 1977, Astrophys. J. Suppl. Series, 33, pp. 495-514.

Huntress, W.T. Jr., and Anicich, V.G.: 1976, Astrophys. J. 208, pp. 237-244.

Huntress, W.T. Jr., and Elleman, D.D.: 1970, J. Amer. Chem. Soc. 92, pp. 3565-3573.

Kim, J.K., and Huntress, W.T. Jr.: 1975, J. Chem. Phys. 62, pp. 2820-2826.

Kim, J.K., Theard, L.P., and Huntress, W.T. Jr.: 1974, Int. J. Mass Spectrom. Ion Phys. 15, pp. 223-244.

Lindinger, W., and Smith, D.: 1983, In "Reactions of Small Transient Species: Kinetics and Energetics" (A. Fontijn, ed.), in press, Academic, New York.

McMahon, T.B., Miosek, P.G., and Beauchamp, J.L.: 1976, Int. J. Mass Spectrom. Ion Phys. 21, pp. 63-71.

Pellerite, M.J., and Brauman, J.I.: 1980, J. Amer. Chem. Soc. 102, pp. 5993-5999.

Pople, J.A.: 1976, Int. J. Mass Spectrom. Ion Phys. 19, pp. 89-106.

Porter, R.N.: 1977, In "State to State Chemistry" (P.R. Brooks
     and E.F. Hayes, eds.) Amer. Chem. Soc. Symp. Series. 56,
     pp. 236-238.
Rosenstock, H.M., Draxl, K., Steiner, B.W., and Herron, J.T.:
     1977, J. Phys. Chem. Ref. Data. 6, Suppl. 1.
Smith, D.: 1981, Phil. Trans. Roy. Soc., A 303, pp. 535-542.
Smith, D., and Adams, N.G.: 1979, In "Gas Phase Ion Chemistry,
     Vol 1" (M.T. Bowers, ed.) pp. 1-44, Academic, New York.
Smith, D., and Adams, N.G.: 1980a, Chem. Phys. Letts. 76,
     pp. 418-423.
Smith, D., and Adams, N.G.: 1980b, Astrophys. J. 242, pp. 424-431.
Smith, D., and Adams, N.G.: 1980c, In "Topics in Current Chemistry,
     Vol 89" (S. Veprek and M. Venugopalan, eds.) pp. 1-43,
     Springer, Berlin.
Smith, D., and Adams, N.G.: 1981, Int. Rev. Phys. Chem., 1,
     pp. 271-307.
Smith, D., and Adams, N.G.: 1983, in preparation.
Smith, D., Adams, N.G. and Henchman, M.J. 1980. J. Chem. Phys.
     72,pp.4951-4957.
Smith, D., Adams, N.G., and Alge, E.: 1982a, Astrophys. J.,
     263, in press.
Smith, D., Adams, N.G., and Alge, E.: 1982b, J. Chem. Phys.,
     77, pp. 1261-1268.
Smith, D.L., and Futrell, J.H.: 1975, J. Phys. B 8, pp. 803-815.
Smith, D.L., and Futrell, J.H.: 1976, Chem. Phys. Letts. 40,
     pp. 229-232.
Smith, R.D., and Futrell, J.H.: 1977, Int. J. Mass Spectrom.
     Ion Phys. 24, pp. 173-179.
Stewart, J.H., Shapiro, R.H., DePuy, C.H., and Bierbaum, V.M.:
     1977, J. Amer. Chem. Soc. 99, pp. 7650-7653.
Su, T., and Bowers, M.T.: 1979, In "Gas Phase Ion Chemistry"
     Vol. 1, (M.T. Bowers, ed.) pp. 83-118, Academic, New York.
Talrose, V.L., Vinogradov, P.S., and Larin, I.K.: 1979, In
     "Gas Phase Ion Chemistry, Vol. 1" (M.T. Bowers, ed.)
     pp. 305-347, Academic, New York.
Villinger, H., Henchman, M.J., and Lindinger, W.: 1982, J. Chem.
     Phys. 76, pp. 1590-1591.
Watson, W.D.: 1976, Rev. Mod. Phys., 48, pp. 513-552.
Watson, W.D.: 1977, In "Topics in Interstellar Matter" (H. van
     Woerden, ed.) pp. 135-147 Reidel, Dordrecht, Holland.
Watson, W.D., Anicich, V.G., and Huntress, W.T. Jr.: 1976,
     Astrophys. J. (Letters), 205, pp. L165-L168.
Winnewisser, G., Churchwell, E., and Walmsley, C.M.: 1979, In
     "Modern Aspects of Microwave Spectroscopy" (G.W. Chantry,
     ed.) pp. 313-503, Academic, New York.

ENERGETICS AND DYNAMICS OF ION-MOLECULE REACTIONS
AT THERMAL ENERGIES

R.Marx

Laboratoire de Résonance Electronique et Ionique
(associé au CNRS) Université de Paris-Sud ,
Centre d'Orsay, 91405 Orsay, France.

Energetics and dynamics of non dissociative charge transfer
reactions at thermal energies are discussed in the general
framework of state selected ion chemistry. Experimental results
including kinetic energy, emission spectra and laser induced
fluorescence excitation spectra of the product ions are
presented. It is shown that in most cases charge transfer
reactions at (near)thermal energies do not proceed via a long
distance electron jump. The energy deposited in the product ion
is not equal to the recombination energy of the reactant ion,
part of this energy being converted into kinetic energy of the
products. Vibrational population distributions in the product
ions are narrow in diatomics (one or two levels) and Franck-
Condon populations have never been observed. In contrast whith
symmetric charge tranfer, no simple model can account for the
experimental results and trajectory calculations on potential
energy surfaces are needed.

INTRODUCTION

Over the past years subtantial progress has been made in
understanding ion-molecule reaction dynamics. This is mainly due
to the increasing number of experimental information on the
kinetic and vibrational energy dependence of reaction rates and
on the partitioning of reactant energy between vibrational (rotational) and translational degrees of freedom of the products.

Beam Experiments

Most of this information has been obtained in beam expe-

M. A. Almoster Ferreira (ed.), Ionic Processes in the Gas Phase, 67–86.
© *1984 by D. Reidel Publishing Company.*

riments at a few eV collision energies with crossed beams or
beam-gas and down to 0.002 eV with merged beams (1).It is outside
the scope of this lecture to review the large amount of
literature published in this field since the last NATO meeting
(2). I will just try to give an idea of the present state of the
art and point out some of the difficulties encountered on the way
of performing "state to state" chemistry. The references cited
have been selected among the more recent papers.

Selection of the reactant ion internal state has been greatly improved by using photoionization (3) and more recently
photoelectron-photoion coincidence techniques (4,5) instead of
electron impact ionization. But increasing selectivity decreases
the ion beam intensity so that only total cross sections could be
measured without any analysis of the product energy distribution.

Kinetic energy and angular distribution of the product ions
have been studied for atomic (6,7) and small molecular (8,9)
reactant ions produced by electron impact. The internal energy of
the reactant is rather well defined for atomic ions but only
average vibrational energies are known for molecular ions.

Internal state of electronically excited products has been
investigated by visible Emission Spectroscopy (ES) mainly in
reactions of atomic ions with diatomic molecules (10-12).

Internal state of ground state products may be investigated
by Infra-Red (IR) emission or Laser Induced Fluorescence (LIF).
To my knowledge there is not yet any result in IR emission and
only two LIF experiments have been reported (13-14) in contrast
with neutral reactions where LIF has been quite extensively used.

Thermal Energy Experiments

There is much less information on thermal energy ion-
molecule reactions. Very little is known about internal energy
dependence of the reaction rates and even less about energy
partitioning in the products. Two methods have been used to get
experimental data:

i) Selected Ion Flow Tubes as described by D.Smith (15)
have been used to determine rate constants and also cross
sections as a function of reactant kinetic energies in Drift SIFT
experiments.

Selection of the reactant internal energy can be performed
rather easily for singly or doubly charged atomic ions using
selective quenching of the different electronic states (16-19).
Indications on the dependence of cross sections and product
distributions on the reactant vibrational energy may be obtained
by Drift SIFT experiments where the reaction cross sections are
measured as a function of the drift potential in various carrier
gases (20-22). However, only the average vibrational energies are
known and it is not straightforward to separate the influence of
kinetic and internal energy of the reactants.

Internal energy of the products has been determined in a few ion-molecule and Charge Transfer (CT) reactions by visible (23,24) and IR (25-27) chemiluminescence.

ii) Ion Cyclotron Resonance (ICR) spectrometers have also been adapted to allow for reactant selection and determination of product energy.

To select reactant internal energy, tandem Dempster-ICR spectrometers have been used (28,29). The reactant ions are produced in a separate source and mass selected before entering the ICR cell to react. Their internal energy depends on the conditions in the source: energy of the ionizing electrons, pressure and composition of the gas, residence time of the ion. Rate constants and product distributions as a function of reactant ion internal energies have been determined this way but there was no information about the product energy distribution.

Measurements of the product ion kinetic energy and spectroscopic investigation of the electronically excited and ground state reaction products have been performed in especially deviced ICR cells (30-32). Although the possibility to select the reactant ion internal energy is very limited, this is to date the only available technique to measure the kinetic energy of product ions in ion-molecule reactions at thermal energies. In addition chemiluminescence and LIF observations can be performed in low pressure conditions where collisional quenching is reduced as compared to flow tubes.

Until now these techniques have been used only to investigate simple systems where detailed experimental information can be obtained and theoretical calculations may be performed, mainly CT reactions of rare gas atomic ions and diatomic molecular ions with small (up to 5 atoms) molecules. Surprisingly enough, although CT is the simplest ion-molecule reaction, its mechanism is still subject to controversy especially at low and thermal energies and there are very few detailed theoretical calculations. I hope that the experimental information reported here will promote some new theoretical investigation and lead to a better understanding of reaction dynamics.

ENERGETICS AND DYNAMICS OF POSITIVE ION CHARGE TRANSFER REACTIONS AT THERMAL ENERGIES

In thermal energy CT reactions: $A^+ + M \longrightarrow M^+ + A$, the reactant collision energy may be neglected and, if M^+ does not dissociate, the energy balance may be approximated by

$$RE(A^+) = IP(M) + IE(M^+) + IE(A) + \Delta T$$

- $RE(A^+)$ is the recombination (i.e., neutralization) energy of A^+ to give ground state A.

- IP(M) is the energy needed to ionize M into the ground electronic and vibrational state of M^+.
- IE(M^+) and IE(A) are the internal energies of M^+ and A with
  respect to their ground electronic and vibrational states.
- $\Delta T = KE(M^+) + KE(A)$ is the translational energy taken away by the
  products.

If the reactant state is known, RE(A^+) and IP(M) may be
found in the literature and a comprehensive picture of the
reaction energetics may be deduced provided two out of the three
quantities IE(M^+), IE(A), ΔT, can be determined experimentally.

Information on the reaction dynamics may be deduced from the
energy partitioning and especially from the kinetic energy
release ΔT:

If $\Delta T = 0$ there is no momentum transfer and:
$$RE(A^+) = IP(M) + IE(M^+) + IE(A)$$
This so called <u>resonant CT</u> can occur only if there is an electronic state of M^+ at the recombination energy of A^+. Further
information on M^+ vibrational level population is needed to
know whether ionization occurs via a vertical Franck-Condon
transition or not. If Franck-Condon population is observed, electron jump probably takes place at large inter-nuclear distance,
if not there must be some perturbation of the reactant internal
state indicating a closer approach.

If $\Delta T \neq 0$, momentum transfer does occur and:
$$RE(A^+) - IE(A) > IP(M) + IE(M^+)$$
Part of the internal energy of the reactants is converted into
kinetic energy of the products indicating that electron exchange
occurs at short distances possibly in an intermediate complex.

To get direct information on the product internal energy,
spectroscopic methods (ES or LIF) are needed. However, internal
energy may be deduced from kinetic energy when the reactant A^+
is an atomic ion. This is because the excited states of A^+ are
too high in energy to fulfill the exothermicity condition for the
CT reaction: RE(A^+) - IE(A) > IP(M). Therefore A must be in its
ground state and IE(A)=0. The energy balance equation becomes:
$$RE(A^+) = IP(M) + IE(M^+) + \Delta T$$
and: $IE(M^+) = \Delta E - \Delta T$ where: $\Delta E = RE(A^+) - IP(M)$

When A^+ is a molecular ion, A may be rotationally and
vibrationally excited and the exothermicity condition still ful-
filled. Then only the sum IE(M^+)+IE(A) may be deduced from ΔT.

1. Experimental

Three different experimental methods have been used to
determine kinetic and internal energy of CT reactions at thermal
energies:
- Kinetic energy measurements of the product ions.

- Emission spectroscopy in the visible and near UV.
- Laser induced fluorescence excitation spectroscopy.

All three experiments make use of an ICR cell as an ion trap.
The experimental techniques have already been described in detail
and I will just recall a few important points.

i) In an ICR cell the reactant gases, A and M, are mixed and
ionized by an electron beam of variable energy (a few eV up to
100 eV) parallel to a magnetic field B. Trapping of the ions is
due both to B and to V_T, the trapping voltage applied to the
plates perpendicular to B. The magnetic field forces the ions
into a cycloidal motion at their cyclotron frequency and prevents
diffusion perpendicular to B, while V_T prevents diffusion along B.
To select the reactant ions A^+, all the other ions produced by
the ionizing electron beam are ejected by RF fields at their
cyclotron frequency. However, both trapping around B and selective ejection are effective only at low pressure ($<10^{-4}$ torr)
i.e., in the normal conditions used for rate constant and kinetic
energy measurements. Because of detection sensitivity problems,
the optical experiments (ES and LIF) have to be performed at
higher pressures ($10^{-3} - 10^{-2}$ torr). As a consequence trapping
times are much shorter (μs instead of ms) and selective ejection
is impossible. This is of course a serious limitation of the
method.

ii) Besides the ion-molecule reactions under investigation
different processes giving the same product may occur in an ICR
cell: ionization by the electron beam, Penning ionization etc..
To separate these processes, the experiments are operated in a
pulsed mode: electron beam, RF field for selective ejection,
trapping potential, laser beam and photon counting are all pulsed
as described in references (30-33). This procedure allows
accumulation and signal averaging which greatly enhances
detection sensitivities.

2. Kinetic Energy Measurements

Kinetic energy of ions is in principle very simple to
determine in an ICR cell. The fraction "f" of M^+ ions having a
given kinetic energy $KE(M^+)$ and trapped in the ICR cell varies
with V_T in the following way:
$f = 1$ when $eV_T > KE(M^+)$ and $f \propto V_T^{1/2}$ when $eV_T < KE(M^+)$.
Thus, if all M^+ had the same KE, f as a function of $V_T^{1/2}$
would show a break at: $eV_T = KE(M^+)$.

The shape of an experimental curve depends on the kinetic
energy distribution of M^+ and as a consequence on the population of its internal energy levels. Information on the energy
distribution may be deduced from these curves either by a fitting
procedure, as described in ref 31, or when the signal to noise
ratio is good enough, by derivation of the curves. Fig.1 gives
two examples of such distributions.

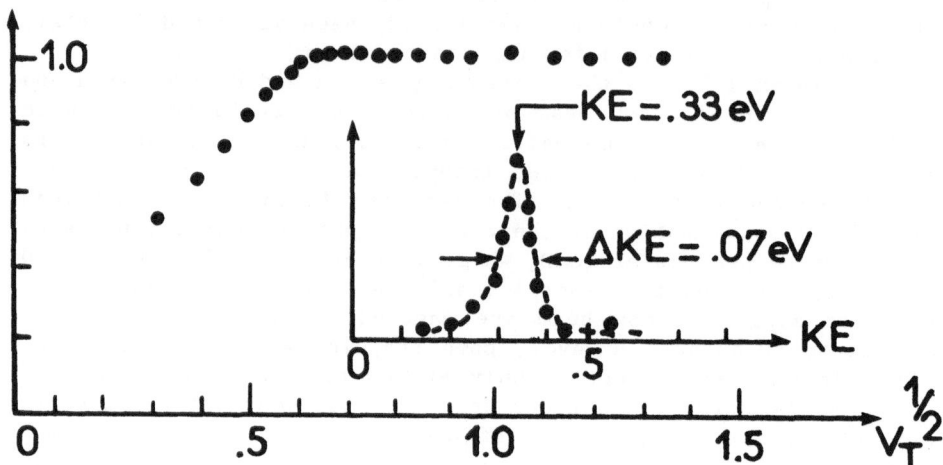

Fig. 1a: ●●● Fraction of trapped CO^+ ions vs. $V_T^{1/2}$ in
reaction $Ar^+ + CO \longrightarrow CO^+ + Ar$
—●— Kinetic energy distribution of the CO^+ ions.

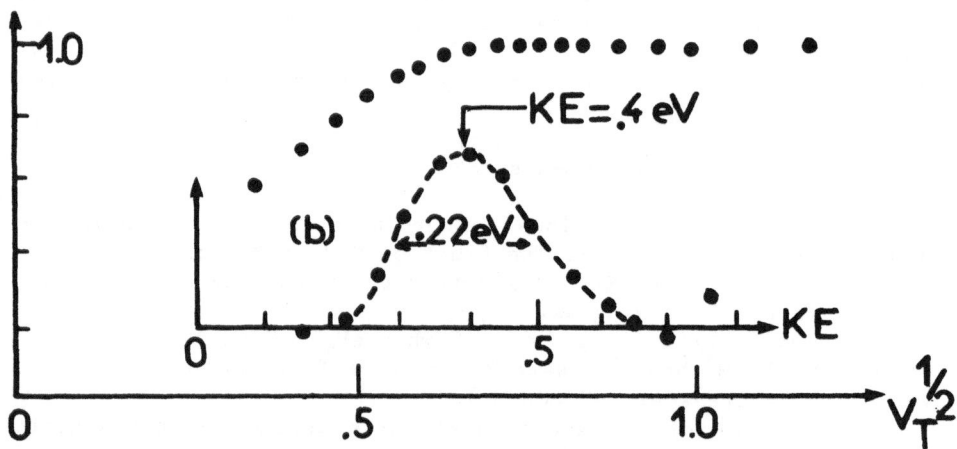

Fig. 1b: ●●● Fraction of trapped CO_2^+ ions vs. $V_T^{1/2}$ in
reaction $Ar^+ + CO_2 \xrightarrow{\hspace{0.8cm}} CO_2^+ + Ar$
—●— Kinetic energy distribution of the CO_2^+ ions

All the experimental results obtained since 1978 are presented in Table I in order to try to draw some general conclusion. The quoted values for $RE(A^+)$ and $IP(M)$ correspond to A^+ and M in their ground state. This is always true for M at room temperature. But if A^+ is produced by 50 eV electrons, both spin states of the rare gas ions are formed and vibrationally excited molecular ions may be formed in some cases. For Ar^+ the $^2P_{3/2}$, $^2P_{1/2}$ energy separation (.18eV) is too small to be resolved in our experiments. For Kr^+ and Xe^+ ions we have good evidence (31) that in our experimental conditions only reaction of the $^2P_{3/2}$ state is observed. To get molecular ions in their ground state, low energy electrons have been used. In addition rapid near resonant CT between A^+ and A removes any residual internal energy.

The total kinetic energy release ΔT is deduced from $KE(M^+)$ by momentum conservation equation:

$$\Delta T = \frac{m_A + m_M}{m_A} \cdot KE(M^+)$$

Finally, when $A = Ar^+$, Kr^+ and Xe^+:

$$IE(M^+) = (\Delta E - \Delta T)$$

and, when $A^+ = N_2^+, O_2^+, CO^+, CO_2^+$

$$IE(M^+) + IE(A) = (\Delta E - \Delta T)$$

2.1. Kinetic energy release. Out of 18 reactions only 3 proceed without momentum transfer:

- $(Xe^+ + O_2)$ which has only one energetically allowed channel giving O_2^+ ($X\ ^2\Pi_g$, v=0)

- $(Ar^+ + NO)$ where there is an electronic state of NO^+ ($a\ ^3\Sigma^+$, v=0) at the recombination energy of Ar^+.

- $(Kr^+ + CH_4$ where there is probably also a vibrational state of CH_4^+ close to the recombination energy of Kr^+.

In all the other cases energy releases ranging from 0.4 to 1.8 eV are observed.

This clearly shows that charge transfer reactions behave quite differently at thermal energies and in medium and high energy beams where resonant CT are observed.

We already pointed out that non resonant CT reactions do occur (34). Now we have enough experimental evidence to say that kinetic energy release is a rule rather than an exception at least for "small systems" (less than 6 or 7 atoms). I will come back to this problem later.

TABLE I: Thermal energy charge transfer reactions:
$$A^+ + M \longrightarrow M^+ + A$$
Kinetic and Internal Energy of reaction products.

A^+ (RE,eV)	M (IP, eV)	ΔE (eV)	$K_{exp} \times 10^9$ cm3s-1	ΔT (eV)	IE(M^+) (eV)	Ref
	NO (9.27)	6.49	0.3	<0.1	6.39 (a $^3\Sigma^+$ v'=0)	
	NH3 (10.16)	5.60	1.84	0.41-1.8	4.52-5.19 (X̃,Ã)	31
Ar$^+$ (15.76)	O2 (12.06)	3.70	0,057	1.80	1.90 (X, v' ∿ 9)	30a
	H2O (12.62)	3.14	1.8	0.3-1	2.14-2.84 (Ã, v'10-16)	38
				∿0.9	∿2.24 (X̃,v')	
	N2O (12.89)	2.87	0.31	0.61-1.41	1.46-2.26 (X̃)	37
	CO2 (13.77)	1.99	0.42	0.65-1.05	0.93-1.33 (X̃)	
	CO (14.01)	1.75	0.05	0.56	1.19 (X̃ v' ∿ 4)	
	NH3 (10.16)	3.84	0.8	0.46-1.20	2.64-3.38 (X̃ v'22-26)	31
Kr$^+$ (14.00)	O2 (12.06)	1.94	0.03	1.17	0.77 (X v' ∿ 3)	30b
	H2O (12.62)	1.38	1.2	∿0.67	∿0.71 (X̃)	
	CH4 (12.99)	1.01	1.03	<0.1	1.0 (X̃)	

TABLE I: continued

A^+ (RE,eV)	M (IP,eV)	E (eV)	$K_{exp} \times 10^9$ cm^{3}s^{-1}	ΔT (eV)	IE(M^+) (eV)	Ref
	NH$_3$ (10.16)	1.97	0.6	0.22-0.54	1.43-1.75 ($\tilde{X}$ v'$_{11-14}$)	31
Xe$^+$ (12.13)	H$_2$S	1.66	0.6	0.32-1	1.34-0.66	
	O$_2$ (12.06)	0.07	0.11	0.1	0. (X v'=0)	30b

					IE(NH$_3^+$) +IE(A)	
N$_2^+$ (15.78)		5.42	2.	0.92±0.2	4.5	
CO$^+$ (14.01)		3.85	1.5	0.76±0.4	3.09	29
	NH$_3$ (10.16)					
CO$_2^+$ (13.77)		3.61	1.86	1.27±0.5	2.34	
O$_2^+$ (12.06)		1.9	2.	0.71±0.1	1.19	

No systematic determination of the kinetic energy distribution has yet been made. The two numbers given in Table I columns 5 and 6 are the lower and upper values of ΔT corresponding to the curved part of "f" vs $V_T^{1/2}$. The general trend seems to be a narrow distribution for diatomic ions: 0.07 eV f.w.h.m in CO$^+$ from Ar$^+$ + CO and broader for polyatomics: 0.22 eV f.w.h.m. for CO$_2^+$ from Ar$^+$+CO$_2$,0.2 to 0.8 eV for NH$_3^+$ from A$^+$+NH$_3$ depending on A$^+$.

2.2. Partitioning of the excess energy between internal and kinetic energy of the products. Partitioning of the excess energy: $\Delta E = RE(A^+)-IP(M)$ is illustrated in Fig.2 for rare gas atomic ions reacting with various molecules and in Fig.3 for atomic and molecular ions reacting with NH$_3$.

One can see in Fig.2 that the fraction of ΔE going into M^+ internal energy, $IE(M^+)/\Delta E$, is minimum when M^+ is a diatomic ion and increases with the number of atoms in M. This trend appears clearly in Kr^+-O_2, Ar^+-CO_2 and Xe^+-NH_3 reactions corresponding to almost equal ΔE. Such a behavior can be accounted for by simple statistical considerations: the fraction of the total energy going into internal energy should increase with the number of available internal states i.e., the number of degrees of freedom. However, although the number of degrees of freedom of the products increases when atomic reactant ions are replaced by molecular ions, Fig.3 shows that the fraction of ΔT going into internal energy of the products does not increase in A^++NH_3 charge transfer reactions.

2.3. <u>Partitioning of the product internal energy between M^+ and A</u>
In CT reactions with molecular ions only the total internal energy: $IE(M^+)$ + $IE(A)$ can be deduced from kinetic energy measurements. Therefore another way to determine $IE(A)$ or $IE(M^+)$ separately has to be found.

The procedure proposed by BOWERS (29) takes advantage of the dependence of M^+ reactivity on its internal energy and makes use of two complementary experiments.
    i) Measurement of M^+ kinetic energy in the CT reaction:
$$A^+ + M \longrightarrow M^+ + A$$
as described above.
    ii) Measurement of the rate constant or product distribution in the reaction: $M^+ + N \longrightarrow$ products
where M^+ is formed   by charge transfer   in the source of a tandem Dempster-ICR spectrometer.

This   method   has   been   used   to   determine   the   energy partitioning in reaction:

$$A^+ + NH_3 \longrightarrow NH_3^+ + A \qquad (1)$$

The diagnostic reaction was

$$NH_3^+ + H_2O \begin{cases} \longrightarrow H_3O^+ + NH_2 \quad + 0.47 \text{ eV} \quad (2a) \\ \longrightarrow NH_4^+ + OH \quad - 0.25 \text{ eV} \quad (2b) \end{cases}$$

where the product distribution $NH_4^+ / (H_3O^+ + NH_4^+)$ strongly depends on NH_3^+ internal energy (34).

To draw the curve - %NH_4^+ vs $IE(NH_3^+)$ - the product distribution for reaction 2 has been measured for NH_3^+ ions resulting from CT of NH_3 with Ar^+, Kr^+ and Xe^+ respectively. The corresponding internal energy of NH_3^+ being known from kinetic energy measurements (Table 1) this gives three

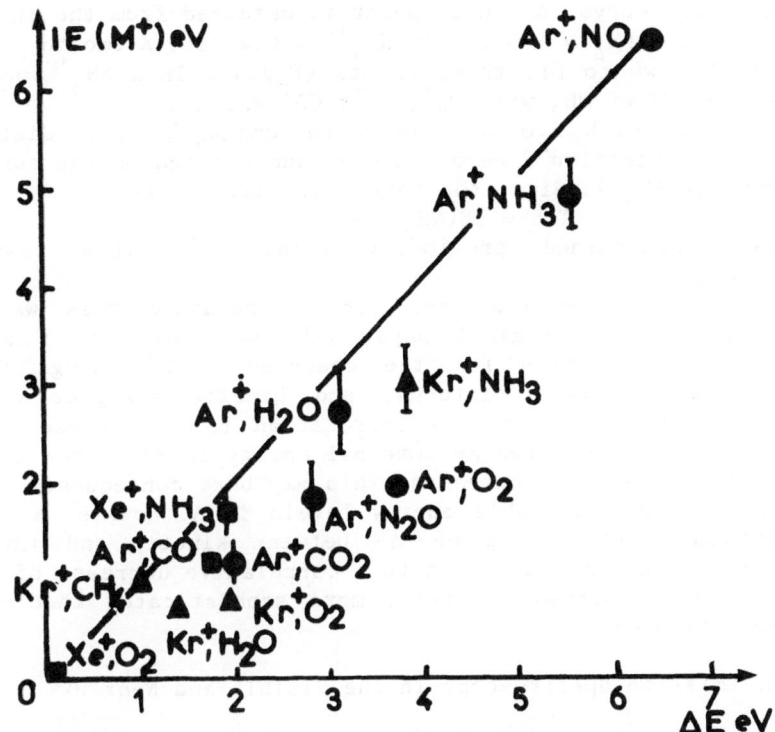

Fig.2. $IE(M^+)$ in charge transfer of Ar^+ (●) Kr^+ (▲) and Xe^+ (■) with various molecules. The full line (—) corresponds to resonant charge transfers ($IE(M^+) = \Delta E$)

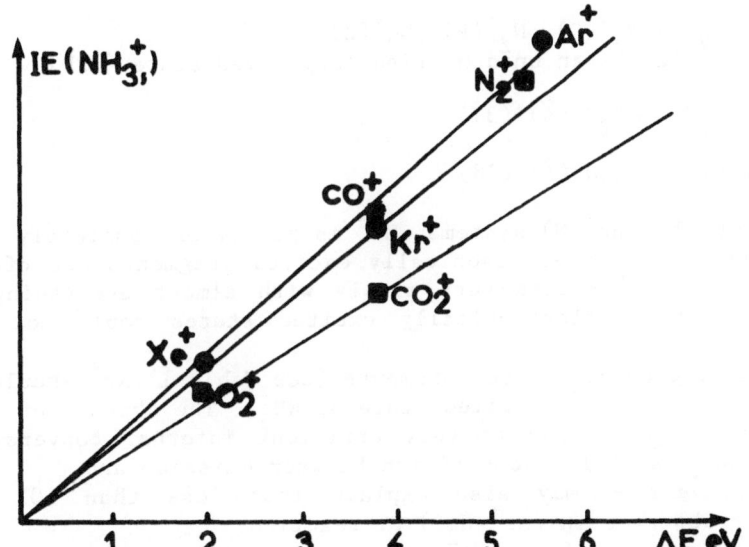

Fig.3. $IE(NH_3^+)$ resulting from thermal energy charge transfer of various atomic (●) and molecular (⊞) ions with NH_3

points on the curve. A fourth point is obtained from the onset of reaction 2a: $\%NH_4^+ = 100$ for $IE(NH_3^+) = 0.47$ eV. A smooth curve was drawn to fit these points (Fig.4). Then NH_3^+ ions produced by CT of NH_3 with O_2^+, N_2^+, CO^+ and CO_2^+ were reacted with H_2O and the corresponding product distributions for reaction 2 were measured and plotted on the curve to determine $IE(NH_3^+)$. Since the total internal energy:

$$TIE = IE(NH_3^+) + IE(A)$$

has been determined previously (Table I), $IE(A)$ can be calculated.

Table II summarizes the results obtained this way. As discussed in the original paper (29) we could not find any obvious interpretation for the observed partitioning of the excess energy between internal and kinetic energies of the products. The only apparent correlation is a decrease in the fraction of ΔE deposited as internal energy in NH_3^+ when the number of atoms in A increases. This may be a consequence of the increase of the available energy levels in A. However a simple statistical partitioning of ΔE between kinetic and internal degrees of freedom would lead to a correlative decrease of $\Delta T/\Delta E$ which is not observed. Clearly, more sophisticated theoretical treatment is needed.

3. Emission Spectroscopy in the Visible and Near UV

Emission spectrum may be expected when $RE(A^+)$ is large enough to populate electronically excited states of M^+. However, among all the energetically allowed CT reactions investigated (36) only three luminescent molecular ions were observed.

$$He^+ + N_2 \longrightarrow N_2^+(C), \ N_2^+(B), \ N_2^+(Z)$$

where $N_2^+(Z)$ is an unidentified long lived state (32b).

$$He^+ + N_2O \longrightarrow N_2O^+(\tilde{A}) \ (37)$$

$$Ar^+ + H_2O \longrightarrow H_2O^+(\tilde{A}) \ (38)$$

In all the (He^+, M) systems, M^{+*} is partly or completely (pre)dissociated and electronically excited fragments are often observed (33). Ne^+ reacts very slowly with almost everything and the possible electronically excited states could not be observed.

According to our KE measurements (see Table 1) Ar^+ should also populate the first excited state of NH_3^+. The absence of emission is probably due to very efficient internal conversion $NH_3^+(\tilde{A}) \longrightarrow NH_3^+(\tilde{X})$ (31). Competition between emission and internal conversion (31) may also explain that less than 40% of $H_2O^+(\tilde{A})$ are observed in (Ar^+-H_2O) CT reaction.

TABLE 2. Energy disposal in: $A^+ + NH_3^+ \rightarrow NH_3^+ + A$ charge transfer reactions at thermal energies. The energies are in eV. The error bars ($\pm$.) include the energy distribution width.

A^+	ΔE	ΔT	TIE	% NH_4	IE(NH_4^+)	IE(A)	k/k_{ADO}
Ar^+	5.6	0.74 $\pm$ .3	4.95$\pm$.3	2$\pm$2	4.95 $\pm$.4	0.0	1.2
N_2^+	5.42	0.92 $\pm$ .2	4.50$\pm$.2	3$\pm$3	3.2 $\pm$.5	1.3$\pm$.5	1.
Kr^+	3.84	0.83 $\pm$ .4	3.$\pm$.4	3$\pm$2	3. $\pm$.4	0.0	0.5
CO^+	3.85	0.76 $\pm$ .4	3.1$\pm$.4	42$\pm$10	1.8 $\pm$.2	1.3$\pm$.6	0.8-1
CO_2^+	3.61	1.3 $\pm$ .5	2.3$\pm$.5	92$\pm$5	1. $\pm$.2	1.3$\pm$.7	1.04
Xe^+	1.97	0.4 $\pm$0.2	1.57$\pm$.2	65$\pm$5	1.6 $\pm$.2	0.0	0.4
O_2^+	1.90	0.7 $\pm$0.1	1.2$\pm$.1	80$\pm$10	1.3$\pm$.2	0	1.06

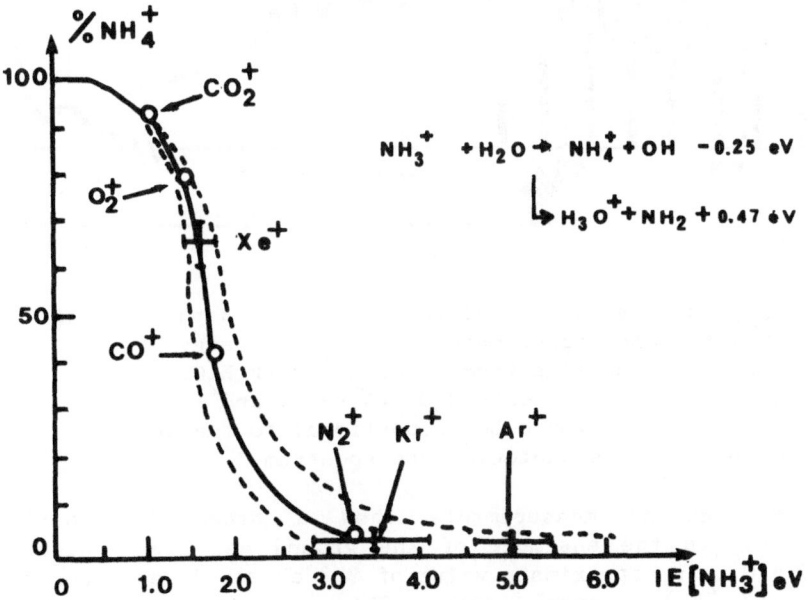

Fig.4: Product distribution vs. NH_4^+ internal energy. NH_3^+ is formed by CT of NH_3 with atomic ions: Ar^+, Kr^+, Xe^+ (⊢┼┤) or molecular ions: CO_2^+, O_2^+, CO^+, N_2^+ (o) (see text)

So it turns out that electronically excited molecular ions are, for various reasons, minor products in thermal energy CT reactions.

Reaction: $Ar^+ + H_2O \longrightarrow H_2O^+ + Ar$, is an interesting case since both spectroscopic investigation and kinetic energy measurements could be performed.

The electronically excited $\widetilde{A}\ ^2A_1$ state accounts for about half of the H_2O^+ ions produced and its vibrational level population could be deduced from the $\widetilde{A} \longrightarrow \widetilde{X}$ emission spectrum. This population is clearly not Franck-Condon as illustrated in Fig.5. The corresponding kinetic energy release: $\Delta T = \Delta E - IE(H_2O^+\ \widetilde{A})$ ranges from $\simeq 0$ to $\simeq 1$ eV.

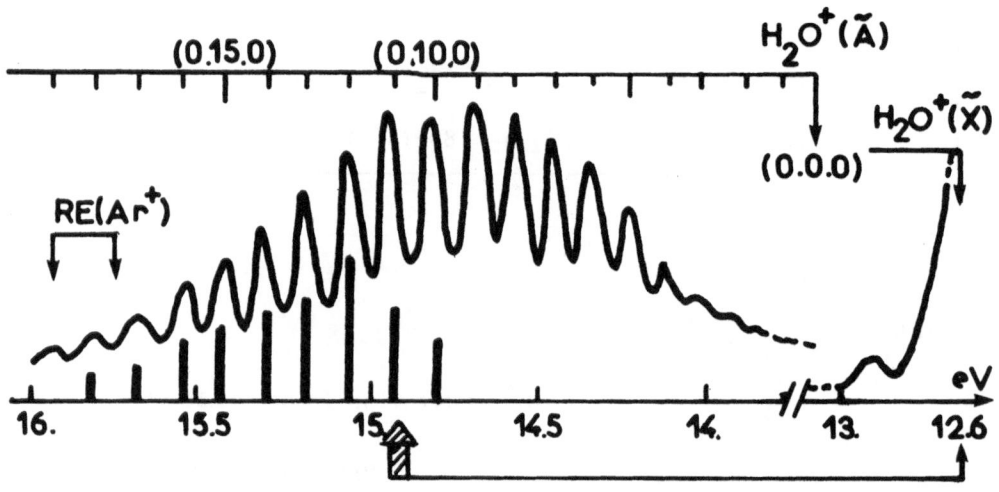

Fig. 5 Histogram of the vibrational level population of H_2O^+ ($\widetilde{A}$) (▌) and approximate vibrational energy of H_2O^+ ($\widetilde{X}$) (▟) resulting from CT of Ar^+ with H_2O. Resonant CT would populate H_2O^+ ($\widetilde{A}$,v=18) and Franck-Condon population would be proportional to the peak intensities of the photoelectron spectrum.

Kinetic energy measurements show a broad distribution corresponding to the ensemble of the ground and electronically excited H_2O^+. An approximate value of 2.3 eV was deduced for the H_2O^+ ($\widetilde{X}\ ^2B_1$) internal energy. This corresponds to very high vibrational levels (v' $\simeq$ 20) lying in an energy range where the $\widetilde{A}$ and $\widetilde{X}$ levels are very close and internal conversion probably occurs.

In conclusion, spectroscopy and kinetic energy measurements both indicate that (Ar^+-H_2O) CT proceeds through an intimate collision and possibly involves an intermediate complex.

The opposite behavior is observed for:

$$He^+ + N_2 \longrightarrow N_2^+ + He$$

one of the most extensively studied CT reactions. In this reaction the main product is $N_2^+(C^2\Sigma_u^+)$ resulting from a <u>resonant CT</u> without momentum transfer. However, vibrational population is not Franck-Condon at thermal and low collision energies (32b and references cited herein) indicating that electron jump does not occur at large distances. We will discuss this point later.

4. Laser Induced Fluorescence of Ion-Molecule Reaction Products

LIF is in principle the best way to probe the internal energy of gas phase reaction products and, since the pioneering experiments of Polanyi and Zare, LIF spectroscopy has been applied to a wide variety of neutral reactions. In contrast, although LIF spectroscopy of ions progressed considerably these last four or five years (39,40), there are very few results on ion-molecule reaction products. The main reason for that is the difficulty to produce and store large enough densities of ions in low pressure conditions where secondary reactions and collisional deactivation are negligible so that nascent product internal states can be determined. To my knowledge there are to date only two published papers on LIF investigation of positive ion-molecule reactions at low energies:

i) Charge transfer of N_2^+ with N_2 and Ar(41) studied in a quadrupole ion trap at collision energies around 0.4 eV. LIF of N_2^+ ions has been recorded as a function of N_2 and Ar pressure. Besides the collisional heating of the rotational temperature, cooling of the vibration due to charge transfer of $N_2^+(X^2\Sigma^+, v>0)$ with Ar and N_2 has been observed. Cross section for vibrational deactivation of $N_2^+(X,v=1)$ was small with Ar ($\sigma_L/20$) and large with N_2 (σ_L). The level v=2 is depleted at almost the same rate as v=1 with Ar and much less with N_2. These changes were qualitatively accounted for by the difference in the Franck-Condon factors involved: smaller for:

$$N_2^+(X,v=1)+Ar \longrightarrow N_2(X,v'=0)+Ar^+$$

than for:

$$N_2^+(X,v=1)+N_2(X,v'=0) \longrightarrow N_2(X,v'=1)+N_2^+(X,v=0)$$

ii) Charge transfer of Ar^+ and CO^+ with CO (42) studied in an ICR cell at (near) thermal energy ($\simeq 0.1$ ev).

According to the kinetic energy measurements (Table I) charge transfer reaction:

$$Ar^+ + CO(X\ ^1\Sigma^+,\ v=0) \longrightarrow Ar + CO^+(X\ ^2\Sigma^+,\ v'')$$

should populate $v''=4$.

An attempt to detect $CO^+(X,\ v''=4)$ exciting the transition $(X,\ v''=4) \rightarrow (A,\ v'=7)$ and observing the transition $(A,\ v''=7) \rightarrow (X,\ v''=5)$ was not conclusive since almost no change of the signal intensity with Ar pressure was observed. In contrast there was an increase in $CO^+(X,\ v''=0)$ LIF signal with Ar pressure. We have good evidence that this increase in ground vibrational level population is not due directly to Ar^+ - CO charge transfer reaction but to the very fast deactivation of pri mary $CO^+(X,\ v''>0)$ into $v''=0$ via the very fast symmetrical CT:

$$CO^+(X,\ v'') + CO(X,\ v=0) \rightarrow CO(X,\ v=v'') + CO^+(X,\ v''=0)$$

The efficiency of such reactions for v'' up to 10 is due to the conjunction of large Franck-Condon factors and small energy excess (42).

The experimental LIF signal of $CO^+(X,\ v''=0)$ as a function of CO pressure fits quite nicely the calculated curve including electron impact ionization, radiative decay of electronically excited CO^+ ions, $(Ar^+ -CO)$ charge transfer and vibrational cooling of $CO^+(X,\ v''\neq0)$.

However we could not obtain quantitative information on the vibrational population of $CO^+(X)$ for two reasons:

i) The detection sensitivity for $CO^+(X,\ v''\neq0)$ was poor.

ii) Deactivation rate being 20 times larger than the production rate, $CO^+(X,\ v''>0)$ concentration is small even at low pressure.

The second point is unfortunately quite a general problem since, at least in diatomics, symmetrical CT is a very efficient deactivating process (43). Therefore, direct information on the internal energy distribution of ground state ionic reaction products may be expected only for fast reactions when symmetrical charge transfer occurs.

5. Conclusion

The experimental results presented here clearly show that except in a few cases CT reactions at (near) thermal energies do not proceed via a long distance electron jump. The energy deposited in the product ion is not in most cases equal to the recombination energy of the reactant ion and the energy partitioning between internal and external modes of the products does not obey simple rules.

This is in contradiction with the conclusions of a recent paper on the effects of collision energy and ion vibrational excitation in CT of H_2^+ with N_2, CO an O_2 (3). The cross sections were found almost independent of collision energy and the reactant ion vibrational effect was accounted for on the basis of a simple model: the charge transfer probability from an intitial $H_2^+(v) + M(v=0)$ state to a given $H_2(v') + M^+(v'')$ final state was assumed to be proportional to the product of the Franck-Condon factors for $H_2^+(v) \rightarrow H_2(v')$ and $M(v=0) \rightarrow M(v'')$ transitions, times an inverse exponential function of the energy difference ΔE between initial and final state ($\sigma \propto exp - \Delta E/E_0$). The total CT probability is then the sum of the probability to all the possible final states.

The authors take argument of the qualitative fit between calculated and experimental cross sections and of the fact that there is no appreciable collision energy dependence to conclude that CT occurs at large reagent separation. They find however quite surprising that Franck-Condon factors are so important in determining CT probability, in spite of the fact that the collisions are slow (0.5 to 3 eV CM) compared to the vibrational period of the molecules. This is indeed surprising and apparently not true in near thermal CT since Franck-Condon population of the product levels was almost never observed.

There may be various reasons for that:

i) The reaction mechanism may change below 0.5 eV (the lowest collision energy used in ref.3), as already observed in some ion-molecule reactions (43).

ii) For some unknown reason H_2^+ reacts in a different way than the other reactant ions investigated so far.

iii) Modelling of the total charge transfer cross section out of various initial vibrational states may not be very sensitive to reaction dynamics. As a consequence the fact that a model, which ignores dynamical effects, reproduces qualitatively the experimental result may not prove that CT is a long range, collision energy independent process.

Anyway product state distributions deduced from trajectory calculations on potential energy surfaces and compared to experimental results will certainly provide more realistic insight into reaction dynamics.

This type of study has been performed (36,44) for the quasi resonant CT reaction:

$$He^+ + N_2 \longrightarrow He + N_2^+(C,v')$$

Potential energy curves and coupling elements for the colinear $(HeN_2)^+$ system have been calculated using an ab-initio

MCSF-CI method and the Bauer Fisher Gilmore multicurve crossing model has been used to determine $N_2^+(C)$ vibrational and rotational population.

The $v'=3/v'=4$ population ratio was well reproduced for both the 14 and 15 N_2 isotopes and as a function of $N_2(X)$ vibrational temperature.

It was also shown that the most efficient transitions between the entrance channel: $He^+ + N_2(X)$
and the outgoing channel: $He + N_2^+(C, v'=3,4)$
take place at $He-N_2$ distances as short as 4-5 a.u.
Calculated cross sections were reproduced within a factor of less than 2.

Although surprisingly good results have been obtained with this simple two dimensional model, more appropriate three dimensional energy surfaces will be calculated in order to get more detailed information on the reaction dynamics.

Calculation of energy surfaces are planned for other triatomic systems ($Ar^+-O_2, Ar^+-CO..$) but ab-initio calculations being very time consuming, DIM (diatomics in molecule) methods will be used.

Of course, such calculations are out of question for systems containing more than three atoms and other methods must be worked out for polyatomics. Phase space treatments which have been used to calculate energy and temperature dependence of ion-molecule reactions and kinetic energy release in unimolecular dissociations may be well suited, at least when long lived intermediate complexes are formed.

ACKNOWLEDGEMENTS

The author wishes to acknowledge the contribution of the members of her group who performed most of the work presented in this lecture : J.Danon, R.Derai, S.Fenistein, M.Gérard-Aîn, T.Govers, G.Mauclaire and the technical assistance of G.Bellec, C.Sourisseau and M.Reyes Pastor. She specially thanks R.Derai and G.Mauclaire for their help in writing the manuscript and P.Collardelle who typed it.

BIBLIOGRAPHY

1. S.L. Anderson, F.A. Houle, D.Gerlich and Y.Lee, J.C.P., 75, (1981)

2. R.W.Gentry in Kinetic of Ion Molecule Reactions, Ed.P.Ausloos Plenum Publ.Corp. 1979 p.81

3. S.L.Anderson, T.Turner, B.H.Mahan and Y.T.Lee, J.C.P., 77,1842 (1982)

4. T.Kato, K.Tanaka, I.Koyano, J.C.P., 77, 834 (1982) and ref. cited herein

5. T.Baer, T.Govers, P.M.Guyon, private communication

6. S.G.Hausen, J.M.Farrah, B.M.Mahan, J.C.P., 73, 3750 (1980)

7. B.Friedrich and Z.Herman, Chem.Phys., 69, 433 (1982)

8. R.M.Billota, J.M.Farrar, J.C.P., 74, 1699 (1981)

9. C.H.Douglass, O.Ringer, N.R.Gentry, J.C.P., 76, 2423 (1982)

10. D.Neuschafer, Ch.Ottinger, S.Zimmermann, Chem.Phys., 55,313 (1981)

11. T.Kusunoki and Ch.Ottinger, J.C.P., 73, 2069 (1980)

12. Ch.Ottinger and J.Reichmuth, J.C.P., 74, 928 (1981)

13. P.J.Dagdigian and J.P.Doering, Chem.Phys.Letters, 64,200(1979)

14. A.Ding, K.Richter and M.Mentzinger, Chem.Phys.Letters, 77,523 (1981)

15. D.Smith, lecture in the present book

16. N.G.Adams, D.Smith and E.Alge, J.Phys.B, 13, 3235 (1980)

17. A.B.Rakshit and P.Warneck, J.C.P., 73, 2673 (1980)

18. H.Stori, E.Alge, H.Villinger, F.Egger and W.Lindinger, Int.J.Mass Spectrom. Ion Phys., 30, 263 (1979)

19. T.T.Jones, K.Birkinshaw, J.D.C.Jones and N.D.Twiddy, J.Phys.B, 15, 2439 (1982)

20. D.L.Albritton in Kinetics of Ion Molecule Reactions, ed. P.Ausloos Plenum Publ.Corp. 1979 p.119

21. E.Alge, H.Villinger and W.Lindinger, S.A.S.P.(1980) and Plasma Chemistry and Plasma Processing, 1, 65 (1981)

22. M.Durup-Ferguson, H.Bohringer, D.W.Fahey, E.E.Ferguson, ASMS annual meeting 1982

23. M.Tsuji, T.Susuki, M.Endoh and Y.Nishimura, Chem.Phys.Lett. 86, 411(1982)

24. M.Tsuji, Y.Nishimura, T.Mizuguchi, Chem.Phys.Lett., 89, 75(1982)

ibid.<u>84</u>, 318(1981) and references cited herein

25. T.S.Zwier, W.M.Bierbaum, G.B.Ellison and S.R.Leone, J.C.P.,
    <u>72</u>, 5426 (1980)

26. J.C.Weisshaar, T.S.Zwier and S.R.Leone, J.C.P.,<u>75</u>,4873 (1981)

27. V.Bierbaum, A.S.M.S. annual meeting 1982

28. D.L.Smith and J.H.Futrell, Int.J.Mass Spectrom. Ion Phys.,
    <u>14</u>, 171 (1974)

29. P.R.Kemper, M.T.Bowers, D.Parent, G.Mauclaire, R.Derai and
    R.Marx, submitted to J.C.P. (1982)

30. a) G.Mauclaire, R.Derai, S.Fenistein and R.Marx, J.C.P.,<u>70</u>,
    4017(1979)
    b) R.Marx, G.Mauclaire, T.R.Govers, M.Gérard-Aïn, R.Derai,
    Journal de Chim.Phys., <u>76</u>, 1077(1979)

31. R.Derai, G.Mauclaire and R.Marx,Chem.Phys.Lett.,<u>86</u>,275 (1982)

32. a) G.Mauclaire, R.Marx, C.Sourisseau, C.A.Van de Runstraat
    and S.Fenistein, Int.J.Mass Spectrom. Ion Phys.,<u>22</u>,339(1976)
    b) T.R.Govers, M.Gérard, G.Mauclaire and R.Marx, Chem.Phys.,
    <u>23</u>, 411 (1977)

33. J.Danon, G.Mauclaire, T.R.Govers and R.Marx, J.C.P.,<u>76</u>,1255
    (1982)

34. R.Marx, in Kinetics of Ion Molecule Reactions ed.P.Ausloos,
    Plenum Publ.Corp 1979 p.103

35. V.G.Anicich, J.K.Kim and W.T.Huntress, Int.J.Mass Spectrom.
    Ion Phys., <u>25</u>, 433 (1977)

36. M.Gérard-Aïn, Thesis, Orsay 1982

37. M.Gérard, T.R.Govers and R.Marx, Chem.Phys.,<u>36</u>, 247 (1979)

38. R.Derai, S.Fenistein, M.Gérard-Ain, T.R.Govers, R.Marx,
    G.Mauclaire, C.Z.Profous and C.Sourisseau, Chem.Phys.<u>44</u>,65
    (1979)

39. T.A.Miller and V.E.Bondibey, Journal de Chimie Physique,<u>77</u>,
    695 (1980)

40. J.Maier's lecture in the present book

41. B.H.Mahan, C.Martner and A.O'Keefe, J.C.P.,76, 4433 (1982)

42. J.Danon and R.Marx, Chem.Phys., <u>68</u>, 255 (1982)

43. D.W.Fahey, I.Dotan, FC.Fehsenfeld and DL.Albritton,
    J.C.P., <u>74</u>, 3321 (1981)

44. M.Gérard-Aïn, T.R.Govers, B.Levy and Ph.Millié, 9th Int.Mass
    Spect. Conference, Vienna (1982)

RECENT STUDIES OF GAS-PHASE REACTIONS OF ANIONS WITH
ORGANIC MOLECULES

S. Ingemann, J.C. Kleingeld and
N.M.M. Nibbering*
Laboratory of Organic Chemistry, University
of Amsterdam, Nieuwe Achtergracht 129,
1018 WS Amsterdam, The Netherlands

INTRODUCTION

The interest in the bimolecular chemistry of anions
in the gas phase has grown steadily (1-4) since the
NATO Advanced Study Institute on Kinetics of Ion-
-Molecule Reactions held in La Baule, France, in
1978 (5). Especially the methods of flowing afterglow
(6,7) and ion cyclotron resonance spectrometry (8)
have been applied to studies of relatively simple anion-molecule reactions in the past few years. A short
description of these methods and the results obtained
for some selected anion-molecule reaction types will
be covered by the following topics to be discussed:
(i)   Methods.
(ii)   General picture of ion-molecule reactions.
(iii) Hydrogen-deuterium exchange reactions.
(iv)   Nucleophilic aromatic substitution.
(v)   Elimination reactions
(vi)   Substitution versus addition-elimination.
(vii) Hydride transfer reactions.

METHODS

        In the flowing afterglow (FA) method (6,7) reac-
tant ions are generated by electron impact from a
reagent gas in the presence of a flowing buffer gas
such as He at a total pressure of 25 to 100 Pa. The
ions rapidly attain thermal equilibrium by many non-
-reactive collisions with the buffer gas. The ther-

M. A. Almoster Ferreira (ed.), Ionic Processes in the Gas Phase, 87–110.
© *1984 by D. Reidel Publishing Company.*

malized ions then react with the neutral reactant
which is added downstream from the ionisation region.
The masses of the ions are measured at the end of the
flow tube by means of a quadrupole mass filter. The
relation between the reactant and product ions can be
established by the selected ion flow tube (SIFT)
method (9,10). In addition to this, the FA method is
considered to be one of the most reliable methods for
obtaining quantitative thermodynamic data and rate
constants of ion-molecule reactions.
The introduction of Fourier transformation in ion cyclotron resonance (ICR) experiments by Comisarow and
Marshall (11) has highly improved this method. In a
Fourier transform ion cyclotron resonance (FT-ICR)
spectrometer (11,12) ions are generated by electron
impact in a cubic cell at total pressures of 1 to 100
µPa. Due to the combined effects of magnetic and
electric fields these ions are trapped in the cell
and start to travel in circular paths. After a certain time -referred to as trapping time- during which
reactions with neutral molecules can take place, the
ions are excited to larger orbits by a fast frequency-
-swept radiofrequency pulse. The coherent motion of
all the excited ions induces image currents in a circuit shunting two receiver plates. These currents are
amplified, transmitted to a computer and Fourier
transformed to generate an interpretable mass spectrum.
Important features of the FT-ICR method are the rapid
acquisition of spectra and the extremely high resolution (13,14). A particularly valuable feature for
ion-molecule reaction studies is the possibility to
excite mass selected ions during the trapping period,
which can be used in three ways:
a) the ions can be excited translationally so that
   they will decompose upon collision with a neutral
   molecule (15);
b) all ions of one specific mass-to-charge ratio can
   be ejected from the cell by "over-exciting" their
   cyclotron motions, so that their reaction products
   can no longer be formed and thus will be absent in
   the spectrum (16);
c) ions of only one specific mass-to-charge ratio can
   be kept in the cell by ejection of all other ions
   so that only their product ions can be formed (17).
As with the SIFT method mentioned before, this enables to identify ion-molecule reaction pathways unequivocally.

GENERAL PICTURE OF ION—MOLECULE REACTIONS

Reactions between ions and molecules occurring under the experimental conditions in an FT-ICR or FA instrument are exothermic or thermoneutral. The rate constants for gas phase ion-molecule reactions can be 10 to 20 orders of magnitude larger than the rate constants for the corresponding reactions in solution. The values are often in the range of 10^{-9} - 10^{-10} cm^3 molecule^{-1} s^{-1} which is close to the collision rate constant (18). The high rates are a consequence of long range attractive ion-dipole/ion-induced dipole interactions between the approaching reactants. These interactions are effective before real chemical reaction occurs, so that the potential energy of the system will decrease initially. The conversion of potential energy to excess internal energy within the reactants could lead to the conjecture that all exothermic gas phase ion-molecule reactions will proceed at the collision rate. Measurements of rate constants for proton transfer reactions to delocalized anions (19) and sterically hindered pyridine bases (20) and for S_N2 substitutions (21,22) have shown, however, that the rates of exothermic reactions can be from almost collision controlled to too slow to be observed. This observation has been rationalized on the basis of the potential energy diagram shown in Fig. 1 for a hypothetical thermoneutral proton transfer reaction (19-23).

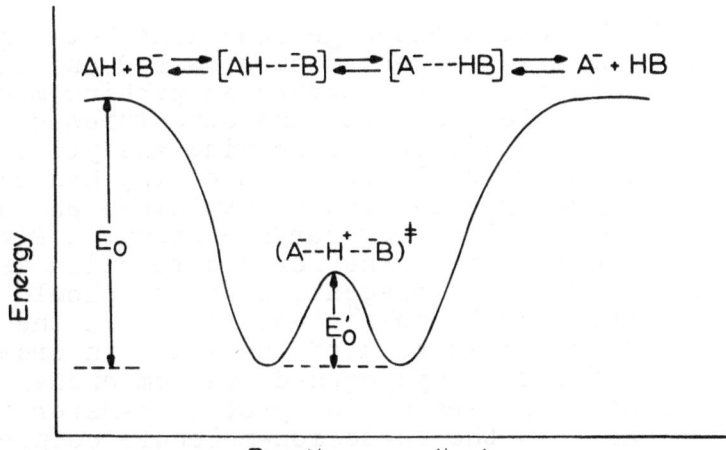

Fig. 1. Schematic potential energy-reaction coordinate diagram for a thermoneutral proton transfer reaction in the gas phase.

The loose ion-molecule complex formed initially corresponds to a minimum in potential energy and is believed to be relatively long lived. It can either
dissociate back to reactants or assume the configuration of the transition state. The transition state
corresponds to the energy maximum of the central barrier and is of lower energy than either the separated
reactants or products. Reaction leads to a second
loose ion-molecule complex which subsequently dissociates. The essential feature of this double-well potential model is that the central barrier is responsible for the wide variation in rate constants. The
height of the barrier will depend on the type of
reaction and on the chemical nature of the ion and
molecule under study.
The double-well potential model can be considered as
qualitative and simple (24). There is increasing evidence that primary reactions can lead to the formation
of loose ion-molecule complexes which are sufficiently
long-lived to allow secondary reactions to occur, as
will be shown below. These observations indicate that
the potential energy surface can be of a more complicated nature than shown in Fig. 1. In such cases the
actual potential energy surface will contain a series
of energy barriers each separating minima corresponding to loose ion-molecule complexes.

HYDROGEN-DEUTERIUM EXCHANGE REACTIONS

    Hydrogen-deuterium exchange reactions between
anions and deuterated reagents, such as D_2O and CH_3OD,
have proven to be a valuable method in probing mechanisms of ion-molecule reactions and structures of ions
in the gas phase (25-29) and in solving analytical
problems (30). The general mechanism of the hydrogen-
-deuterium exchange process can be explained on the
basis of the potential energy surface shown in Fig. 1
with the modification that the forward reaction is
endothermic (25-27). The reagent, however, should not
be more than about 80 kJ mol^{-1} less acidic in the gas
phase than the conjugate acid of the anion in order
to observe an effective hydrogen-deuterium exchange.
The anion formed irreversibly by proton transfer to
the OD^- ion can form the loose ion-molecule complex
1, if it meets a D_2O molecule. Part of the energy of
about 40-80 kJ mol^{-1}, released upon formation of
this complex, may be used to overcome the barrier
towards endothermic proton transfer leading to complex 2. This complex has insufficient energy to dis-

$$R_2CH_2 + OD^- \longrightarrow R_2CH^- + HDO$$

$$R_1CH^- + D_2O \rightleftharpoons [R_2CH^-\cdot D_2O]^* \rightleftharpoons [R_2CHD\cdot OD^-]^* \rightleftharpoons \quad (1)$$
$$\underline{1} \qquad\qquad\qquad \underline{2}$$

$$[R_2CD^-\cdot HDO]^* \rightleftharpoons R_2CD^- + HDO$$
$$\underline{3}$$

sociate, but proton transfer to the OD⁻ ion results in complex $\underline{3}$ capable of dissociation (see equation 1).

Recently, DePuy et al. (27) have reported on the exchange reactions between the 2-phenylallyl anion and D_2O. The allylic hydrogen atoms in 2-phenylpropene are slightly more acidic than those of water, whereas the aryl hydrogen atoms can be expected to be less acidic (31). Nevertheless, exchange of the four allylic hydrogen atoms and at least four aryl hydrogen atoms could be observed when the 2-phenylallyl anion was formed in the presence of D_2O:

$$C_6H_5-C\!\!\begin{array}{c}/CH_2^-\\ \\ \diagdown CH_2\end{array} \xrightarrow{\quad} \xrightarrow{D_2O} \longrightarrow C_6HD_4-C\!\!\begin{array}{c}/CD_2^-\\ \\ \diagdown CD_2\end{array} \qquad (2)$$

The energy gained upon approach of the anion and the D_2O molecule must be large enough to form a complex analogous to complex $\underline{2}$ (eq. 1) and to induce a second endothermic proton transfer from one of the ring positions to the OD⁻ ion in the complex.

It is evident from this result that the potential energy surface, describing the exchange of the hydrogen atoms of the 2-phenylallyl anion with D_2O, is more complicated than the one shown in Fig. 1.

Exchange of different types of hydrogen atoms within a carbanion with D_2O is also observed for the conjugate bases of 2-, 3- and 4-fluoranisole (29). Up to 6 hydrogen atoms in the conjugate bases of 2- and 4--fluoroanisole exchange with deuterium atoms, whereas the conjugate base of the more acidic 3-fluoroanisole exchanges 3 hydrogen atoms. In other words, the conjugate bases of 2- and 4-fluoroanisole exchange the aryl and the methyl hydrogen atoms when formed in the presence of D_2O:

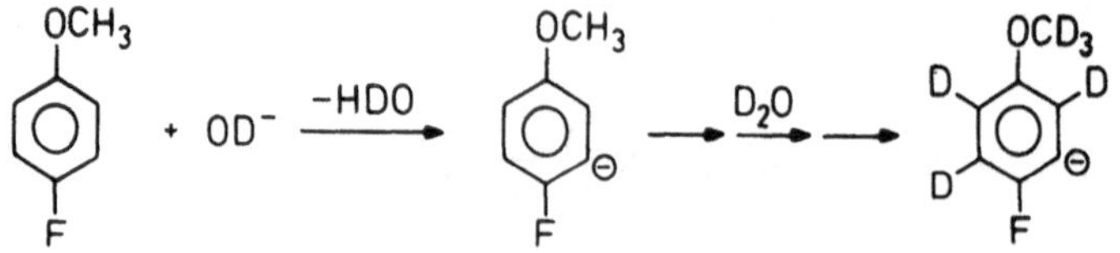

OCH$_3$ ... + OD$^-$ $\xrightarrow{-\text{HDO}}$ OCH$_3$ ... $\xrightarrow{D_2O}$ OCD$_3$ (3)

By following the hydrogen-deuterium exchange reactions of the conjugate base of 4-fluoroanisole as a
function of trapping time, it is observed that the
two types of hydrogen atoms exchange with different
rates. Exchange reactions in the conjugate base of 4-
-FC$_6$H$_4$OCD$_3$ with either H$_2$O or D$_2$O reveal that the
aryl hydrogen atoms exchange with a rate much larger
than the methyl hydrogen atoms. A mechanism, rationa-
lizing the exchange in the conjugate base of 4-fluoro-
anisole, is shown in Scheme 1.
Formation of complex 5 followed by proton abstraction
from one of the ring positions can account for the
exchange of the aryl hydrogen atoms. The essential
feature of the mechanism is that the excess internal
energy arising from the ion-dipole/ion-induced dipole
interactions occasionally can promote an unfavoured
intramolecular proton transfer from the methyl group

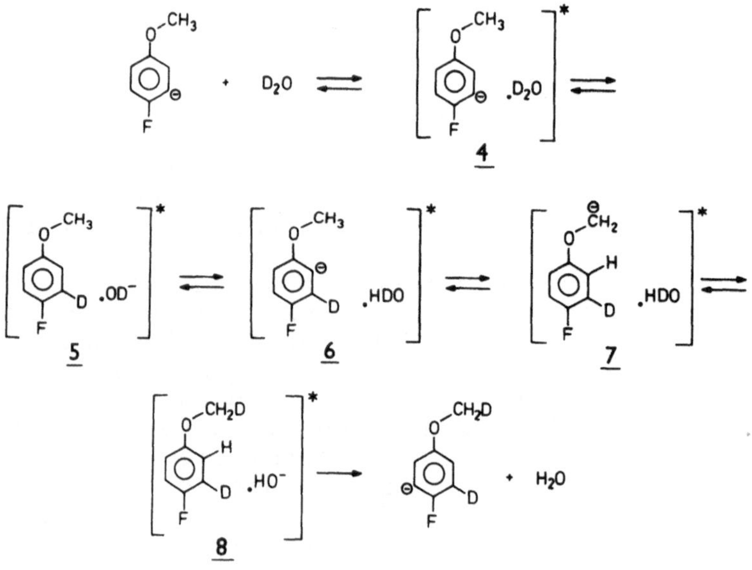

CH$_3$ ... + D$_2$O $\rightleftharpoons$ [CH$_3$ ... .D$_2$O]* $\rightleftharpoons$

4

[CH$_3$ ... D .OD$^-$]* **5** $\rightleftharpoons$ [CH$_3$ ... D .HDO]* **6** $\rightleftharpoons$ [CH$_2$ ... H D .HDO]* **7** $\rightleftharpoons$

[CH$_2$D ... H D .HO$^-$]* **8** $\rightarrow$ CH$_2$D ... D + H$_2$O

Scheme 1. Rationalization of the exchange of the hydrogen atoms in the conjugate base of 4-fluoroanisole
in the presence of D$_2$O.

to the negative charge located at the 2 position with respect to the methoxy group in complex 6. This results in a primary carbanion which rapidly can abstract a deuteron from the HDO molecule present in complex 7.
This simple rationalisation of the exchange of the methyl hydrogen atoms is supported by the observation that reaction between NH_2^- and $4-FC_6H_4OCD_3$ results in the formation of the $[M-H]^-$ ion and the product ions given in eqs. 4a and 4b.

$$
\begin{array}{c}
\text{OCD}_3 \\
\underset{\text{F}}{\bigcirc} + NH_2^-
\end{array}
\longrightarrow
\begin{cases}
\xrightarrow{50\%} C_6DH_3F^- + CD_2O + NH_3 & (4a) \\
\xrightarrow{50\%} C_6D_2H_2F^- + CDHO + NH_3 & (4b)
\end{cases}
$$

The same result is obtained, if NH_2^- reacts with $3-FC_6H_4OCD_3$.
In the $2-FC_6H_4OCD_3$ case only the $C_6DH_3F^-$ ion and the $[M-H]^-$ ion are formed.
The important observation here is that the ratio between $[C_6DH_3F^-]$ and $[C_6D_2H_2F^-]$ is unity, which means that there is a complete equilibration between the three deuterium atoms at the methyl group and the ortho hydrogen atom prior to loss of formaldehyde. An intermediate primary carbanion must be formed and must live sufficiently long to allow this equilibration before fragmentation occurs.

NUCLEOPHILIC AROMATIC SUBSTITUTION

Gas phase nucleophilic aromatic substitutions have recieved rather limited attention (29, 32-35). The known reactions can best be described as proceeding by the addition-elimination pathway, that is the S_NAr mechanism (36).
Particularly interesting examples are the reactions of methyl pentafluorophenyl ether (35) and 2-fluoroanisole (29) with nucleophiles to form long lived F^- ion-molecule complexes (Scheme 2).
Nucleophilic attack on the fluorine substituted carbon atoms leads to formation of a σ-anion complex, which might be on either a local minimum or maximum on the potential energy surface. Simple dissociation of the F^- ion-molecule complex is not observed in any of the methyl pentafluorophenyl ether systems studied (35). The F^- reattacks instead the newly formed molecule under formation of new ion-molecule complexes

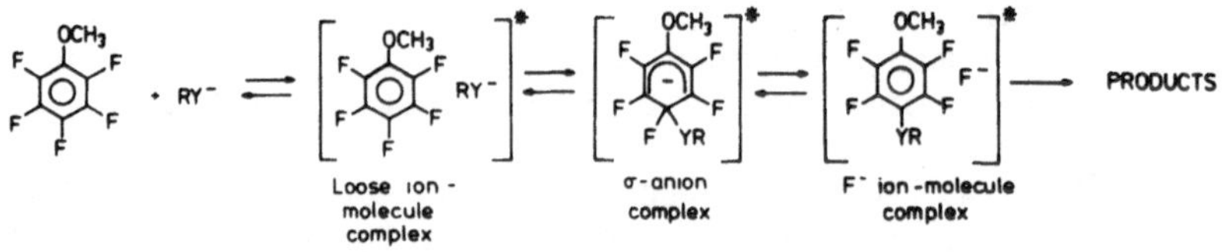

Scheme 2. Formation of the F^- ion-molecule complex.

which subsequently dissociate. Although the lifetime
of the F^- ion-molecule complexes is not known, they
must live long enough to allow secondary reactions to
occur. These can be classified as: proton transfer,
S_N2 substitution and E2 elimination.
Proton transfer to the displaced F^- ion is the dominant reaction, if the neutral in the complex is more
acidic than HF. This is the case when the primary
reactant ions are NH_2^-, OH^-, SH^-:

$$C_6F_5OCH_3 + YH^- \longrightarrow [HYC_6F_4OCH_3 \cdot F^-]^* \longrightarrow CH_3OC_6F_4Y^- + HF$$

$$Y = O, S, NH \tag{5}$$

The major part of the even electron ions formed is
observed to undergo a fast unimolecular reaction to
give a radical anion, if Y = NH, O. With the assumption that the OH^- ion has attacked the para position
with respect to the methoxy group, the radical anion
of para-tetrafluorobenzoquinone (EA = 2.92 eV (37))
will be formed:

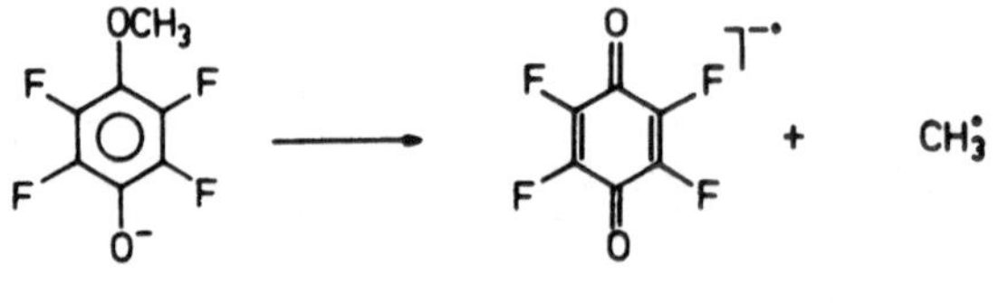

$$\tag{6}$$

The formation of relatively long-lived F^- ion-molecule
complexes is proven by reaction between CD_3O^- and
$C_6F_5OCH_3$:

$$C_6F_5OCH_3 + CD_3O^- \longrightarrow [CD_3OC_6F_4OCH_3 \cdot F^-]^* \xrightarrow{S_N2}$$

$$\xrightarrow{S_N2} \begin{array}{l} \xrightarrow{50\%} CD_3OC_6F_4O^- + CH_3F \tag{7a} \\ \\ \xrightarrow{50\%} CH_3OC_6F_4O^- + CD_3F \tag{7b} \end{array}$$

This result shows that a concerted formation and breakdown of the σ-anion complex in analogy with the mechanism, proposed to account for formation of fluorinated enolate anions in gas phase systems composed of alkoxide ions and either CF_2CF_2 (38) or CH_2CF_2 (38,39), can be excluded.

In the case of reaction between CH_3S^- and $C_6F_5OCD_3$ reattack by F^- occurs nearly exclusively on the methyl group bonded to the oxygen atom:

$$C_6F_5OCD_3 + CH_3S^- \longrightarrow [CH_3SC_6F_4OCD_3 \cdot F^-]^* \xrightarrow{S_N2}$$

$$\xrightarrow{S_N2} \begin{cases} \xrightarrow{> 99\%} CH_3SC_6F_4O^- + CD_3F \quad\quad (8) \\ \\ \xrightarrow{< 1\%} CD_3OC_6F_4S^- + CH_3F \quad\quad (9) \end{cases}$$

We have rationalized this selectivity in reattack as mainly being due to the higher electropositive character of the carbon atom bonded to the oxygen atom (35).

These results clearly show that the potential energy surface can contain a series of minima. The fact, that selectivity in reattack by the F^- ion can be observed, indicates that the differences between the energy barriers for the secondary reactions control the distribution of the final products.

The multistep character of these processes is further illustrated by the reactions observed when enolate anions are used as reactant ions.

The ambient enolate anions may react with methyl pentafluorophenyl ether at the carbon or the oxygen site. If they react with the carbon site at the fluorine bearing carbon atoms, then the molecule in the formed F^- ion-molecule complex contains relatively acidic hydrogen atoms.

The enolate anion 9 formed by proton transfer to the F^- ion is not observed. An intramolecular nucleophilic aromatic substitution occurs instead and leads to a second F^- ion-molecule complex. The F^- ion in this complex reattacks then the formed substituted benzofuran molecule by either proton transfer or S_N2 substitution.

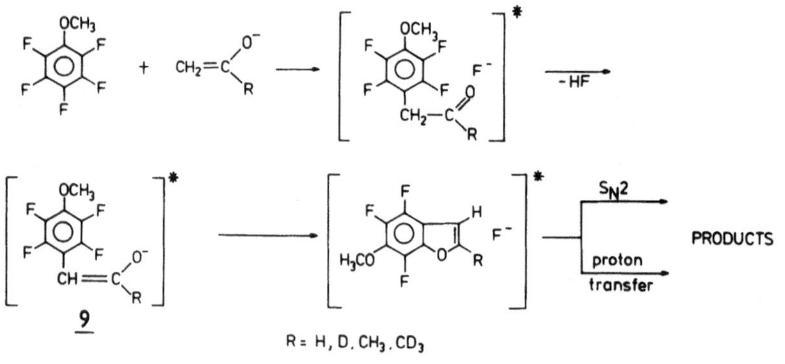

Scheme 3. Carbon site attack in the enolate anions/
$C_6F_5OCH_3$ systems.

The displaced F^- ion can in principle induce an E2
elimination, if the original reactant ion contains
β-hydrogen atoms. For example:

$$C_6F_5OCH_3 + CH_3CH_2O^- \longrightarrow [CH_3CH_2OC_6F_4OCH_3 \cdot F^-]^* \longrightarrow$$

$$\xrightarrow{\text{E2}} CH_3OC_6F_4O^- + CH_2CH_2 + HF \tag{10}$$

$$\xrightarrow{S_N2} CH_3OC_6F_4O^- + CH_3CH_2F \tag{11}$$

$$\xrightarrow{S_N2} CH_3CH_2OC_6F_4O + CH_3F \tag{12}$$

$$[CH_3CH_2OC_6F_4O^-] : [CH_3OC_6F_4O^-] = 1 : 9$$

The observed ratio between the abundances of the two
product ions indicates that the $CH_3OC_6F_4O^-$ ion is
formed by an E2 elimination. Increasing the number of
β-hydrogen atoms in the reactant anion favours formation of $CH_3OC_6F_4O^-$ even more:

$$C_6F_5OCH_3 + (CH_3)_3CO^- \longrightarrow [(CH_3)_3COC_6F_4OCH_3 \cdot F^-]^* \longrightarrow$$

$$\xrightarrow[4\%]{S_N2} (CH_3)_3COC_6F_4O^- + CH_3F \tag{13}$$

$$\xrightarrow[96\%]{E2} CH_3OC_6F_4O^- + (CH_3)_2C=CH_2 + HF \tag{14}$$

Changing of the reactant ion to $(CH_3)_3CS^-$ alters the competition between E2 elimination and S_N2 substitution:

$$C_6F_5OCH_3 + (CH_3)_3CS^- \longrightarrow [(CH_3)_3CSC_6F_4OCH_3 \cdot F^-]^* \longrightarrow$$

$$\underbrace{S_N2}_{29\%} \blacktriangleright (CH_3)_3CSC_6F_4O^- + CH_3F \tag{15}$$

$$\underbrace{E2}_{71\%} \blacktriangleright CH_3OC_6F_4S^- + (CH_3)_2C{=}CH_2 + HF \tag{16}$$

The effect of changing $(CH_3)_3CO^-$ to $(CH_3)_3CS^-$ apparently is to raise the height of the local energy barrier towards E2 elimination, thereby making it less favoured over S_N2 substitution. This can be explained on the basis of the difference in electronegativity of the oxygen and the sulphur atom as follows: in the transition state for the E2 elimination the developing negative charge will be accommodated less effectively by the sulphur atom than by the oxygen atom.

ELIMINATION REACTIONS

     Although elimination reactions have been shown to be facile processes in the gas phase (vide infra), it is often difficult to distinguish them from substitution reactions: both reactions mostly lead to the same product ions, but not to the same neutral products, which however are not known in most experiments (40). The reactions of OH^- with diethyl ether, for example, leads to $C_2H_5O^-$ ions (41, 42), which can be explained by both an S_N2 and an E2 mechanism (reactions 17 and 18, respectively).

$$OH^- + C_2H_5OC_2H_5 \longrightarrow$$

$$\xrightarrow{S_N2} C_2H_5OH + C_2H_5O^- \tag{17}$$

$$\xrightarrow{E2} H_2O + C_2H_4 + C_2H_5O^- \tag{18}$$

$$\longrightarrow C_2H_4 + C_2H_5O^- \cdot H_2O \tag{19}$$

In this case the S_N2 reaction is exothermic by about 38 kJ mol^{-1}, whereas the E2 reaction is slightly endothermic, but it may be that elimination is favoured because of the large increase in entropy (reaction 18: $\Delta H = +8$ kJ mol^{-1}, $\Delta G = -39$ kJ mol^{-1}) (41, 42). The additional formation of hydrated ethoxide ions $(C_2H_5O^- \cdot H_2O)$, which can only arise from an elimina-

tion process (reaction 19), supports this view. It
has been shown that the abundance of such solvated
alkoxide ions increases with decreasing exothermicity
of the E2 reaction, even though the S_N2 reaction is
still highly exothermic in some cases (41, 42).
Further support for the elimination mechanism is derived from the observation that the reaction of OH^-
with diethyl ether is more than 2000 times faster
than the reaction of OH^- with dimethyl ether (42).
This large difference indicates that the two reactions proceed via different mechanisms. Substitution
is the only pathway open for the reaction of OH^- with
dimethyl ether, so that it is clear that the E2 channel dominates in the case of diethyl ether.
The reactions of cyclic ethers are a good probe to
study elimination reactions, because the "leaving
group" remains with the anion. For example, the reaction of NH_2^- with tetrahydrofuran (THF, 10) leads to
$(M-H)^-$ ions. Deuterium labelling has shown that the
proton is abstracted exclusively from the β-position
(42, 43). This points to the elimination process
shown in reaction 20:

$$NH_2^- + \quad \beta\!\!\underset{\alpha}{\diagdown}\!\!\underset{O}{\Box} \quad \longrightarrow \quad CH_2=CH-CH_2-CH_2-O^- + NH_3 \qquad (20)$$

$$\underline{10}$$

In some of the cyclic ethers the product ions may be
formed with excess internal energy because of the
strain relieved upon ring opening. An example is seen
in the reaction of NH_2^- with 2-methyloxetane 11. Although the initial product ion is the same as in the
reaction of NH_2^- with THF, it has enough energy because of loss of ring strain to allow further fragmentation (43):

$$NH_2^- + \quad \overset{CH_3}{\underset{O}{\Box}} \quad \longrightarrow \quad CH_2=CH-CH_2-CH_2-O^- + NH_3$$

$$\underline{11} \qquad\qquad\qquad \downarrow \qquad\qquad\qquad\qquad (21)$$

$$C_3H_5^- + CH_2O$$

Furthermore, in the reaction of OH^- with 11 almost no
(< 2%) hydrated alkoxide ions are formed, whereas in
the reaction of OH^- with THF these ions make up 54%
of the spectrum (43). This is in line with the observation that the abundance of solvated alkoxide ions

is dependent on the exothermicity of the corresponding reaction.
From the reactions in cyclic ethers indications have
been found to support the view that gas phase E2
reactions of ethers prefer transition states in which
the oxygen atom and the β-hydrogen are antiperiplanar, like those in solution. For example, oxetane and
ethylene oxide, which both lack the possibility of
antiperiplanar stereochemistry, do not react via an
E2 mechanism, whereas THF and tetrahydropyran, which
can assume antiperiplanar configurations, do (43).
Interesting examples of elimination reactions are also seen in the reactions of cyclic sulphides (44).
1,3-Dithiane 12 and 1,3-dithiolane 13 react with several bases in two ways, as has been confirmed by deuterium labelling:

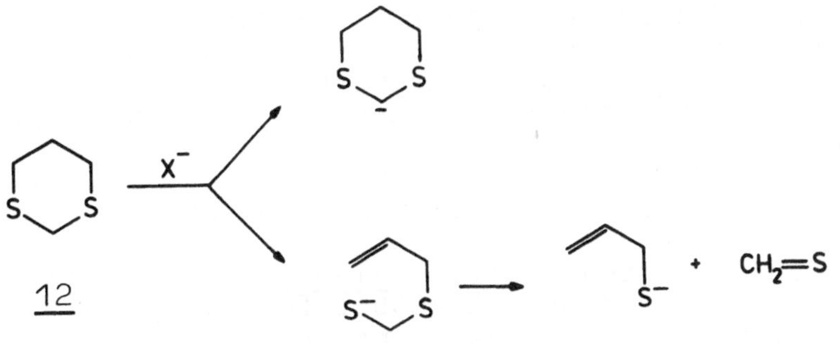

(22)

(23)

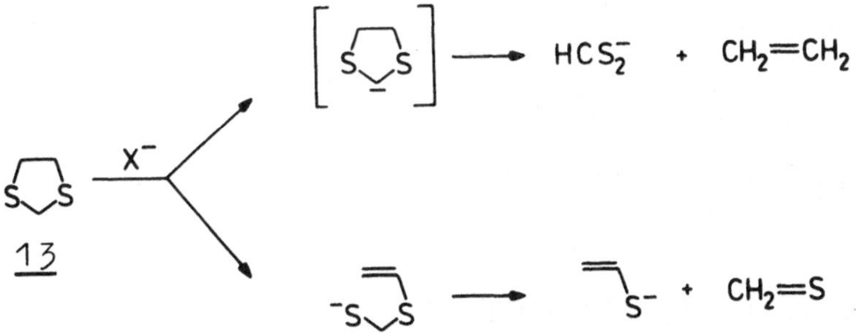

(24)

(25)

The actual product ion distribution is dependent on
the base strength of the bases used. On the basis of
the occurrence/non-occurrence of the reaction channels thermochemical information has been derived
(44).

For 2,2-dimethylthioxolane $\underline{14}$ there are two possible
elimination pathways:

$$\left[\begin{array}{c} O \diagdown S^- \\ \times \end{array} \right] \longrightarrow O^- + \overset{S}{\diagdown} \qquad (26)$$

$$\underline{15}$$

$$\underline{14}$$

$$\left[\begin{array}{c} ^-O \diagdown S \\ \times \end{array} \right] \longrightarrow \diagup S^- + \overset{O}{\diagdown} \qquad (27)$$

$$\underline{16}$$

No CH_2CHO^- ions are observed, so that it can be concluded that elimination proceeds via abstraction of
the proton adjacent to sulphur. This results in the
more stable product ion (CH_2CHS^-), although the intermediate ion $\underline{16}$ is less stable than ion $\underline{15}$. This
might point to a concerted mechanism in which no such
intermediate ion is formed, but it might also be that
the higher acidity of the proton adjacent to sulphur
compared to that adjacent to oxygen is the direction
controlling factor.
As has been discussed earlier (vide supra), ion-molecule complexes can remain together for sufficient
time to allow secondary reactions to occur. This is
also shown by the reactions of ethylene glycol methyl
alkyl ethers. In addition to other product ions, the
enolate anion of acetaldehyde is formed (42). This
ion may arise from a second E2 reaction within the
initially formed complex $\underline{17}$.

$$B^- + CH_3OCH_2CH_2OCH_2CH_2R \xrightarrow{E2}$$

$$BH + [CH_3O^- + CH_2=CHOCH_2CH_2R]^* \xrightarrow{E2}$$

$$\underline{17}$$

$$CH_2CHO^- + CH_2=CHR + CH_3OH \qquad (28)$$

The direction of elimination has been investigated by
using mixed ethers. It has been shown that NH_2^- preferentially abstracts a proton from the larger of the
two alkyl groups, forming the smaller alkoxide ion
(42). The most dramatic example is given in the reaction of NH_2^- with ethyl tert-butyl ether which yields

almost exclusively $C_2H_5O^-$ and $C_2H_5O^-.NH_3$. This has
been interpreted by means of an E1cB type mechanism
in which the breaking of the carbon-hydrogen bond is
most important, so that the stability of the intermediate carbanion plays a decisive role. Proton abstraction from the ethyl group would lead to ion 18,
which will be less stable than ion 19, generated by
proton abstraction from the tert-butyl group (42).

$$H_2\overset{..}{N}{}^-...{}^{\delta+}H...{}^{\delta-}CH_2CH_2OC(CH_3)_3 \qquad C_2H_5O-\underset{\underset{CH_3}{|}}{\overset{\overset{CH_3}{|}}{C}}-CH_2^{\delta-}...{}^{\delta+}H...{}^-\overset{..}{N}H_2$$

18   19

The direction of elimination in the reactions of OH^-
with mixed ethers is less pronounced, probably because the results are obscured by the high amount of
hydrated alkoxide ions formed (42).
Elimination reactions have also been observed in aromatic ethers (45). The anions NH_2^-, $C_2H_5O^-$ and
n-$C_4H_9O^-$, which hardly react or do not react at all
with methyl phenyl ether generate $C_6H_5O^-$ ions in high
abundance from ethyl phenyl ether. This indicates
that these bases react with ethyl phenyl ether by
elimination, whereas they react with methyl phenyl
ether mainly by S_N2 substitution. This S_N2 process,
however, may become so slow that an ipso substitution
can start to compete, as has been shown by the
reaction of $^{18}OH^-$ with methyl phenyl ether which
yields 85% $C_6H_5O^-$ (S_N2 reaction) and 15% $C_6H_5^{18}O^-$
(ipso substitution) (45).
The reactions of ethyl phenyl ether show that the
elimination reaction needs not to be as straightforward as one might think. For example, reaction of the
side-chain d_5-labelled compound $(C_6H_5OC_2D_5)$ with NH_2^-
yields mainly $C_6H_4DO^-$ (46). This shows that the elimination occurs after abstraction of a proton from
the aromatic ring:

A similar conclusion can be drawn from the observation that reaction of OD⁻ with ethyl phenyl ether leads to a 1:1 ratio of $C_6H_5O^-$ and $C_6H_4DO^-$ (46). Obviously, exchange of a ring proton with OD⁻ occurs prior to the elimination reaction.

SUBSTITUTION VERSUS ADDITION-ELIMINATION

Important conclusions concerning the potential energy surface of gas phase reactions can be drawn, if two or more reactions are in competition. Examples can be seen in the previous section. In many cases substitution is far more exothermic than elimination. However, from the observation that elimination quite often is favoured over substitution it can be concluded that the central barrier in the potential energy surface of substitution reactions can be much higher than that of elimination reactions. In this section the competition between substitution and addition-elimination reactions will be discussed.

For example, the reaction of OH⁻ with phenyl trifluoroacetate yields -in addition to other products- $C_6H_5O^-$ and CF_3COO^- ions (47). Use of $^{18}OH^-$ has shown that the latter are formed via a $B_{AC}2$ mechanism:

$$OH^- + CF_3-\overset{O}{\overset{\|}{C}}-OC_6H_5 \xrightarrow{B_{AC}2} \left[CF_3-\overset{O^-}{\underset{OH}{\overset{|}{\underset{|}{C}}}}-OC_6H_5 \right]^* \longrightarrow$$

$$\underline{20}$$

$$CF_3COO^- + C_6H_5OH \qquad\qquad\qquad (30)$$

However, it is unlikely that the intermediate $\underline{20}$ will dissociate into the phenoxide anion and neutral trifluoroacetic acid, which is less exothermic than reaction 30 by about 112 kJ mol⁻¹. Therefore, the phenoxide anions are probably formed via the S_N2 reaction 31:

$$OH^- + CF_3-\overset{O}{\overset{\|}{C}}-OC_6H_5 \xrightarrow{S_N2} CF_3OH + CO + C_6H_5O^- \quad (31)$$

This is supported by the observation that the abundance ratio of the trifluoroacetate and phenoxide ions decreases from 3 to almost 1 (47) upon ejection of the collision complex between OH⁻ and $C_6H_5OCOCF_3$, having a nominal mass of 207 daltons, but not seen in the normal ICR mass spectra. This observation has been explained as follows:

The $B_{AC}2$ reaction leading to CF_3COO^- ions must involve
the tetrahedral type of ion 20, which most probably
corresponds with a minimum in the potential energy
surface. The $B_{AC}2$ reaction should be described, therefore, by a triple-well potential model, where the
outer minima will correspond with the clustered reactant and product ions and the central minimum with the
tetrahedral intermediate ion 20.
For an S_N2 reaction, however, a double-well potential
model has been proposed (21,22). The two minima in
this model correspond again with the clustered reactant and product ions, but now the species essential
for the reaction (ion 21) will lie on a potential maximum.

$$\left[HO \ldots \overset{\overset{F}{|}}{\underset{/\backslash}{C}} \ldots \overset{\overset{O}{||}}{C} \ldots OC_6H_5 \right]^{*-} \qquad \underline{21}$$

$$\underset{F \quad F}{}$$

The differences between reactions 30 and 31 is therefore
an additional minimum in the potential energy surface
for the former reaction. On the basis of this difference the ion-molecule complex in the $B_{AC}2$ reaction is
expected to live longer than that in the S_N2 reaction.
The former complex will thus be more sensitive to radiofrequency pulses applied to eject it than the latter, as is observed.
The competition between $B_{AC}2$ and S_N2 reactions is also
demonstrated by the reactions of alkoxide ions with
dialkyl carbonates (48,49). The products of these
reactions can be explained by the following reactions:

$$R^1O^- + (R^2O)_2CO \xrightarrow{B_{AC}2} \left[\begin{array}{c} O^- \\ | \\ R^2O-C-OR^2 \\ | \\ OR^1 \end{array} \right]^* \begin{array}{c} \nearrow R^2OCO_2^- \quad (32a) \\ \searrow R^1OCO_2^- \quad (32b) \end{array}$$

$$R^1O^- + (R^2O)_2CO \xrightarrow{S_N2/E2} R^2OCO_2^- \qquad (33)$$

For R^2 = ethyl or propyl no distinction can be made
whether reaction 33 proceeds via an S_N2 or an E2 mechanism. Based on the high amount of $R^2OCO_2^-$ formed
in these cases compared to R^2 = methyl the E2 channel
is most likely. Therefore, the best probe to study the
S_N2 versus $B_{AC}2$ reactions is the reaction of methoxide
with dimethyl carbonate. Deuterium labelling has shown
that in this case 70% of the product ions are formed
via the $B_{AC}2$ channel (reactions 32a and 32b), whereas
the remaining 30% is formed via reaction 33 (48). Use
of $CH_3{}^{18}O^-$ has shown that the $B_{AC}2$ reaction proceeds

through a symmetrical decomposing ion, as would be expected (48,49).
Similar observations have been made for some methyl
esters ($RCOOCH_3$, R = CF_3, C_6H_5). Reaction of CD_3O^-
ions with these compounds only yields $RCOO^-$ ions,
which at first sight would suggest that they proceed
purely via an S_N2 mechanism (50). However, use of
$CH_3{}^{18}O^-$ has revealed substantial ^{18}O incorporation in
the product ions (51). This points to a mechanism in
which a tetrahedral intermediate (ion 22) is formed.
The S_N2 reaction then occurs after breaking of this
intermediate in either $[CH_3{}^{18}O^- + RCOOCH_3]^*$ or
$[CH_3O^- + RCO^{18}OCH_3]^*$.

$$
\begin{array}{c}
O^- \\
| \\
R\text{-}C\text{-}OCH_3 \qquad \underline{22} \\
{}^{18}| \\
{}^{18}OCH_3
\end{array}
$$

However, it has been shown (51) that a direct S_N2 mechanism still competes with the formation of the tetrahedral intermediate.
This is also shown by the reaction of allyl anion
($C_3H_5{}^-$) with methyl trifluoroacetate (52). The major
reaction channel appears to be an S_N2 reaction leading
to CF_3COO^- ions, but the product ions $CF_3COCHCHCH_2{}^-$
and $CH_3OCOCHCHCH_2{}^-$ point to a $B_{AC}2$ reaction:

$$
C_3H_5{}^- + CF_3COOCH_3 \longrightarrow \left[\begin{array}{c} O^- \\ | \\ CF_3\text{-}C\text{-}OCH_3 \\ | \\ CH_2CH=CH_2 \end{array} \right]^* \longrightarrow
$$

$$
\nearrow \quad \begin{array}{c} O^- \\ | \\ CF_3C=CH\text{-}CH=CH_2 \end{array} + CH_3OH \qquad\qquad (34a)
$$

$$
\longrightarrow \quad \begin{array}{c} O^- \\ | \\ CH_3OC=CH\text{-}CH=CH_2 \end{array} + CF_3H \qquad\qquad (34b)
$$

The additional formation of $CF_3{}^-$ is probably not due
to a $B_{AC}2$ reaction, but to decomposition of excited
CF_3COO^- formed via the S_N2 reaction (52). However, it
might be that $CF_3{}^-$ is formed by electron transfer from
$C_3H_5{}^-$ to neutral CF_3COOCH_3. The resulting $CF_3COOCH_3{}^{-\cdot}$
would then decompose to give $CF_3{}^-$. The electron affinity of $C_3H_5{}^-$ is about 0.55 eV, whereas fluorinated
compounds usually have much larger electron affinities
(53).
Interesting examples of addition-elimination reactions

are also seen in the reactions of some anions with
carboxylic esters (54). For example, enolate anions
react with their carbon site via the $B_{AC}2$ mechanism to
eliminate the alkoxide ion which then abstracts a proton from the generated 1,3-diketone species. With a
triple-well potential model it has been possible to
account for this sequence of steps in a semi-quantita-
tive way (54).
Addition-elimination reactions in competition with S_N2
have also been observed for reactions of various anions with aliphatic nitrites (55,56). A nice example
is the addition-elimination of enolate ions to nitrites which results mainly in three product ions, as
has been measured by the FA method (56):

$$X-\overset{O}{\overset{\|}{C}}-CHR^- + R^1ONO \longrightarrow \left[X-\overset{O}{\overset{\|}{C}}-CHR-\underset{\underset{OR^1}{|}}{N}-O^- \right]^* \longrightarrow$$

$$\left[\overset{O}{\overset{\|}{X-\overset{}{C}}}-CR-N=O \right]^*$$

$$\xrightarrow{a} \overset{O}{\overset{\|}{X}}C-CR^--NO + R^1OH \qquad (35a)$$

$$\underset{23}{} \longrightarrow XCOO^- + RCN \qquad (35b)$$

$$\xrightarrow{c} {}^-CHRNO + XCOOR^1 \qquad (35c)$$

Reaction 35c involves two consecutive addition-elimination steps. It has been proposed that reaction 35b
can only occur in the syn form of ion 23 (56).
These reactions have been used to distinguish isomeric
enolate anions generated by reaction of fluoride, amide or hydroxide with the corresponding isomeric trimethylsilyl enol ethers (57).

HYDRIDE TRANSFER REACTIONS

Only a few examples of gas phase hydride transfer
reactions have been reported. The earliest examples
are the transfer of a hydride from HCO^- to formaldehyde (58) and of a deuteride from $DNO^-•$ to $(CH_3)_3B$
(59). Other examples are the hydride transfer from the
conjugate base of 1,4-cyclohexadiene to benzaldehyde
(60) and from alkoxide ions to singlet oxygen (61).
One of the reactions involving hydride transfer, which
has synthetic importance in solution chemistry, is the
Meerwein-Ponndorf-Verley reduction (62). In this reaction an alkoxide donates a hydride to a carbonyl

group. Such a reaction can also occur in the gas phase
as shown by the hydride transfer from methoxide to
acrolein (63). However, because in this case methyl
nitrite was used to generate CH_3O^- ions, reaction of
HNO^-· having the same nominal mass is difficult to ex-
clude. This problem can be overcome by the FT-ICR
method. Methoxide ions can now be generated by reaction of OH^- with methanol, after which residual OH^-
and its other reaction products can be ejected. It is
then possible to study the reactions of CH_3O^- or
CD_3O^-, if CD_3OD is used, with the substrate of interest. In this way it has been shown that methoxide is
capable of reducing a number of aldehydes lacking α-
-hydrogen atoms (RCHO, R=H, C_6H_5, $(CH_3)_3C$,1-adamantyl)
(64), but it can also reduce a compound like carbon
disulphide (65).

The reaction between CD_3O^- and formaldehyde gives some
information on the complex involved in this reaction.
From the observation that the abundance of the CDH_2O^-
ions is much higher than that of the CD_2HO^- ions, it
can be concluded that k_1 is much greater than k_2 in
reaction 36 (64).

$$CD_3O^- + CH_2O \rightleftharpoons [CD_3O^-.CH_2O]^*$$

$$[CD_2O.CDH_2O^-]^* \xrightarrow{k_1} CD_2O + CDH_2O^- \quad (36)$$
$$\underline{24}$$

$$k_2 \updownarrow$$

$$[CD_2HO^-.CDHO]^* \longrightarrow CDHO + CD_2HO^-$$
$$\underline{25}$$

The ions $\underline{24}$ and $\underline{25}$ can transfer a hydride to another
formaldehyde molecule to generate CH_3O^- ions. All
product ions can react with neutral CD_3OD to generate
CD_3O^- again. The system is thus predicted to reach
equilibrium, as is observed (64). It can be concluded
that the hydride transfer reaction is competitive with
the proton transfer reaction, both reactions being
nearly thermoneutral.

Hydride transfer reactions from alkoxides other than
methoxide to aldehydes have also been observed (64).
As mentioned before, the product ions of the reactions
of OH^- with the substrates have been ejected during
the hydride transfer study (64). However, one of these
reactions generated ions which are unexpected at the
low pressures used in an FT-ICR instrument, which is

the reaction of OH^- with formaldehyde yielding $OH^- \cdot H_2O$ and $CH_3O_2^-$ ions. It is hard to believe that these ions are formed by direct addition of OH^- to H_2O and CH_2O, respectively, because third-body collisions to stabilize the adducts are very unlikely in FT-ICR. Further investigations have shown that the ions are formed via H_3O^- (66).

Deuterium and ^{18}O labelling studies have shown that the structure of this ion most likely resembles a hydrated hydride: $H^- \cdot H_2O$. This is formed via a two-step process in the collision complex of OH^- and formaldehyde:

$$OH^- + CH_2O \rightleftharpoons [OH^- \cdot CH_2O]^* \rightleftharpoons [H_2O \cdot HCO^-]^* \rightleftharpoons$$
$$\underset{26}{} $$
$$[H^- \cdot H_2O \cdot CO]^* \longrightarrow H^- \cdot H_2O + CO \tag{37}$$

The first step is proton transfer from formaldehyde to OH^-, but the resulting complex 26 cannot dissociate, because water is more acidic than formaldehyde (58). In the second step HCO^- transfers a hydride to water, resulting in H_2O^- and CO.

On the basis of the observed reactions the heat of formation of H_3O^- has been bracketed (66).

ACKNOWLEDGEMENT

The authors wish to thank the Netherlands Organisation for Pure Research (SON/ZWO) for financial support.

REFERENCES

1. Bowie, J.H.: 1980, Acc. Chem. Res. 13, pp. 76-82.
2. DePuy, C.H., and Bierbaum, V.M.: 1981, Acc. Chem. Res. 14, pp. 146-153.
3. Budzikiewicz, H.: 1981, Angew. Chem. Int. Ed. Engl. 20, pp. 624-637.
4. Nibbering, N.M.M.: 1981, Recl. Trav. Chim. Pays--Bas 100, pp. 297-306.
5. Ausloos, P. (Ed.): 1979, "Kinetics of Ion-Molecule Reactions" NATO ASI Vol. B40, Plenum Press, New York.
6. Smith, D., and Adams, N.G.: 1979, in "Gas Phase Ion Chemistry" (Bowers, M.T., Ed.) Vol. 1, Academic Press, New York, Chapter 1, pp. 1-44.
7. Bierbaum, V.M., DePuy, C.H., Shapiro, R.H., and Stewart, J.H.: 1976, J. Am. Chem. Soc. 98, pp. 4229-4235.

8. Lehman, T.A., and Bursey, M.M.: 1976, "Ion Cyclo-
    tron Resonance Spectrometry", Wiley-Interscience,
    New York.
9. Adams, N.G., and Smith, D.: 1976, Int. J. Mass
    Spectrom. Ion Phys. 21, pp. 349-359.
10. Adams, N.G., and Smith, D.: 1976, J. Phys. B 9,
    pp. 1439-1451.
11. Comisarow, M.B.: 1978, in "Transform Techniques
    in Chemistry" (Griffiths, P.R., Ed.), Plenum
    Press, New York, Chapter 10, pp. 257-284 and re-
    ferences cited therein.
12. Wilkins, C.L., and Gross, M.L.: 1981, Anal. Chem.
    53, pp. 1661A-1676A.
13. White, R.L., Ledford, E.B., Jr., Ghaderi, S.,
    Wilkins, C.L., and Gross, M.L.: 1980, Anal. Chem.
    52, pp. 1527-1529.
14. Alleman, M., Kellerhals, Hp., and Wanczek, K.-P.:
    1980, Chem. Phys. Lett. 75, pp. 328-331.
15. Cody, R.B., Burnier, R.C., and Freiser, B.S.:
    1982, Anal. Chem. 54, pp. 96-101.
16. Comisarow, M.B., Grassi, V., and Parisod, G.:
    1978, Chem. Phys. Lett. 57, pp. 413-416.
17. Kleingeld, J.C., and Nibbering, N.M.M.: 1982,
    Org. Mass Spectrom. 17, pp. 136-139.
18. Su, T., and Bowers, M.T.: 1979, in "Gas Phase Ion
    Chemistry" (Bowers, M.T., Ed.), Vols 1 and 2,
    Academic Press, New York, Chapter 3, pp. 83-118.
19. Farneth, W.E., and Brauman, J.I.: 1976, J. Am.
    Chem. Soc. 98, pp. 7891-7898.
20. Jasinski, J.M., and Brauman, J.I.: 1980, J. Am.
    Chem. Soc. 102, pp. 2906-2913.
21. Olmstead, W.N., and Brauman, J.I.: 1977, J. Am.
    Chem. Soc. 99, pp. 4219-4228.
22. Pellerite, M.J., and Brauman, J.I.: 1980, J. Am.
    Chem. Soc. 102, pp. 5993-5999.
23. Asubiojo, O.I., and Brauman, J.I.: 1979, J. Am.
    Chem. Soc. 101, pp. 3715-3724.
24. Brauman, J.I.: 1979, in "Kinetics of Ion-Molecule
    Reactions" NATO ASI Vol. B40 (Ausloos, P., Ed.),
    Plenum Press, New York, pp. 153-164.
25. Stewart, J.H., Shapiro, R.H., DePuy, C.H., and
    Bierbaum, V.M.: 1977, J. Am. Chem. Soc. 99, pp.
    7650-7653.
26. DePuy, C.H., Bierbaum, V.M., King, G.K., Shapiro,
    R.H.: 1978, J. Am. Chem. Soc. 100, pp. 2921-2922.
27. Squires, R.R., DePuy, C.H., and Bierbaum, V.M.:
    1981, J. Am. Chem. Soc. 103, pp. 4256-4258.
28. Noest, A.J., and Nibbering, N.M.M.: 1980, J. Am.
    Chem. Soc. 102, pp. 6427-6429.

29. Ingemann, S., and Nibbering, N.M.M.: J. Org. Chem., in press.
30. Hunt, D.F., and Sethi, S.K.: 1980, J. Am. Chem. Soc. 102, pp. 6953-6963.
31. Bartmess, J.E., and McIver, R.T.: 1979, in "Gas Phase Ion Chemistry" (Bowers, M.T., Ed.), Vols 1 and 2, Academic Press, New York, Chapter 11, pp. 87-121.
32. Briscesse, S.M.J., and Riveros, J.M.: 1975, J. Am. Chem. Soc. 97, pp. 230-231.
33. Bowie, J.H., and Stapleton, B.J.: 1977, Aust. J. Chem. 30, pp. 295-300.
34. Bruins, A.P., Ferrer-Correia, A.J., Harrison, A. G., Jennings, K.R., and Mitchum, R.K.: 1977, Advan. Mass Spectrom. 7, pp. 355-358.
35. Ingemann, S., Nibbering, N.M.M., Sullivan, S.A. and DePuy, C.H.: J. Am. Chem. Soc., in press.
36. March, J.: 1977, "Advanced Organic Chemistry", 2nd Ed., McGraw Hill, New York.
37. Cooper, C.D., Frey, W.F., and Compton, R.N.: 1978, J. Chem. Phys. 69, pp. 2367-2374.
38. Sullivan, S.A., and Beauchamp, J.I.: 1977, J. Am. Chem. Soc. 99, pp. 5017-5022.
39. Riveros, J.M., and Takashima, K.: 1976, Can. J. Chem. 54, pp. 1839-1840.
40. Smith, M.A., Barkley, R.M., and Ellison, G.B.: 1980, J. Am. Chem. Soc. 102, pp. 6851-6852 and references cited therein.
41. Van Doorn, R., and Jennings, K.R.: 1981, Org. Mass Spectrom. 16, pp. 397-399.
42. DePuy, C.H., and Bierbaum, V.M.: 1981, J. Am. Chem. Soc. 103, pp. 5034-5038.
43. DePuy, C.H., Beedle, E.C., and Bierbaum, V.M.: submitted.
We gratefully acknowledge Professor DePuy for sending us a preprint of this paper.
44. Bartmess, J.E., Hays, R.L., Khatri, H.N., Misra, R.N., and Wilson, S.R.: 1981, J. Am. Chem. Soc. 103, pp. 4746-4751.
45. Kleingeld, J.C., and Nibbering, N.M.M.: 1980, Tetrahedron Lett. 21, pp. 1687-1690.
46. Kleingeld, J.C., and Nibbering, N.M.M.: Unpublished results.
47. Kleingeld, J.C., and Nibbering, N.M.M.: in "Lecture Notes in Chemistry" (Hartmann, H., and Wanczek, K.-P., Eds.), Springer Verlag, Berlin, in press.
48. Dottore, M.F., and Bowie, J.H.: 1982, J. Chem. Soc. Perkin II, pp. 283-286.

49. Comisarow, M.B.: personal communication as quoted in: Nibbering, N.M.M.: 1981, Ann. di Chim. 71, pp. 3-28.
50. Comisarow, M.B.: 1977, Can. J. Chem. 55, pp. 171-173.
51. Comisarow, M.B.: personal communication as quoted in ref. 4.
52. McDonald, R.N., and Chowdhury, A.K.: 1982, J. Am. Chem. Soc. 104, pp. 901-902.
53. Janousek, J.I., and Brauman, J.I.: 1979, in "Gas Phase Ion Chemistry" (Bowers, M.T., Ed.), Vol. 2, Academic Press, New York, chapter 10, pp. 53-86.
54. Bartmess, J.E., Hays, R.L., and Caldwell, G.: 1981, J. Am. Chem. Soc. 103, pp. 1338-1344.
55. Noest, A.J., and Nibbering, N.M.M.: 1980, Adv. Mass Spectrom. 8A, pp. 227-237.
56. King, G.K., Maricq, M.M., Bierbaum, V.M., and DePuy, C.H.: 1981, J. Am. Chem. Soc. 103, pp. 7133-7140.
57. Squires, R.R., and DePuy, C.H.: 1982, Org. Mass Spectrom. 17, pp. 187-191.
58. Karpas, Z., and Klein, F.S.: 1975, Int. J. Mass Spectrom. Ion Phys. 18, pp. 65-68.
59. Murphy, M.K., and Beauchamp, J.L.: 1976, J. Am. Chem. Soc. 98, pp. 1433-1440.
60. DePuy, C.H., Bierbaum, V.M., Schmitt, R.J., and Shapiro, R.H.: 1978, J. Am. Chem. Soc. 100, pp. 2920-2921.
61. Schmitt, R.J., Bierbaum, V.M., and DePuy, C.H.: 1979, J. Am. Chem. Soc. 101, pp. 6443-6445.
62. Wilds, A.L.: 1944, "Organic Reactions", Vol. 2, Wiley and Sons, New York, pp. 178-223 and references cited therein.
63. Bartmess, J.E.: 1980, J. Am. Chem. Soc. 102, pp. 2483-2484.
64. Ingemann, S., Kleingeld, J.C., and Nibbering, N.M.M.: J. Chem. Soc. Chem. Commun., in press.
65. DePuy, C.H.: personal communication.
66. Kleingeld, J.C., and Nibbering, N.M.M.: submitted to Int. J. Mass Spectrom. Ion Phys.

GAS-PHASE STUDIES OF THE INFLUENCE OF SOLVATION ON ION
REACTIVITY

D.K. Bohme

Department of Chemistry and C.R.E.S.S.
York University
Downsview, Ontario M3J 1P3, Canada

The absolute influence of step-wise solvation on ion
reactivity is viewed from the perspective provided by recent
gas-phase measurements of rate constants and product distributions for binary reactions of unsolvated and solvated positive
and negative ions with neutral substrates. Available data is
reviewed for solvated proton transfer (acid-base), solvent
exchange, charge transfer, associative detachment and nucleophilic displacement reactions.

1. INTRODUCTION

A variety of experimental techniques are now available
which allow the production, chemical manipulation, and detection
of both unsolvated and solvated ions in the gas phase. As a
result it has become feasible:
1. to examine ion stability and the influence of solvation on
this stability in the gas phase,
2. to execute in the complete absence of solvent molecules
numerous ionic reactions which occur in liquid solutions, and to
measure their reaction efficiencies, and
3. to scrutinize the reactivity of an ion as a function of step-
wise solvation, in some instances all the way from the intrinsic
(unsolvated) reactivity to reactivities for degrees of solvation
at which the distinction between the gas phase and solution
begins to dissolve.

The intrinsic stabilities (heats of formation) of a large
number of ions are now firmly established, as are their intrinsic
reactivities in a large variety of ion-molecule reactions. Also,

111

M. A. Almoster Ferreira (ed.), Ionic Processes in the Gas Phase, 111–134.
© *1984 by D. Reidel Publishing Company.*

a considerable body of data now exists on the rates of ternary
solvent (ligand) – molecule addition reactions to a variety of
positive and negative core ions, and much progress has been made
in establishing, primarily from equilibrium constant measurements,
the thermodynamics associated with successive solvent molecule
additions. Here we address gas-phase laboratory studies which
have begun to explore the absolute influence of solvent molecules
on intrinsic ion reactivity. Available data is gathered and considered for a variety of types of binary positive and negative
ion-molecule reactions, but no attempt has been made to be completely comprehensive. Two aspects of the influence of solvation
will be emphasized (see Fig. 1):
1. the absolute degree to which step-wise solvation influences
ion reactivity as revealed by measurements of rate constants, and
2. the nature of the participation of solvent molecules in the
mechanism of reaction as inferred from measurements of product
distributions.

BRIDGING THE GAP BETWEEN THE GAS PHASE AND SOLUTION

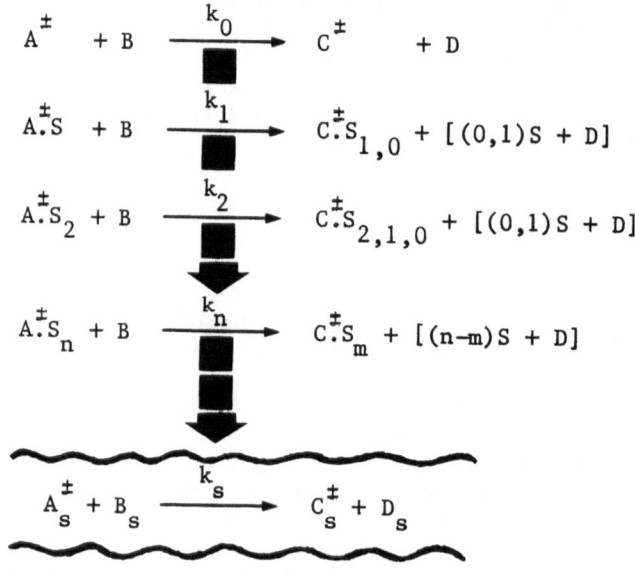

Fig. 1. An illustration of the transition in the kinetics (rate
constants, k_n) and product distributions of ion-molecule reactions
between the gas phase and solution. In the case of the solvated
reactions in the gas phase allowance has been made for the in-
complete retention of solvent molecules by the product ions and for
solvation of the product neutrals.

The reactions which shall be considered have been studied up to only three or four additions of solvent molecules which corresponds to a solvent environment still far removed from that found in solution. Nevertheless, we can at least begin to build a bridge between our knowledge of the reactivity of naked ions in the gas phase and that of solvated ion reactivity in solution.

2. POSITIVE ION CHEMISTRY

2.1. Reactions of Hydrated Hydronium Ions.

The gas-phase reactions of hydrated hydronium ions have received the most attention by the atmospheric chemist, particularly from the time when these ions were shown by in-situ mass spectrometer measurements to dominate the positive ion content of the lower D-region of the earth's ionosphere (1). Water cluster ions appear principally as terminal ions in this region but not at lower altitudes where the atmosphere contains a larger number of neutrals with high proton affinities such as ammonia, formaldehyde, methanol, and nitric and sulfuric acid. Hydrated hydronium ions appear as reactive intermediates in recent models of stratospheric and tropospheric ion chemistry (2). Laboratory measurements are now available for gas-phase reactions of these ions with $CFCl_3$ (3); H_2SO_4 (4); CH_3CN (5); NH_3 (6,7); CH_2O and CH_3OH (7,8); CH_3CHO, H_2S, $HCOOH$, CH_3COOH, $HCOOCH_3$, $(CH_3)_2O$, and $(CH_3)_2CO$ (7); and $(C_2H_5)_2S$ (8).

Observations of product ions indicate a mode of reaction for hydrated hydronium ions which may be represented by equation (1):

$$H_3O^+.(H_2O)_n + B \rightarrow BH^+.(H_2O)_m + (n-m+1)H_2O \qquad (1)$$

Here m may take on values between 0 and n depending on the degree to which water molecules are retained by the product ions. The water may be produced as individual molecules or conceivably even as hydrogen-bonded dimers or polymers (7). The overall standard enthalpy change for reaction (1) is determined by three factors: the difference in the proton affinities of H_2O and B, ΔPA, the difference in the standard enthalpies of hydration of H_3O^+ and BH^+, and the standard enthalpy associated with the possible hydration of the product water molecules. Clearly hydration will alter the exothermicity of the unsolvated reaction and if the sum of intrinsic exothermicity, viz. ΔPA, and the enthalpy of hydration of the product ions is insufficient to offset the enthalpy of hydration of the hydronium ions, hydration will render reaction (1) endothermic. In terms of standard free energies, it is conceivable that hydration shifts the position of equilibrium from right to left in favour of the hydrated hydronium ions to a degree which renders the overall standard free energy change positive and so reverses the relative basicity of H_2O and B.

The gas-phase reactivity of the naked H_3O^+ ion is now well established (10). The observed reaction channel has invariably been proton transfer. For exothermic proton transfer the measured specific reaction rates are in the range from approximately 1 to 5×10^{-9} cm^3 molecule^{-1} s^{-1}. Comparisons with collision rate constants derived from currently available theories suggest that H_3O^+ transfers a proton at essentially every collision over a range in exothermicity from approximately 4 to 35 kcal mole^{-1}.

Two extreme types of behaviour have been observed for the influence of hydration on the reactivity of H_3O^+ in the gas phase. In Fig. 2 hydration is seen to lead to a quite uniform and only

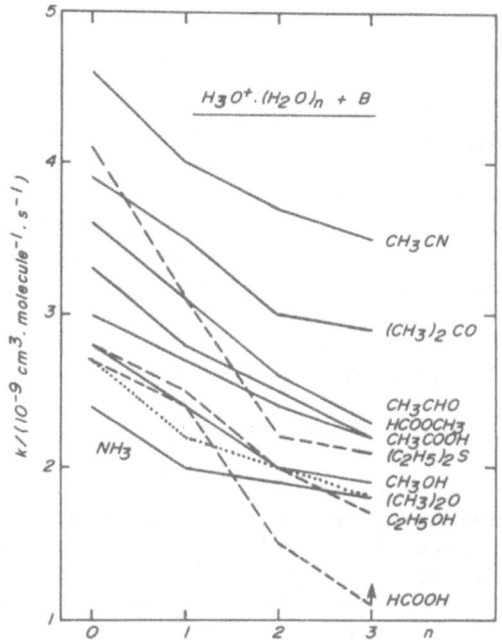

Fig. 2. Observed variation in the rate constant with extent of hydration for gas-phase reactions of hydrated hydronium ions with a variety of bases at 298 K.

slight depression in the specific reaction rate for up to three molecules of hydration. This trend follows the decrease in collision rate expected from the increase in the reduced mass of the reacting pair. In sharp contrast to this behaviour, observations of the reactions with H_2S, H_2SO_4, and $CHC\ell_3$ indicate that the addition of just one molecule of hydration reduces the reaction rate constant precipitously from 1.9 to <0.001, from 0.26 to <0.06, and from 0.4 to <0.001 x 10^{-9} cm^3 molecule^{-1} s^{-1}, respectively. Clearly the former group of reactions remains exothermic and proceeds without activation energy and steric constraint while the latter group presumably becomes endoergic. Indeed, for B = H_2S ΔPA is small and ΔG° has been shown to change from -3.8 for the unsolvated reaction to +7.1 and +6.3 kcal mole^{-1} for n = 1 and 2, respectively, at 300 K (7). The reactions with H_2SO_4 and $CFC\ell_3$ are also likely to immediately become endoergic with hydration since the proton affinities of these two molecules are likely to be close to that of water (3,4). The reactions shown in Fig. 2 are intrinsically more exothermic and the hydrated product ions are quite stable so that the preferred direction of reaction is preserved upon hydration for all of these reactions. The reaction (1) with formaldehyde exhibits an intermediate behaviour in that hydration results in a decrease in the equilibrium constant which has been observed to converge to a value of 1 for hydration with three water molecules, i.e. this reaction appears to become isoergic upon solvation.

Product ion distributions associated with specifically hydrated hydronium ions have generally not been determined. This would be difficult with the flowing afterglow technique employed in most of these measurements. However the product spectra observed with this technique do suggest a preference for hydrated product ions with m < n. One recent study of the very exothermic reaction with NH_3 (11), carried out with a SIFT apparatus which allows the study of specifically hydrated hydronium ions in isolation, has shown that the less exothermic channel is followed exclusively: $H_3O^+ \cdot H_2O$ was found to produce only NH_4^+ and not $NH_4^+ \cdot H_2O$ while $H_3O^+ \cdot (H_2O)_2$ was found to produce only $NH_4^+ \cdot H_2O$ and not $NH_4^+ \cdot (H_2O)_2$ (formation of NH_4^+ is endothermic in this case). These results taken together demonstrate that in sufficiently exothermic reactions some of the excess energy can appear in the product ion and lead to the 'boil-off' of water of hydration.

Some further insight into reaction mechanism has been provided by the SIFT studies with NH_3 (11). Reactions with $D_3O^+ \cdot (D_2O)_{0,1,2}$ were observed not to lead to H/D scrambling in the product ions. This result favours a mechanism which proceeds via the discrete transfer of a proton together with the minimum number of water molecules which is energetically allowed (see Fig. 3) rather than via the formation of an intermediate complex which always decomposes by ejecting the maximum number of water molecules which is energetically allowed. In contrast, the thermoneutral isotopic exchange reactions of $H_3O^+ \cdot (H_2O)_{0,1}$ and

$D_3O^+ \cdot (D_2O)_{0,1}$ with D_2O and H_2O were observed to occur rapidly and lead to a distribution of H and D amongst the product ions and neutrals which was purely statistical. This result implies that these thermoneutral reactions proceed via the formation of an intermediate long-lived association ion in which total randomization of the H and D atoms takes place prior to unimolecular decomposition.

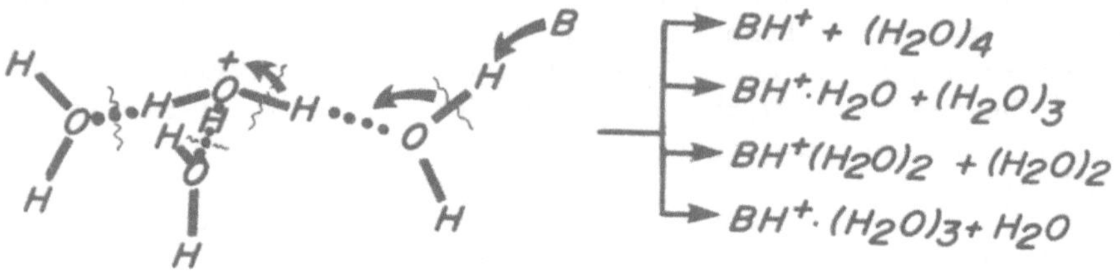

Fig. 3. Possible 'proton-jump' mechanism for gas-phase reactions of hydrated hydronium ions with molecules B. The base can attack at any edge of the hydrated hydronium ion and discretely pick up a proton along with one or more water molecules. Energy redisposition may result in hydrogen-bond scission in the ion at one of several sites and so lead to geometrically favourable formation of polymeric water molecules which retain the extra stabilization provided by hydrogen bonding (7). No solvent reorganization is required for the total retention of the water of hydration but initially the new core ion is likely to be at the 'edge' rather than the 'center' of the product water aggregate. A more stable hydration shell may be achieved by a subsequent reorganization of water molecules.

2.2. Solvent Exchange Reactions.

Gas-phase measurements have shown that solvated cluster ions can evolve rapidly at room temperature to achieve a more stable solvent shell simply by exchange of solvent molecules in binary solvent 'switching' reactions of the following general type:

$$A^+ \cdot S_n + S' \rightarrow A^+ \cdot S_{n-1} \cdot S' + S \tag{2}$$

Many examples of such processes, both inorganic and organic, have been reported in the literature, but no attempt is made here to document them. It has been generally observed that solvent exchange proceeds rapidly, $k \gtrsim 1 \times 10^{-10}$ cm^3 $molecule^{-1}$ s^{-1}, when exothermic.

Sequential solvent exchange reactions of type (2) can ultimately lead to the complete replacement of one solvent by another. Alternatively, depending on the energetics of solvation, the sequential solvent exchange may also be arrested   at some intermediate stage of 'mixed' solvation. For example, in the replacement of water in the hydrated ion $CH_3CNH^+.(H_2O)_n$ by methyl cyanide, SIFT measurements (5) have indicated the following progressions and terminations:

$$CH_3CNH^+.(H_2O)_2 \xrightarrow{CH_3CN} CH_3CNH^+.CH_3CN.H_2O \xrightarrow{CH_3CN} CH_3CNH^+.(CH_3CN)_2 \quad (3)$$

$$CH_3CNH^+.(H_2O)_3 \xrightarrow{CH_3CN} CH_3CNH^+.CH_3CN.(H_2O)_2 \xrightarrow{CH_3CN} \quad (4)$$

$$CH_3CNH^+.(CH_3CN)_2.H_2O \xrightarrow{CH_3CN} CH_3CNH^+.(CH_3CN)_3$$

The terminal steps are believed to be endothermic due to a cross-over in the relative binding energy of H_2O and CH_3CN for ligand attachment to $CH_3CNH^+.CH_3CN$ and $CH_3CNH^+.(CH_3CN)_2$, respectively. Crossovers of this type are not uncommon and can be expected to govern the overall energetics for some solvent exchange reactions particularly at higher cluster sizes (12).

2.3. Anomalous Solvent Participation.

A few isolated examples have been reported which suggest that the solvent molecule may not always play a passive role in solvated ion-molecule reactions. Sequential build-up of a solvation shell by ternary association reactions with O_2^+ and NO^+ is arrested by the following rapid binary reactions which appear to lead to unusual transformations of the original core ion (13):

$$O_2^+.H_2O + H_2O \quad \rightarrow \quad H_3O^+.OH + O_2 \qquad\qquad (5)$$

$$\rightarrow \quad H_3O^+ + (OH + O_2)$$

$$NO^+.(H_3O)_3 + H_2O \rightarrow H_3O^+.(H_2O)_2 + HNO_2 \qquad (6)$$

$$NO^+.NH_3 + NH_3 \quad \rightarrow \quad NH_4^+ + ONNH_2 \qquad\qquad (7)$$

A similar transformation also seems to be indicated for n = 1 to 3 in the following binary reactions (13):

$$NO^+.(H_2O)_n + NH_3 \rightarrow NH_4^+.(H_2O)_{n-1} + HNO_2 \qquad (8)$$

A consistent mechanism has been suggested for these reactions (13). The reactant solvent molecule is viewed to attach to the solvated core ion whereupon a proton is transferred from one solvent molecule to another leaving the original core ion neutralized as a result of the transfer of an electron, as in reaction (5), or the addition of a negative ion (OH^- or NH_2^-), as in

reactions (6) to (8). Another plausible mechanism suggests itself if the solvated reactant ions are viewed as isomeric
protonated neutral molecules: $O_2^+ \cdot H_2O$ as $(HO_3)H^+$, $NO^+ \cdot (H_2O)_n$ as
$(HNO_2)H^+ \cdot (H_2O)_{n-1}$, and $NO^+ \cdot NH_3$ as $(ONNH_2)H^+$. These isomeric
forms can be viewed to react simply by unsolvated or solvated
proton transfer to yield the observed products. This interpretation would require that $PA(H_2O) < PA(HNO_2) < PA(NH_3)$ and that
two molecules of hydration are sufficient to promote the transfer of a proton from $(HNO_2)H^+$ to H_2O.

3. NEGATIVE ION CHEMISTRY

3.1. Solvated Proton-Transfer Reactions.

Gas-phase measurements have been carried out for solvated
proton-transfer reactions between negative ions and both inorganic
and organic acids. The observed behaviour of these reactions
can be discussed in terms of the following general equation:

$$B^- \cdot (BH)_n + AH \rightarrow A^- \cdot (BH)_m + (n-m+1) \ BH \qquad (9)$$

The overall standard free energy change, $\Delta G°$, of reaction (9) is
determined by the difference in the gas-phase acidities of AH and
BH, the difference in the standard free energies of solvation of
B^- and A^-, and the standard free energy associated with the
possible solvation of the neutral products. The free energy of
solvation of the reactant ion acts to decrease the reaction
exoergicity and can lead to a change in the sign of $\Delta G°$, viz. a
reversal in relative acidity, and therefore a change in the
preferred direction of the reaction, unless it is offset by the
solvation energy of the conjugate base produced.

Fig. 4 exemplifies the pattern of reactivity observed for
totally unsolvated B^- when the negative charge in B^- is localized.
Proton transfer occurs at essentially every gas-kinetic collision except when the reaction becomes nearly thermoneutral or
endothermic. Charge delocalization and steric factors may also
act to reduce reaction probability (14,15,16).

A controlling influence of the overall energetics on reaction kinetics is also evident in the observed reactions with
solvated B^- ions. Consider, for example, the results presented
in Fig. 5 for systematic measurements of the following two sets
of reaction (17,18):

$$OH^- \cdot (H_2O)_n + AH \rightarrow A^- \cdot (H_2O)_m + (n-m+1) \ H_2O \qquad (10)$$

$$CH_3O^- \cdot (CH_3OH)_n + AH \rightarrow A^- \cdot (CH_3OH)_m + (n-m+1) \ CH_3OH \qquad (11)$$

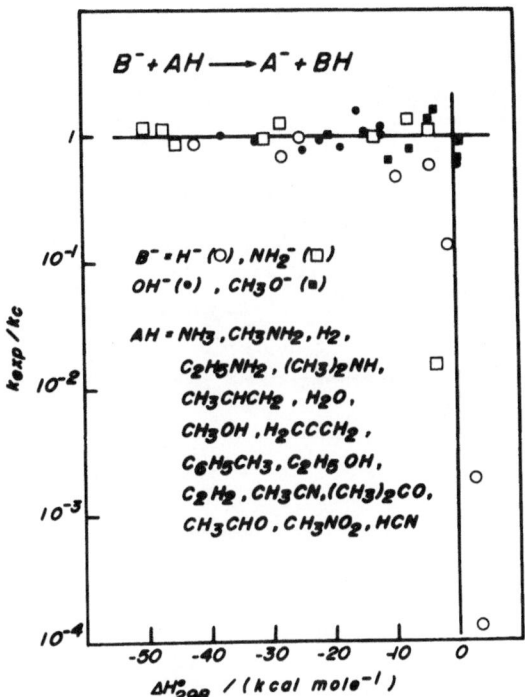

Fig. 4. A correlation between the efficiency of proton transfer, k_{exp}/k_c, and the overall change in enthalpy, $\Delta H°$, in the total absence of solvation at 298 K.

The reactions included in Fig. 5 span a large variety of organic acids AH and a wide range in intrinsic exoergicity from 0.2 to 37 kcal mole^{-1}. Solvation is seen to result either in a slight or a precipitous reduction in the specific rate of reaction, the latter occuring in some instances already after the addition of just one molecule of solvent. As was the case for the analogous positive ion proton-transfer reactions of $H_3O^+ \cdot (H_2O)_n$ discussed earlier, this divergence in behaviour can be accounted for by a consideration of the overall energetics of reaction, even though these have not been established quantitatively in most cases.

Of the acids shown in Fig. 5, allene and toluene are closest in acidity to water and methanol. The carbanions produced by their deprotonation are expected to have free energies of solvation insufficient to compensate for the high solvation energies of OH^- and CH_3O^-. The sharp drop by three orders of magnitude in the specific rate of deprotonation

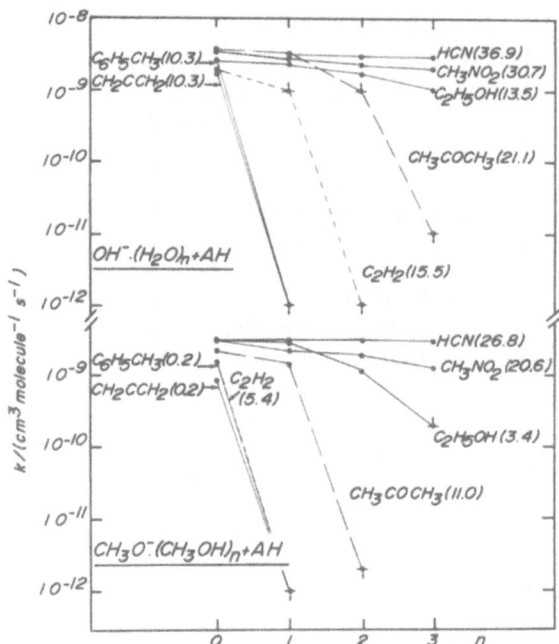

Fig. 5. Observed variations in the rate constants at 298 K for acid-base reactions in the gas phase between a variety of acids AH and hydrated hydroxide ions (top) and methoxide ions solvated with methanol (bottom). The values in parentheses correspond to the relative intrinsic acidities, viz. − ΔG° for the solvent-free reactions at 298 K.

observed with the addition of just one molecule of solvent is therefore a manifestation of an immediate reversal in the relative acidity of these two carbon acids and water or methanol. a A similar reversal is apparent with acetylene but in this case two water molecules are required to drive endoergic the intrinsically more exoergic reaction with OH^-. For the deprotonation of acetone by OH^- and CH_3O^-, both reversals also occur, but they are delayed by one additional molecule of solvent: the sharp drop in the rate occurs at n = 3 and 2, respectively. No reversals are observed for the deprotonations of HCN, CH_3NO_2, and C_2H_5OH. In the former two cases a high intrinsic acidity difference serves to preserve the order, while in the latter case the order of acidity is preserved by the strong solvation of $C_2H_5O^-$. The general trends in the rates shown in Fig. 5 are consistent with rate measurements made in solution and with known solution acidities. For example, the reaction between OH^- and HCN is rapid in aqueous solution while its reaction with acetone is 10 orders of magnitude slower (19)!

Flowing afterglow measurements have also been reported for the following interesting inorganic acid-base reaction (4):

$$NO_3^-.(HNO_3)_n + H_2SO_4 \rightarrow HSO_4^-.(HNO_3)_n + HNO_3 \qquad (12)$$

The reactions with n = 0 and 1 occur on essentially each colli- sion but for n = 2 a slightly reduced efficiency is observed: k_n = 9.7, 8.6 and 4.0 x 10^{-9} cm^3 $molecule^{-1}$ s^{-1}. A consideration of the difference in intrinsic acidity between HNO_3 and H_2SO_4 and the relative HNO_3 solvation energy of NO_3^- and HSO_4^- leads to the implication that reaction (12) is exothermic for n = 0 and 1 but nearly thermoneutral or even endothermic for n = 2. Con- sequently the transfer of 2 molecules of nitric acid along with the proton may involve a slight barrier to reaction (4).

Little data is available on the product distributions which apply for reactions of type (9). For the reactions (10) and (11) formation of product ions of the type $A^-.S_n$ was observed for all of the reactions with k $\geq 10^{-10}$ cm^3 $molecule^{-1}$ s^{-1} (17). SIFT measurements of exact product distributions are only now becoming available. These will be important in elucidating features of the reaction mechanisms which apply for such reac- tions as the one proposed in Fig. 6.

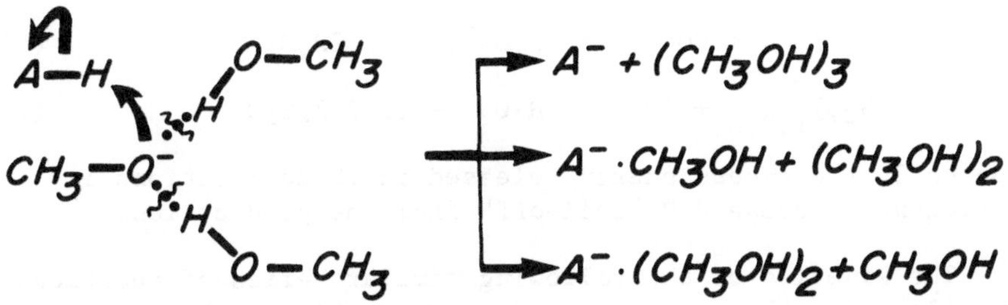

Fig. 6. Possible mechanism for the gas-phase reaction of solvated methoxide ions with acids AH. Hydrogen-bond scission in the ion adduct may lead to product A^- ions with or without retention of methanol solvent and to hydrogen-bonded dimers or trimers of methanol as possible neutral products.

3.2. Solvent Exchange Reactions.

As was the case for positive cluster ions, gas-phase

measurements have shown that negative cluster ions can evolve
rapidly at room temperature to achieve a more stable solvent
shell by exchange of solvent molecules in binary 'switching'
reactions of the following general type:

$$A^-.S_n + S' \rightarrow A^-.S_{n-1}.S' + S \tag{13}$$

Again, many examples of such processes have been reported and
they are generally observed to proceed rapidly, $k \geq 1 \times 10^{-10}$
cm^3 molecule^{-1} s^{-1}, when exoergic. Sequential reactions can
lead to further evolution of the solvent shell and the complete
replacement of one solvent by another as in the following example
(18):

$$CN^-.(CH_3OH)_2 + HCN \rightarrow CN^-.CH_3OH.HCN + CH_3OH \tag{14}$$

$$CN^-.CH_3OH.HCN + HCN \rightarrow CN^-.(HCN)_2 + CH_3OH \tag{15}$$

In some instances ligand exchange can lead from relatively
weak electrostatic or hydrogen bonding in the solvated reactant
ion to bonding in the product ion which is much more covalent in
nature. For example, flowing afterglow measurements have shown
that $OH^-.(H_2O)_n$ reacts rapidly with CO_2 and SO_2 for values of n
from 1 to 3 to form product ions of the type $HCO_3^-.(H_2O)_m$ and
$HSO_3^-.(H_2O)_m$ (20). The predominant products of these reactions
have recently been elucidated in tandem mass spectrometer studies
at collision energies less than 1 eV(CM) and are as follows (21):

$$OH^-.(H_2O)_{1,2} + CO_2 \rightarrow HCO_3^- + (1,2)H_2O \tag{16}$$

$$OH^-.(H_2O)_3 + CO_2 \rightarrow HCO_3^-.H_2O + 2H_2O \tag{17}$$

$$OH^-.(H_2O)_{1,2,3} + SO_2 \rightarrow HSO_3^- + (1,2,3)H_2O \tag{18}$$

Apparently the excess energy released in these reactions is
sufficient to cause H_2O "boil-off" from the product ions.

The kinetics of the following similar series of reactions
have also recently been followed as a function of hydration (22):

$$O_2^-.(H_2O)_n + SO_2 \rightarrow SO_4^-.(H_2O)_m + (n-m)H_2O \tag{19}$$

$$O_2^-.(H_2O)_n + CO_2 \rightarrow CO_4^-.(H_2O)_m + (n-m)H_2O \tag{20}$$

$$O_2^-.(H_2O)_n + NO \rightarrow NO_3^-.(H_2O)_{n-1} + H_2O \tag{21}$$

The measured rate constants for these reactions are presented
graphically in Fig. 7. For the reactions with NO, one water
molecule was observed to be exchanged rapidly for the NO molecule for values of n from 1 to 5. Also it was proposed that the

NO_3^- is likely to be in the peroxy nitrite form, $OONO^-$, rather than the more stable symmetric nitrate form. Secondary reactions prevented the determination of the principal product ions for the reactions with SO_2. Only $O_2^-\cdot(H_2O)_{1,2}$ were observed to react with CO_2. The reaction with n = 2, m = 1 is slightly endothermic and its rate constant was found to be an increasing function of temperature.

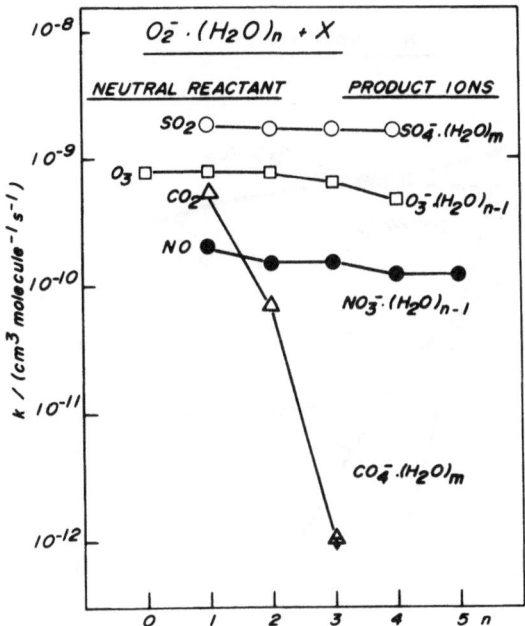

Fig. 7. Observed variation in the rate constant with extent of hydration for gas-phase ligand exchange reactions of $O_2^-\cdot(H_2O)_n$ with SO_2, CO_2 and NO, and the charge transfer reactions of $O_2^-\cdot(H_2O)_n$ with O_3.

3.3. Charge Transfer Reactions.

A few gas-phase measurements have been reported which provide evidence for the simultaneous transfer of an electron and one or more water molecules in ion-molecule reactions with hydrated anions which may be represented by the following equation:

$$A^-\cdot(H_2O)_n + B \rightarrow B^-\cdot(H_2O)_m + [(n-m)H_2O + A] \qquad (22)$$

The rate constants which have been reported for reactions of
type (22) are presented graphically in Fig. 8 (20,22). The
overall energetics are governed by the intrinsic energy change
determined by the difference in the electron affinities of A and
B, the difference in the hydration energies of A^- and B^-, and
the energy associated with the possible hydration of the neutral
products.

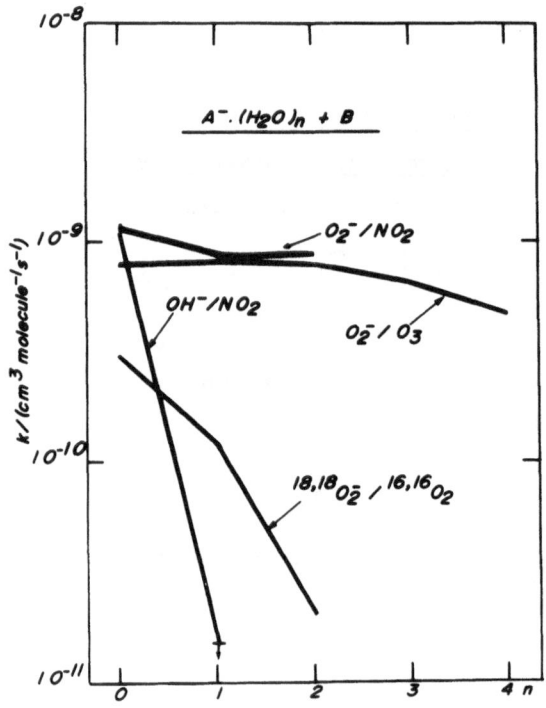

Fig. 8. Observed variation in the rate constant with extent of
hydration for gas-phase charge transfer reactions of negative
ions.

The reaction of hydrated O_2^- with NO_2 is intrinsically the
most exothermic (~ 46 kcal $mole^{-1}$). The unhydrated charge
transfer proceeds rapidly. Hydration of O_2^- with up to two water
molecules was found not to significantly reduce the rate of
reaction. Although the hydration energies of NO_2^- are less than
those of O_2^-, the reaction exothermicity is maintained well beyond n = 2 for complete transfer of water because of the high
intrinsic exothermicity of the reaction. In fact, the reactions
are exothermic even if no water molecules are transferred along
with the electron. However products have not been elucidated for

these reactions so that we have no information on the degree of solvent retention.

The intrinsic exothermicity for the charge transfer between O_2^- and O_3 is somewhat less than the previous case but still quite high (~ 37 kcal mole^{-1}). Charge transfer was distinguished from oxygen-atom transfer with isotope labeling experiments. It was found to proceed rapidly for values of n from 0 to 4 in the temperature range from 335 to 181 K, the higher hydrates being studied at lower temperatures. The observed product spectra indicated a preferential ejection of one molecule of H_2O in each case:

$$O_2^-\cdot(H_2O)_n + O_3 \rightarrow O_3^-\cdot(H_2O)_{n-1} + H_2O + O_2 \qquad (23)$$

If the hydration energies of O_3^- are estimated to be equal to the O_2^- hydration energies, reaction (23) remains exothermic for $m \geq n - 2$ at all values of n corresponding to the ejection of up to two water molecules in $O_3^-\cdot(H_2O)_n$. However, the reactions for $m = n - 2$ are nearly thermoneutral and are not expected to be rapid since a constraint arises from the difference in the equilibrium angle in O_3 and O_3^-. All of the reactions for $m = n$ are nearly 37 kcal mole^{-1} exothermic so that product $O_3^-\cdot(H_2O)_n$ ions can be expected to contain too much energy to be stable against ejection of one molecule of H_2O.

We now turn to the charge transfer reaction of OH^- with NO_2 which is of a still lower exothermicity (~ 14 kcal mole^{-1}). For this system a fast reaction is observed only for the unsolvated ion. Since the hydration energies of NO_2^- are substantially less than those of OH^-, the reactions of $OH^-\cdot(H_2O)_n$ become endothermic and therefore not allowed for $n > 1$. The reaction for $n = m = 1$ is only slightly exothermic but is constrained by the energy required to bend NO_2 to the equilibrium angle of NO_2^- (116 to 134°).

Finally, initial results have been reported for the following thermoneutral charge transfer reaction

$$^{18\,18}O_2^-\cdot(H_2O)_n + {}^{16\,16}O_2 \rightarrow {}^{16\,16}O_2^-\cdot(H_2O)_n + {}^{18\,18}O_2 \qquad (24)$$

for $n = 0, 1,$ and 2. No oxygen-atom scrambling was observed to occur. Room temperature measurements provided the rate constant ratios $15/6/1$ for $k_0/k_1/k_2$ and so indicate a significant decrease in the reaction probability for charge transfer with reactant ion hydration. The measured absolute value for k_0, $(3 \pm 1) \times 10^{-10}$ cm^3 molecule^{-1} s^{-1}, is nearly one half of the collision limited value and is understood in terms of an O_4^- reaction complex in which each O_2 molecule has an equal probability of being charged when the complex decomposes. Hydration can be expected to delocalize the charge in the hydrated reactant ion so that the two O_2 charge centers may no longer be equal in the reaction complex. It has been suggested that this could lead to a statistical decomposition of the hydrated reaction complex back to reactants which becomes increasingly preferred with increasing hydration and thus account for the observed drop in reaction probability (22).

3.4. Associative Detachment Reactions.

One study has been reported (23) which deals with the absolute influence of solvation on the gas-phase kinetics of associative detachment reactions of the following type:

$$A^- \cdot (H_2O)_n + H \rightarrow [AH + n\ H_2O] + e \qquad (25)$$

Rate constants were determined at 296 K for reactions of atomic hydrogen with $A^- = C\ell^-$, OH^- and O_2^- for values of n up to 3. They are presented graphically in Fig. 9. In all cases hydration was observed to produce a decrease in the reactivity of the anion.

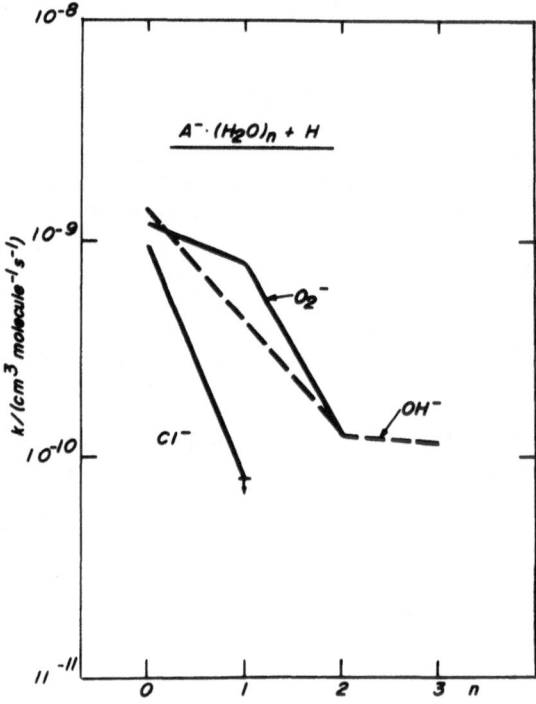

Fig. 9. Observed variation in the rate constant with extent of hydration for gas-phase associative detachment reactions of negative ions with atomic hydrogen at 296 K.

Although hydration leads to increasingly less favourable energetics for associative detachment, all of the reactions in Fig. 9 have exothermic associative detachment channels with the exception of the reaction with $Cl^-.(H_2O)_2$ which is therefore not expected to proceed with a measurable rate. Several alternate reaction channels leading to ion formation are possible for the reactions with $O_2^-.(H_2O)_n$, as, for example, in reaction (26), but branching ratios for the various channels were not ascertained.

$$O_2^-.H_2O + H \rightarrow HO_2 + H_2O + e \qquad\qquad (26a)$$

$$\rightarrow OH^-.H_2O + O \qquad\qquad (26b)$$

Several possible explanations have been offered for the observed decrease in the probability of associative detachment upon hydration (23). It has been pointed out that the reduced reactivities might result from:
1. a potential barrier due to the presence of H_2O which inhibits the formation of the autodetaching complex,
2. a steric constraint due to the presence of H_2O which prevents the H atom from approaching the A^- ion,
3. an increase upon hydration in the autodetachment lifetime of the negative ion complex relative to the lifetime for decomposition to reform the reactants.
No additional measurements have since been reported which provide further insight into the reaction mechanism. Clearly experiments as a function of temperature of the reacting system or energy of the reacting ions would be highly desirable in this regard.

The specific rates measured for the hydrated hydroxide reactions correlate well with the value of 3×10^{-14} cm^3 $molecule^{-1}$ s^{-1} which has been reported for the specific rate for the conversion of atomic hydrogen to hydrated electrons by hydroxide ions in water (24). It is just conceivable that this reaction with $n \geq 6$ or 7 may also be a potential source of hydrated electrons in the gas phase. Free hydrated electrons have now been observed in the gas phase with $n \geq 8$ (25) and calculations have shown that at least six water molecules are required to bind an electron in the gas phase (26).

3.5. Nucleophilic Displacement Reactions.

We now turn to reactions of anions which, in contrast to those considered previously, involve the breaking and making of strong chemical bonds between heavy atoms. Nucleophilic displacement reactions between anions and methyl halides were first executed in the gas phase in 1970 in an early application of the flowing afterglow technique to gas-phase organic anion chemistry (27). The intrinsic reactivity patterns of these and other related reactions have since been thoroughly characterized

primarily with flowing afterglow (28) and ICR measurements
(29,30,31).

For exothermic unsolvated nucleophilic displacement reactions of the following general type:

$$A^- + RB \rightarrow B^- + AR \tag{27}$$

gas-phase measurements have shown that variations in the
nucleophile (A^-), leaving group (B^-), and alkyl substrate (R)
lead to a wide range in specific rate, but generally the
specific rates are much higher (up to 20 orders of magnitude!)
in the gas phase than in solution. In contrast to the behaviour
generally observed for proton transfer, reaction exothermicity
does not appear to be the only determinant of reaction efficiency. Although S_N2 efficiency is generally observed to
decrease with decreasing exothermicity, particularly within
certain homologous series of exothermic reactions, reactions
having similar exothermicities may proceed with quite different
efficiencies (28,30). Variations in reaction efficiency have
also been found to correlate with the degree of charge delocalization in the nucleophile A^- (27,30).

Solution measurements have clearly demonstrated the influence of the medium on the efficiency of S_N2 reactions which can
be as much as a million times faster in dipolar aprotic than in
protic solvents (32). The early flowing afterglow studies (27)
provided, for the first time, a method of measuring the absolute
influence of solvation on this efficiency, albeit in the gas
phase in the absence of bulk solvent effects. Reactions of the
type:

$$RO^-.(ROH)_n + CH_3C\ell \rightarrow C\ell^-.(ROH)_m + [(n-m)ROH + ROCH_3] \tag{28}$$

were executed in the gas phase as a function of step-wise
solvation and a significant decrease in reactivity was observed
already for the addition of just one solvent molecule to the
nucleophile. Since then more quantitative rate measurements have
been carried out both at room temperature with a flowing afterglow
(33,9) and at collision energies less than 1 eV(CM) with a tandem
mass spectrometer (21). The flowing afterglow results obtained
for reactions of CH_3Br with hydrated hydroxide ions and methoxide
ions solvated with methanol are summarized in Fig. 10. Similar
results have been obtained with $CH_3C\ell$ and comparative measurements are now in progress for reactions of the more general
type (29) with other anions solvated with either protic or
dipolar aprotic solvents:

$$A^-.S_n + CH_3B \rightarrow B^-.S_m + (n-m)S + CH_3A \tag{29}$$

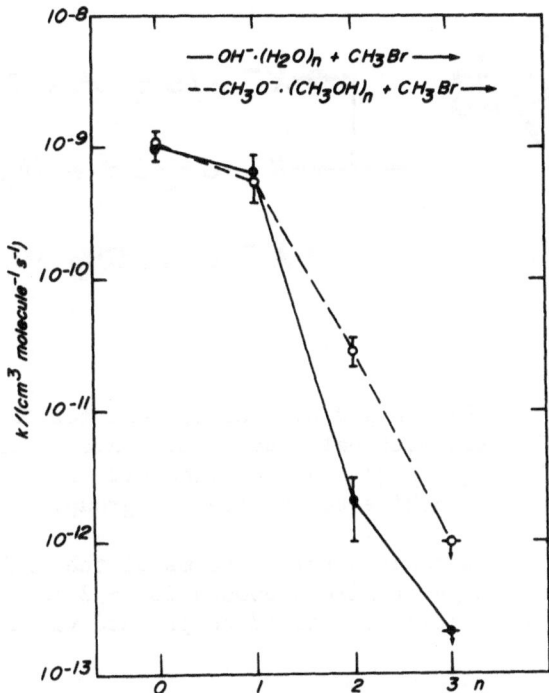

Fig. 10. Observed variation in the rate constant with extent of
solvation for gas-phase nucleophilic displacement reactions of
solvated anions with methyl bromide at 298 K.

As regards product distributions, the available flowing afterglow
and tandem mass spectrometer measurements generally indicate in-
complete solvent retention by the leaving group, viz. a preference for for product ions $B^-.S_m$ with values of $m < n$. This has
been clearly established in a recent SIFT study for the reaction
of $OH^-.H_2O$ with CH_3Br. 95% of this reaction proceeds
to produce the unsolvated Br^- product at room temperature (34).
Fig. 11 expresses the various options for the similar reactions
of $CH_3O^-.CH_3OH$. For $X = Br$ and Cl, the product ion $X^-.CH_3OCH_3$
was not observed in the flowing afterglow experiments and
formation of X^- was seen to be preferred over $X^-.CH_3OH$.

     The overall decrease in the specific rate for displacement
shown in Fig. 10 amounts to at least 3 orders of magnitude with
the addition of just 3 molecules of solvent to the nucleophile.
The addition of the first solvent molecule leads to only a slight
decrease but further solvent addition reduces the reactivity
precipitously. Such a reduction in reactivity with solvation

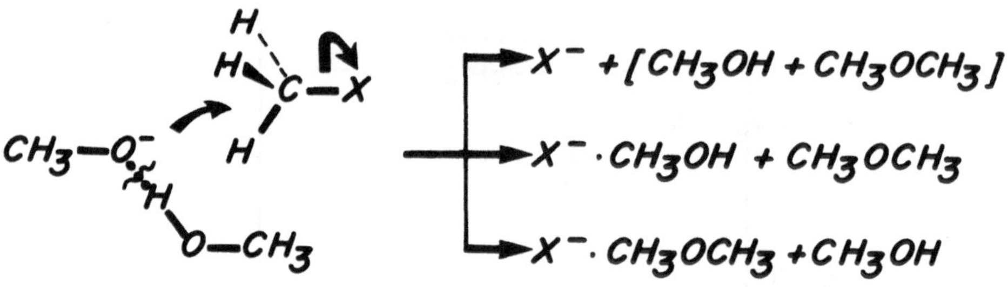

Fig. 11. Possible mechanism and product options for the gas-phase
reaction of $CH_3O^-.CH_3OH$ with halogenated methanes. Formation of
the product ion $X^-.CH_3OH$ requires the transport of a methanol
molecule from the nucleophile to the leaving group.

can be qualitatively accounted for in terms of the double
minimum potential energy profile proposed for S_N2 reactions by
Brauman (33). The reaction is viewed to proceed via a 3-step
mechanism (33):

1. formation of a loosely bound cluster:

$$A^-S_n + CH_3B = A^-.S_n...CH_3B \tag{30a}$$

2. isomerization of the cluster (CH_3 transfer with inversion):

$$A^-.S_n...CH_3B = ACH_3...B^-.S_n \tag{30b}$$

3. final dissociation of the loosely bound cluster to products:

$$ACH_3...B^-.S_n = ACH_3 + B^-.S_n \tag{30c}$$

As the nucleophile becomes more solvated the differential
solvation of the reactants and the cluster leads to a transition
state which becomes relatively less stable and so to an
increasingly large central barrier to reaction. The energy
difference between the reactants and the central barrier can be
correlated with the efficiency of reaction (30). While the
energy of the central barrier may remain below the energy of the
reactants at low levels of solvation and so account for the high
reaction efficiencies observed at these levels, further solvation is expected eventually to lead to a central barrier with an
energy greater than the reactants and thus to a dramatic decrease
in reaction efficiency (33).

Very recently Morokuma has actually attempted with ab initio SCF calculations to follow the intermediate features of the potential energy profiles at various degrees of solvation (35). The calculations were carried out for the isoergic reactions:

$$Cl^-.(H_2O)_n + CH_3Cl \rightarrow ClCH_3 + Cl^-.(H_2O)_n \qquad (31)$$

for n = 0,1 and 2. Some of the interesting results which were derived from these calculations are reproduced in Fig. 12. This figure is quite instructive since it allows us to follow various possible paths for the migration of the solvent molecules along the reaction coordinate alluded to in Fig. 11. For the un-hydrated reaction the calculations produce a double minimum potential. Hydration with one molecule leads to a higher barrier whether the reaction proceeds by the simultaneous migration of H_2O and CH_3 transfer with inversion or sequentially by CH_3 transfer with inversion followed by migration of H_2O. The doubly hydrated reaction involves two H_2O migrations and a CH_3 transfer with inversion. These may take place one by one, two by one or all three simultaneously. Fig. 12 (C) shows that the last of these paths appears to be the least favourable. In the most favourable path one molecule of H_2O is transferred first with little or no barrier. This is followed by CH_3 transfer with inversion and then the transfer of the second H_2O molecule. The calculated geometries show that the Cl^- moves out of the C_{3v} axis to assist the migration of the H_2O to the leaving group. Finally, Morokuma also makes the important point that for large n, when the first solvation shell of Cl^- is completed, the initial interaction between the reagents involves dehydration and so a new feature is expected to appear in the potential surface. This will apply generally to reactions of type (29) at higher levels of solvation. Unfortunately, given the dynamic range in rate constant measurements of ion molecule reactions available with present experimental techniques, we can expect the solvated S_N2 reactions to become immeasurably slow at these levels of solvation in the gas phase. This is somewhat ironical as it was once supposed that unsolvated S_N2 reactions would be too fast to be measurable in the gas phase (36).

4. CONCLUDING REMARKS

An attempt has been made here to gather together, for the first time, the results of experimental investigations which provide insight into the influence of solvation on the kinetics of binary ion-molecule reactions in the gas phase. Results have been reviewed and discussed for a variety of positive and negative ion-molecule reactions. Many of the underlying measurements were instigated by an interest in atmospheric ion chemistry and were carried out primarily with the flowing afterglow technique at the NOAA Laboratories in Boulder, Colorado, by E.E. Ferguson and

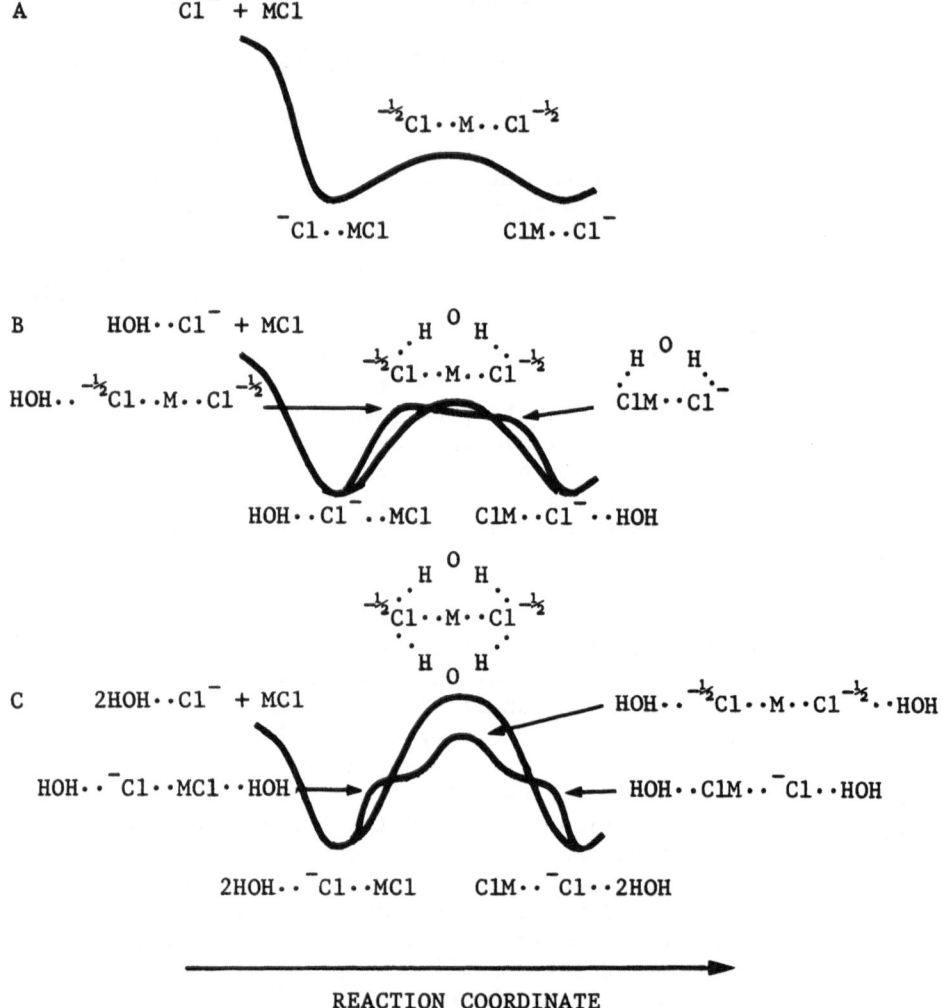

REACTION COORDINATE

Fig. 12. Calculated profiles for the potential energy surfaces of the unhydrated (A), singly hydrated (B), and doubly hydrated (C) reaction of $Cl^-\cdot(H_2O)_n$ with CH_3Cl (35).

F.C. Fehsenfeld and their colleagues. Studies of a similar nature and using other techniques are only now beginning to be extended to more traditional areas of chemistry so that we are still very much at the beginning of our understanding of the kinetics of ion-molecule reactions at the frontier between the gas phase and solution.

5. REFERENCES

(1) Narcisi, R.S. and Bailey, A.D.: 1965, J. Geophys. Res. 70, p. 3687.

(2) Ferguson, E.E., and Arnold, F.: 1981, Acc. Chem. Res. 14, p. 327.

(3) Fehsenfeld, F.C., Crutzen, P.J., Schmeltekopf, A.L., Howard, C.J., Albritton, D.L., Ferguson, E.E., Davidson, J.A. and Schiff, H.I.: 1976, J. Geophys. Res. 81, p. 4454.

(4) Viggiano, A.A., Perry, R.A., Albritton, D.L., Ferguson, E.E. and Fehsenfeld, F.C.: 1980, J. Geophys. Res. 85, p. 4551.

(5) Smith, D., Adams, N.G. and Alge, E.: 1981, Planet. Space Sci. 29, p. 449.

(6) Fehsenfeld, F.C. and Ferguson, E.E.: 1973, J. Chem. Phys. 59, p. 6272.

(7) Bohme, D.K., Mackay, G.I. and Tanner, S.D.: 1979, J. Am. Chem. Soc. 101, p. 3724.

(8) Fehsenfeld, F.C., Dotan, I., Albritton, D.L., Howard, C.J. and Ferguson, E.E.: 1978, J. Geophys. Res. 83, p. 1333.

(9) Bohme, D.K., and Rakshit, A.B.: 1982, unpublished results.

(10) Mackay, G.I., Tanner, S.D., Hopkinson, A.C. and Bohme, D.K.: 1979, Can. J. Chem. 57, p. 1518.

(11) Smith, D., Adams, N.G. and Henchman, M.J.: 1980, J. Chem. Phys. 72, p. 4951.

(12) Castleman, A.W.: 1979, in "Kinetics of Ion-Molecule Reactions," NATO ASI Vol. B40 (Ausloos, P., Ed.), Plenum Press, New York.

(13) Fehsenfeld, F.C., and Ferguson, E.E.: 1971, J. Chem. Phys. 54, p. 439.

(14) Mackay, G.I., Tanaka, K., and Bohme, D.K.: 1977, Int. J. Mass Spectrom. Ion Phys. 24, p. 125.

(15) Farneth, W.E., and Brauman, J.I.: 1976, J. Am. Chem. Soc. 98, p. 7891.

(16) Bohme, D.K., Lee-Ruff, E., and Young, L.B.: 1972, J. Am. Chem. Soc. 94, p. 5153.

(17) Bohme, D.K., Rakshit, A.B., and Mackay, G.I.: 1982, J. Am. Chem. Soc. 104, p. 1100.

(18) Mackay, G.I., Rakshit, A.B, and Bohme, D.K.: 1982, Can. J. Chem., in press.

(19) Eigen, M.: 1964, Angew. Chem. Int. Ed. Engl. 3, p. 1.

(20) Fehsenfeld, F.C., and Ferguson, E.E.: 1974, J. Chem. Phys. 61, p. 3181.

(21) Hierl, P.M., Henchman, M.J., and Paulson, J.F.: 1981, presented at the 29th Ann. Conf. on Mass Spectrom. and Allied Topics, Minneapolis, MN, May 24 - 29.

(22) Fahey, D.W., Bohringer, H., Fehsenfeld, F.C., and Ferguson, E.E.: 1982, J. Chem. Phys. 76, p. 1799.

(23) Howard, C.J., Fehsenfeld, F.C., and McFarland, M.: 1974, J. Chem. Phys. 60, p. 5086.

(24) Hart, E.J.: 1966, in "Actions Chimiques et Biologiques des

Radiations" (Haissinsky, M., Ed.), Dixieme Serie, Masson and Co., Paris, Chapter 1, p. 34.
(25) Armbruster, M., Haberland, H., and Schindler, H.-G.: 1981, Phys. Rev. Letters 47, p. 323.
(26) Jortner, J., and Gaathon, A.: 1977, Can. J. Chem. 55, p. 1801.
(27) Bohme, D.K., and Young, L.B.: 1970, J. Am. Chem. Soc. 92, p. 7374.
(28) Tanaka, K., Mackay, G.I., Payzant, J.D., and Bohme, D.K.: 1976, Can. J. Chem. 54, p. 1643.
(29) Brauman, J.I., Olmstead, W.N., and Lieder, C.A.: 1974, J. Am. Chem. Soc. 96, p. 4030.
(30) Olmstead, W.N., and Brauman, J.I.: 1977, J. Am. Chem. Soc. 99, p. 4219.
(31) Asubiojo, O.I., and Brauman, J.I.: 1979, J. Am. Chem. Soc. 101, p. 3715.
(32) Parker, A.J.: 1969, Chem. Rev. 69, p. 1.
(33) Bohme, D.K., and Mackay, G.I.: 1981, J. Am. Chem. Soc. 103, p. 978.
(34) Paulson, J.F., Henchman, M.J. and Hierl, P.M.: private communication.
(35) Morokuma, K.: 1982, J. Am. Chem. Soc. 104, p. 3732.
(36) Bathgate, R.H., and Moelwyn-Hughes, E.A.: 1959, J. Chem. Soc., p. 2642.

EXOTHERMIC BIMOLECULAR ION MOLECULE REACTIONS WITH NEGATIVE TEMPERATURE DEPENDENCE

T. F. Magnera and P. Kebarle
Chemistry Department, University of Alberta,
Edmonton, Alberta, Canada T6G 2G2

INTRODUCTION, ION MOLECULE COLLISION RATES

The majority of exothermic bimolecular ion molecule reactions (1), whose rates have been measured, proceed at near

$$A^+ + B = C^+ + D \tag{1}$$

collision rates. The collision rate constants of ion molecule
reactions are very large due to the long range attractive forces
between the ion and the molecule. Adequate theoretical expres-
sions for the collision rates have been developed, starting with
a transition state treatment of the orbiting ion molecule complex by Eyring (1), followed by the expression of Giomousis and
Stevenson (2) for molecules without a permanent dipole and the
more recently developed theory for molecules with permanent dipoles, Duggan, Bowers, Ridge, Bates (3-7). Collision theory has
been recently summarized by Su, Chesnavitch and Bowers (5). A
convenient equation for the collision rate constant k_c is given
in eq. 2 where q is the electronic charge, μ the reduced mass, α
the polarizability, D the dipole moment and c a factor between 0
and 1, the actual value being dependent on the reactants (5).

$$k_c = \frac{2\pi q}{\mu^{\frac{1}{2}}} \left[\alpha^{\frac{1}{2}} + cD \left(\frac{2}{\pi kT} \right)^{\frac{1}{2}} \right] \tag{2}$$

k_c is temperature independent for molecules without permanent
dipole and has a very weak negative temperature dependence for
molecules with a permanent dipole. The reactant parameters have
only a weak influence on k_c, so that this rate constant is found

135

M. A. Almoster Ferreira (ed.), Ionic Processes in the Gas Phase, 135–157.
© 1984 by D. Reidel Publishing Company.

to be close to 10^{-9} molecules^{-1} cm^3 sec^{-1} for most reactants over a wide temperature range. Thus, reactions that proceed at collision rates are chemically featureless i.e. the rates reveal little about the details of the reaction.

EXPERIMENTALLY OBSERVED REACTIONS WITH LOW COLLISION EFFICIENCIES AND NEGATIVE TEMPERATURE DEPENDENCE

There have been several classes of reactions reported which proceed at rates significantly lower than collision rates. The temperature dependence of some of them has been also investigated. In most cases the temperature dependence was found to be negative i.e. the rate constant increased with decrease of temperature. Listed below are reactions which have been shown to have negative temperature dependence and also reactions which were found to have low collision efficiency at room temperature. These reactions, as will be shown later, probably also have negative temperature dependence.

a. Hydride transfer reactions: $R^+ + R'H = RH + R'^+$ example:

$$sec-C_3H_7^+ + HC(CH_3)_3 = C_3H_8 + (CH_3)_3C^+$$

Ausloos and Lias (8) were the first to show that many hydride transfer reactions proceed with low collision efficiencies. Subsequently Meot-Ner and Field (9,10) made a comprehensive study of these reactions and showed that they have negative temperature dependence.

b. Alkylation of bases B by chloronium ions: $Me_2Cl^+ + B =$
    $MeB^+ + MeCl$ $(B = NH_3, Me_3N, Me_2O, Et_2O, i-C_3H_7 - C_6H_5$
    $MeC_6H_5, C_6H_6)$.

The above reactions were studied by Sen Sharma and Kebarle (11), some were found to proceed at collision rates (NH_3, Me_3N), others below collision rates and with negative temperature dependence.

c. $C_6H_5CH_2^+ + C_6H_5CH_3 = CH_3C_6H_4CH_2^+ + C_6H_6$

Negative temperature dependence was observed for this and other related reactions by Sen Sharma and Kebarle (12).

d. Nucleophilic substitition, example: $Cl^- + CH_3Br = ClCH_3 + Br^-$.

This class of reactions was first shown to proceed at low collision efficiencies by Bohme et al (13,14) in a comprehensive series of experimental investigations. Later Brauman and co-workers (15) made nucleophilic displacement reactions a subject of theoretical and experimental investigations. The temperature dependence of these reactions has not been studied experimentally, it is probably negative as will be shown further on.

e. Claisen, Aldol, Dieckman condensations, example:

$$R_1-\underset{\underset{CH_2}{|}}{\overset{\overset{O}{||}}{C}} + R-\underset{X}{\overset{\overset{O}{||}}{C}} \longrightarrow R_1-\underset{\underset{CH_2}{|}}{\overset{\overset{O}{||}}{C}}-\underset{R}{\overset{\overset{O^{\cdot}}{|}}{C}}-X \longrightarrow R_1-\underset{\underset{CH_2}{|}}{\overset{\overset{O}{||}}{C}}-\underset{R}{\overset{\overset{O}{||}}{C}} + X^-$$

These reactions bear some similarity to nucleophilic substitution, but involve a tetrahedral intermediate complex. They have been studied by Brauman and Bartmess (16) and shown to have low collision efficiencies.

f. Proton transfer involving delocalized anions, example:

$$Me-\langle\rangle-O^{\cdot} + \langle\rangle-OH = Me-\langle\rangle-OH + \langle\rangle-O^-$$

Several proton transfers involving charge delocalized anions (negative charge on phenoxide anion is delocalized to the ortho and para positions) were investigated by Farneth and Brauman (17) and found to have low collision efficiencies. It is important to note that most of the reactions that were examined were only weakly exothermic.

g. Proton transfer involving steric hindrance, example:

(diagram of substituted pyridinium/pyridine proton transfer with Et and Bu substituents)

Reactions of this type were shown to have low collision efficiencies by Jasinski and Brauman (18).

From the above examples it is possible to extract some empirical rules concerning the factors that lead to low collision efficiency and negative temperature dependence. However it is more profitable to postpone the classification of the reaction types until after one has examined low collision efficiency and negative temperature dependence on basis of theory.

REACTIONS WITH LOW COLLISION EFFICIENCY. REACTION COORDINATE, ENERGY AND ANGULAR MOMENTUM DEPENDENCE.

A clear discussion of the factors leading to low collision efficiency was given first by Olmsted and Brauman (15). The treatment is statistical and thus presupposes persistent reaction complexes. The argument is summarized in Figure 1. The reactants A and B form a long lived complex AB* at a collision rate given by k_c. AB* can either lead to products at a rate k_p or back decompose with rate k_b. Applying the steady state assumption: $d[AB*]/dt = 0$, one obtains eq. 3, while the reaction efficiency per

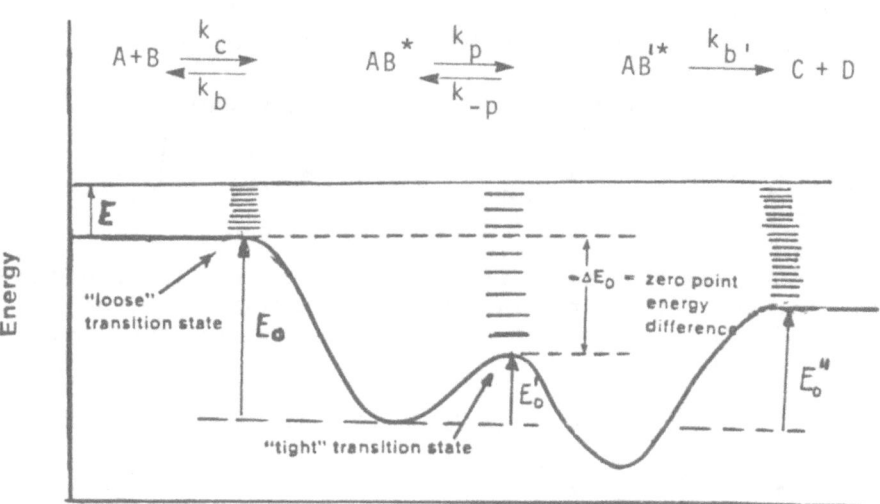

FIGURE 1.   Potential energy of reaction coordinate for reactions
with low collision efficiency and negative temperature dependence. The collision (orbiting) transition states occurring over
E_0 (and E_0'') are loose transition states. The product forming
transition state (over E_0') is a tight transition state. A
bottle neck generally develops at that point if $-\Delta E_0 = E_0 - E_0'$ is
small i.e. does not exceed   10 to 15 kcal/mol.

$$k = \frac{k_c k_p}{k_b + k_p} \qquad (3)$$

$$eff = \frac{k}{k_c} = \frac{k_p}{k_b + k_p} \qquad (3a)$$

collision is given by (3a). It should be noted that AB* is
chemically activated i.e. internally excited by the exothermicity of the reaction A + B → AB. At pressures up to a few torr,
generally, a given reacting pair does not exchange energy with
the surroundings by collisions or radiation, it is thus characterized by a constant total energy E and constant total angular
momentum J.

It is clear that the reaction will be very efficient
for reactants for which the zero point energy difference between
the two complexes $-\Delta E_0$, is very large. For such reactions $k_p \gg k_b$
such that $k = k_c$.   Reactions which have moderate or small
$-\Delta E_0$ may be expected to proceed with low collision efficiencies.
For the moment we will assume that the internal barrier is a
chemical barrier i.e. it is caused by an increase in the electronic energy due to the specific bonding changes required by the
reactions. Later, other less evident causes of internal barriers
will be examined. The complex corresponding to the top of the
internal barrier and relating to k_p will be a tight complex while

the collision complex corresponding to k_c is a loose complex. In other words, the density of states or microcanonical crossings at the collision complex is much larger than that at the product complex. This difference is mostly due to the presence of six external free rotations of the reactants in the loose collision complex while in the tight product complex generally some four of these rotations will not be present. Additional tightening of the product complex associated with other specific constraints due to the reactive change may occur also. Crossing of the internal barrier is therefore not so probable, and the reaction efficiency may be low even if $-\Delta E_o = E_o - E_o'$ is moderately large say, 10-15 kcal/mol. It is to be expected that the collision efficiency will be very low for cases where $-\Delta E_o$ is quite small.

The considerations leading to eq. 3 are based on the assumption that $k_b' \gg k_{-p}$ i.e. that every complex AB* that crosses the internal barrier leads to reaction. This assumption is justified so long as the reaction is exothermic by a few kcal/mol. Since the complex at E_o'' is a loose complex, passage over this barrier will be fast compared with passage over E_o'.

Results obtained by Brauman (15) for a nucleophilic displacement reaction whose reaction coordinate is assumed to have an internal barrier as in Fig. 1 are shown in Fig. 2. The calculations employed the RRKM formalism. The results illustrate that the reaction efficiency becomes appreciably less than unity for $-\Delta E_o$ which are smaller than ~8 kcal/mol. The (CH_3O^-, CH_3Br) reaction has a lower efficiency than (Cl^-, CH_3Br) (at a given ΔE_o), a result to be expected from the larger rotational loss of freedom on formation of the tight complex for CH_3O^- as compared to the monoatomic Cl^-. Since the actual value of ΔE_o for the given systems is unknown Brauman used the experimentally determined efficiencies and the results in Fig. 2 to predict an approximate ΔE_o. The $-\Delta E_o$ for Cl^-, CH_3Br, a reaction that is only ~1% efficient at room temperature, is ~1 kcal/mol.

Surprisingly Olmstead and Brauman (15) did not calculate reaction efficiences at temperatures other than room temperature, nor did they comment on the temperature dependence of the reactions. It is quite clear that reactions with a reaction coordinate as in Fig. 1 will have negative temperature dependence. At higher temperature, the average internal energy E of a colliding pair will be larger. This will make it easier for the complex to backdecompose to the reactants since as E increases the large density of states of the loose transition state becomes more important.

The requirement for conservation of angular momentum in the complex also leads to a decrease of reaction efficiency at high temperatures. This effect can be demonstrated with the aid of

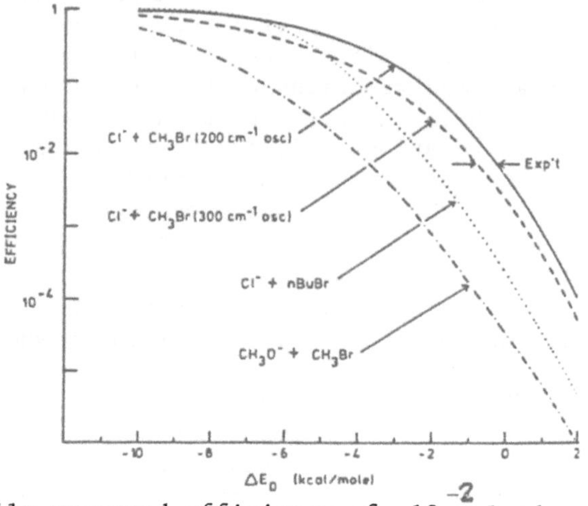

FIGURE 2. Reaction ef-
ficiencies at 300K in
function of ΔE_o from
Olmstead and Brauman
(15) based on RRKM
calculations. ΔE_o is
defined in Fig. 1.
Reaction efficiency is
seen to decrease
rapidly as $-\Delta E_o$ decreases. The two curves
given for reaction,
$Cl^- + CH_3Br = ClCH_3 +$
$Br-$, illustrate effect
of changing the three
softest frequencies
in the tight transi-
tion state from 200
to 300 cm^{-1}. Experimentally measured efficiency of $\sim 10^{-2}$ leads
to a $\Delta E_o \sim -1$ kcal/mol.

Fig. 3. In an orbiting collision, as A and B are pulled together by the attractive potential and the distance r_{AB} decreases,
conservation of angular momentum of the orbiting, quasidiatomic
molecule AB requires an increase of the rotational energy E_J.
This follows from equation for a diatomic rotor. This additional

$$E_J = \frac{h^2 J(J+1)}{8\pi^2 \mu r_{AB}^2} \qquad (J = const)$$

energy must be supplied from the attractive potential. This means
that not all of the potential drop is available as internal
excitation (non-fixed energy) in the complex AB.

The potential energy tied up in rotation is shown schematically in Fig. 3. We see that a high quantum number J increases
the effective barriers for both crossings but the increase of the
barrier of the tight complex is much larger than that of the loose
complex since r_{AB} for the tight complex is smaller.

The effect of increasing the energy E and the angular momentum J is illustrated in Fig. 4. This figure is based on calculations from this laboratory for the hydride transfer reaction:
$C_3H_7^+ + iso\text{-}C_4H_{10} = C_3H_8 + t\text{-}C_4H_9^+$. (The results for the reaction
$(C_3H_7^+, C_4H_{10})$ are only used here as an illustration of the effect
of E and J. This reaction will be discussed in greater detail
later). The calculations are for microscopic rates at a given
energy E above the centrifugal barrier and angular momentum J.

FIGURE 3. Schematic representation of effect of
angular momentum J in the
quasidiatomic molecule AB.
Angular momentum conser-
vation (J=const) requires
increase of rotational
energy as distance between
AB decreases. This fixed
rotational energy leads to
a larger rotational
barrier at the complex

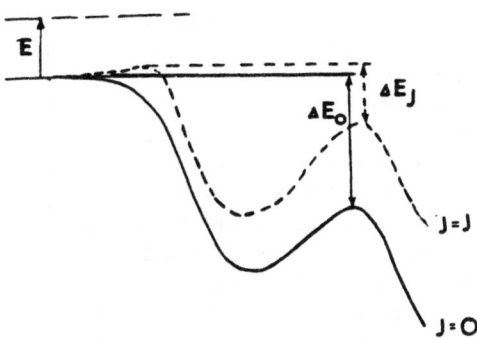

with the smaller A...B distance i.e. at the tight complex.

FIGURE 4. Effect of energy
E and angular momentum J
of quasidiatomic molecule
AB on reaction efficiency.
RRKM calculations for re-
action: sec-$C_3H_7^+$ +
$HC(CH_3)_3$ = C_3H_8 + $(CH_3)_3C^+$.
Reaction assumed to have an
internal barrier as in Fig.
1. Results illustrate that
efficiency decreases
rapidly with increase of
E or J.

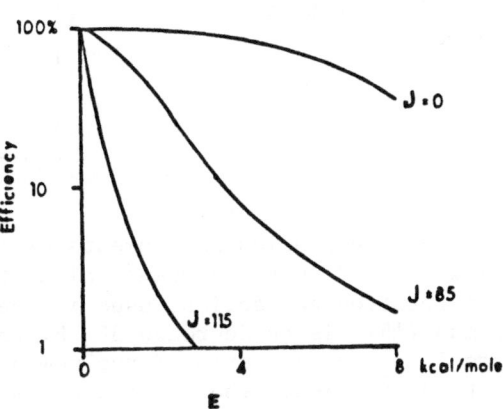

At constant J the efficiency decreases with E, this is due to the
fact that at high E the complex AB* can cross more easily the
high barrier of the back reaction to A + B. At constant E, the
efficiency decreases rapidly with increasing J because of the
more rapid increase of the fixed rotational energy in the tight
complex where r_{AB} is smaller.

Calculations from our laboratory of the temperature dependence of the nucleophilic displacement reaction (4):

$$Cl^- + n\text{-}BuBr = Cl\text{-}nBu + Br^- \qquad (4)$$

are given in Fig. 5. ΔE_o is used as a parameter because accurate
potential energy diagrams are not available. Results in which the
need to conserve angular momentum is not considered are shown
together with calculations where corrections for the conservation of angular momentum were made.

FIGURE 5.   Arrhenius plot of
calculated reaction efficiencies for reaction:
$Cl^- + n\text{-}BuBr = n\text{-}BuCl + Br^-$.
Unknown ΔE_o is treated as a
variable parameter. Reaction
is predicted to have negative temperature dependence.
Straight line plots are not
obtained. Average slope of
curves increases as $-\Delta E_o$ increases. —— conservation
of angular momentum was
neglected. ---- conservation
of angular momentum was
included.

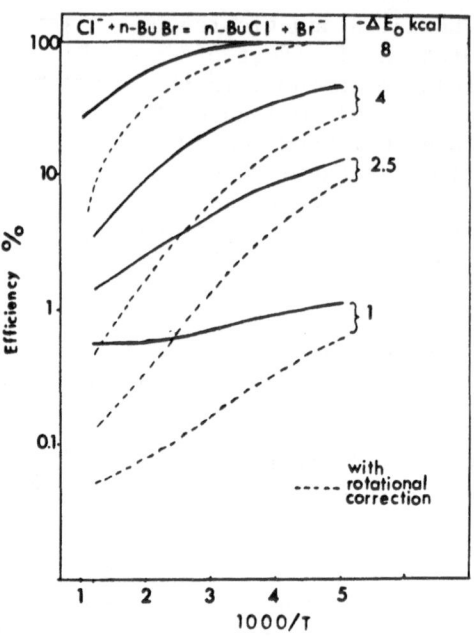

The frequencies and moments of inertia provided by Brauman
(15) were used in the present calculation. The usual RRKM adiabatic rotation correction, used in the present calculation and by
Brauman (15), is to decrease ΔE_o by the quantity $kT(I_L/I_T - 1)$,
where I_L/I_T is the ratio of the moments of inertia of the loose
and tight complex.   The correction is obtained from the assumption that the complexes crossing the   barrier have an
average orbiting energy equal to kT i.e. the average energy of a
two dimensional rotor.

The reaction is predicted to have negative temperature dependence for the ΔE_o values shown. Inclusion of the rotational
correction increases the fall off in efficiency with increase of
temperature, an effect expected from Figures 3 and 4.   The
Arrhenius plots do not lead to straight lines. Plots of log
(efficiency)vs log T (not shown) also don't lead to straight
lines.   The results in Fig. 5 show that the reaction efficiency
at a given temperature increases rapidly as $-\Delta E_o$ is increased.
The average slope also increases with increase of $-\Delta E_o$. Similar
calculations of the temperature dependence were made for other
nucleophilic displacement reactions (Cl^-,CH_3Br); (CH_3O^-, CH_3Br).
These reactions also showed negative temperature dependence for
$-\Delta E_o$ in the range 2.5 - 8 kcal/mol. Again neither the Arrhenius
nor the log k vs log T plot gave straight lines.

Since experimental measurements of the temperature dependence
of the rate constants for nucleophilic displacement reactions are
not available, further discussion of the above results or attempts

to refine the calculations are not really profitable. Our experi-
ence with these calculations suggests that matching the efficiency
at a single temperature is not a very reliable means of testing
the theoretical model and determining a value of ΔE_o. We shall
return to theoretical predictions when discussing the hydride
transfer reactions for which experimental results over a wide
temperature range are available.

EXPERIMENTAL VERIFICATIONS OF THE PRESENCE OF INTERNAL BARRIERS
AND THEIR EFFECT ON REACTION EFFICIENCY AND TEMPERATURE DEPEND-
ENCE

   A very direct method to demonstrate the presence of internal
barriers is illustrated in Fig. 6. We envisage a reaction series:
A^+ + B → products, in which one reactant (B) is varied. The
changes of B are aimed to gradually increase the internal barrier,

FIGURE 6. Potential
energy of reaction
coordinate for a re-
action series: A^+ + B =
C^+ + D in which re-
lated compounds B (B,
B',B") are used. The
changes of the nature
of B raise the height
of the internal
barrier. The rates
with B and B' have
negative temperature

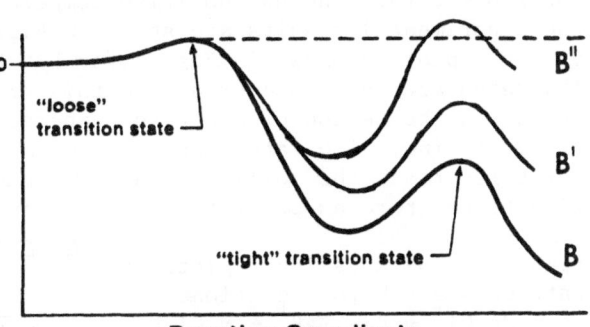

dependence, with B" positive temperature dependence.

as is the case from B to B' (Fig. 6). At some point (B") the
internal barrier becomes higher than the energy level of A + B
and then the reaction is not only the slowest but, what is more
important, the temperature dependence switches from negative to
positive.

   The existence of such a reaction series was found in the
alkylation reactions of bases B by the dimethyl chloronium ion
as shown in eq. (5).

$$Me_2Cl^+ + B = MeB^+ + MeCl \tag{5}$$

   The structure of the reactants, where B = dimethyl ether is
shown in scheme I. The reaction is closely related to nucleophilic displacement. The nucleophilic attack is by the oxygen

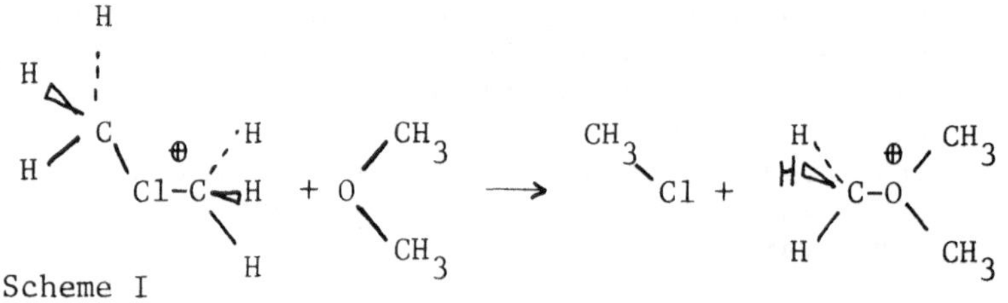

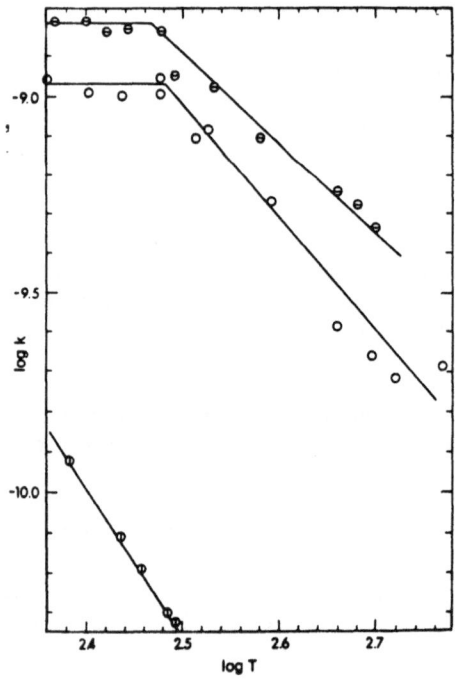

Scheme I

of dimethyl ether on the methyl group carbon. In a series of
B's studied in our laboratory (11), it was found that for the
bases B of decreasing basicity: Me_3N, H_3N, Me_2O, Et_2O, $i-C_3H_7-$
benzene, toluene, benzene, the rate constants also decreased
with decreasing basicity. Thus, slightly above room temperature,
Me_3N and NH_3 proceeded at collisional efficiency, Me_2O and Et_2O
were below collision efficiency and had negative temperature dependence.   iso-C_3H_7 benzene and toluene were still slower and had
negative temperature dependence. The weakest base, benzene had
the slowest rate and had positive temperature dependence. Some of
the experimental results are shown in Figures 7 and 8. A log k
vs log T plot is shown in Fig. 7 for B = Et_2O, Me_2O and toluene.
The rates are seen to decrease in this order. The experimental
points in the region of negative temperature dependence fit on a
straight line.   The rate constants for Me_2O and Et_2O at low temperatures reach the collision rate and become essentially constant
with temperature as predicted by equation 1. Figure 8 gives

FIGURE 7. log k vs log T plots of
rate constant k for reactions
(5): Me_2Cl^+ + B = MeB^+ + MeCl.B =
Et_2O, ⊖; B = Me_2O, 0; B = toluene,
Ⓓ . Rate constants have negative
temperature dependence. Collision
rates are reached with Me_2O and
Et_2O at T ~310°K. From Sen
Sharma, Kebarle (11).

Arrhenius plots for B = isopropyl benzene, toluene and benzene.
Benzene which has the lowest rate constant has positive temperature dependence. The slopes of the three Arrhenius plots lead to

FIGURE 8. Arrhenius type plots
of rate constants k for reaction (b) with B = isopropyl
benzene, toluene and benzene.
Rate constants for isopropyl
benzene, and toluene have
negative temperature dependence while the much lower
rate constants for benzene
have positive temperature
dependence. Thus, this series
of reactions fits scheme
represented in Fig. 6 with
B and B' being isopropyl
benzene and toluene and
B" = benzene. From Sen
Sharma and Kebarle (11).

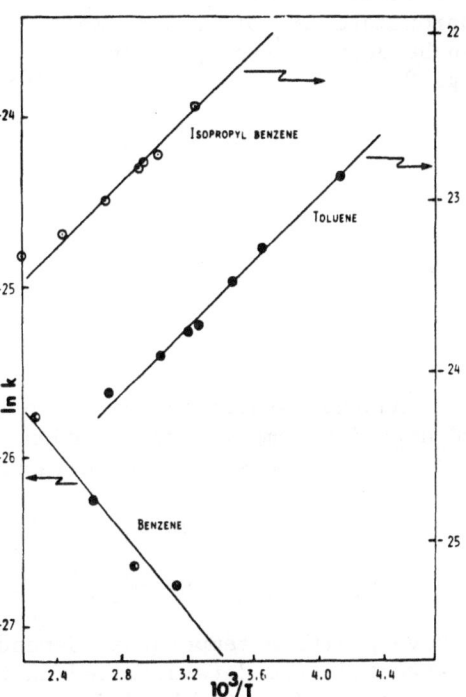

"activation energies". For benzene this value is positive and
should correspond to $\Delta E_o \simeq E_A$, where E_A is the Arrhenius activation energy. For toluene and isopropyl benzene the values of
E_A are negative. Since we deal with chemically activated systems,
the E_A for these systems can not be identified with ΔE_o, however
considering that the $-E_A$ are quite small one might assume that
the ΔE_o are not too far from the observed negative activation
energies.

        A reaction coordinate diagram (Fig. 9) summarizes the
approximate information available for toluene. At very low temperatures an adduct, $Me_2Cl.toluene^+$ was observed (11) and the
equilibrium constant K_6^o led to a $\Delta H_6^o = -11$ kcal/mol which is shown

$$Me_2Cl + toluene = Me_2Cl.toluene^+ \qquad\qquad (6)$$

in Fig. 9. For B = benzene, the diagram should be similar but
with one important difference, namely that the internal barrier
leading to the tight complex protrudes a bit above the energy
level of the reactants.

FIGURE 9. Probable potential
energy of reaction coordinate
for reaction: Me_2Cl^+ + toluene =
$Me.toluene^+$ + MeCl. An
approximate ΔE_o of -2 kcal/mol
can be deduced from plot in
Fig. 8.

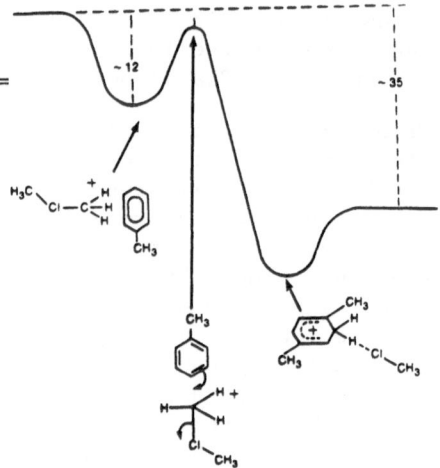

Another system in which two related reactions have positive
and negative temperature dependence are reactions (7) and (8).
Reaction (7) was studied by Hiraoka and Kebarle (20) and found

$$C_2H_5^+ + CH_4 = sec\text{-}C_3H_7^+ + H_2 \tag{7}$$

$$C_2H_5^+ + C_2H_6 = sec\text{-}C_4H_9^+ + H_2 \tag{8}$$

to have positive temperature dependence. A reaction coordinate
constructed from data detailed in that work is shown in Figure
10. The three centre bonded structures indicated in the Figure
are only suggestions. The complex   H   is really

$$CH_3CH_2 \overset{H}{\diagup\diagdown} CH_3^+$$

loosely associated $C_2H_5^+.CH_4$ whose formation from $C_2H_5^+$ and CH_4
could be observed and the equilibrium measured. Available data
for reaction (8) are also given in Figure 10. Negative temperature dependence was observed for reaction (8), by Hiraoka and
Kebarle (20. The origin of other data given in the diagram are
detailed in the publication (20). The lower energy of the tight
complex for reaction (8) observed experimentally is in line with
the expected better stabilization of the transition state of that
system since the carbon loosing the H_2 molecule is flanked by
methyl and ethyl rather than by two methyls as in the case in
reaction (7).

The reactions with positive temperature dependence
(examples: (5) B = benzene and (7)) have transition states which
require thermal activation. These reactions can be treated by
conventional transition state theory (see eq. 9)

$$k = \frac{k_B T}{h} \frac{Q^{\ddagger}_{AB}}{Q_A Q_B} e^{-\Delta E_o^{\ddagger}/RT} \approx A e^{-E_A/RT} \tag{9}$$

where k_B is the Boltzman constant, h is the Planck constant and Q are partition functions. The preexponential factor for such reactions is much smaller than that predicted by collision

FIGURE 10. Example of potential energy diagrams for related reactions (7) and (8) deduced from experimental measurements. Reaction (7) had positive, while (8) had negative temperature dependence. Hiraoka and Kebarle (20).

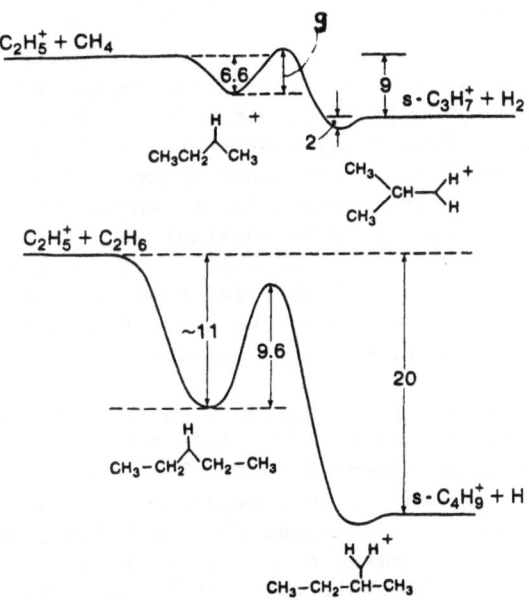

theory when the transition state is a tight transition state as is the case here. We expect therefore preexponential factors for reaction (5, benzene) and (7) which are much smaller than the collision rate i.e. much smaller than 10^{-9} molecules cm^3 sec. The preexponential factors, observed experimentally (11,20), given below are in agreement with this expectation.

$$A_5 = 9 \times 10^{-11}; \quad A_7 = 6 \times 10^{-13} \text{ (molecules}^{-1} \text{ cm}^3 \text{ sec}^{-1})$$

REACTION INEFFICIENCY AND NEGATIVE TEMPERATURE DEPENDENCE NOT CAUSED BY A CHEMICAL INTERNAL BARRIER: LOCKED ROTOR. HYDRIDE TRANSFER REACTIONS.

A very comprehensive experimental study of the temperature dependence of rate constants of hydride transfer reactions has been made by Meot-Ner, Field and coworkers (9,10). The results for several reactions studied by them are shown in Figure 11. All reactions shown have negative temperature dependence and reach collision rates at low temperature. From the many hydride transfer reactions studied by Meot-Ner and Field, we selected reaction (10) for a theoretical study. This reaction was followed over the widest temperature range by Meot-Ner and Field. Also, it involves reactants of only moderate complexity so that the required parameters for the RRKM calculations (vibrational frequencies

$$\begin{array}{c} CH_3 \\ \diagdown \\ CH^+ \\ \diagup \\ CH_3 \end{array} + \begin{array}{c} CH_3 \\ \diagup \\ H-C \!-\! CH_3 \\ \diagdown \\ CH_3 \end{array} = \begin{array}{c} CH_3 \\ \diagdown \\ CH_2 \\ \diagup \\ CH_3 \end{array} + \begin{array}{c} CH_3 \\ +\diagup \\ C \!-\! CH_3 \\ \diagdown \\ CH_3 \end{array} \qquad (10)$$

and moments of inertia of the reactants and the loose and tight
complex) could be obtained from the literature or estimated. The
results from a calculation using the Brauman model (see Fig. 1
and eq. (3) and (3a)), are shown in Fig. 12. Proper account of
the need for conservation of angular momentum was also taken (19).
The microscopic rate constant and the steady state distribution
of AB* were evaluated explicitly for each J as well as for E.
This requires a double integration over J and E but avoids the
uncertainty of using the usual RRKM correction with which the
results of Fig. 4 were obtained.

Since ΔE_0 is unknown, calculations for various ΔE_0 were made.
The ΔE_0 = -9.8 kcal/mol led to a curve which gave the closest fit
to the experimental data. This curve is the one shown in Fig. 12.
The agreement between experimental and calculation is only qualitative. In particular the sharp break observed in the experimental data between the rates proceeding at collision efficiency
and the fallen off rates at higher temperature is not reproduced
in the calculation. Calculations performed with different vibrational frequencies in the complexes and sec-$C_3H_7^+$, moved and
changed somewhat the shape of the calculated curve but did not
produce a sharper break.

FIGURE 11. Temperature dependence of rate constants
for hydride transfer re-
actions: sec-$C_3H_7^+$ + RH =
C_3H_8 + R^+ from Meot-Ner
and Field (9,10). These
authors were the first to
demonstrate that the
limiting, collision rates
are reached at low tem-
peratures.

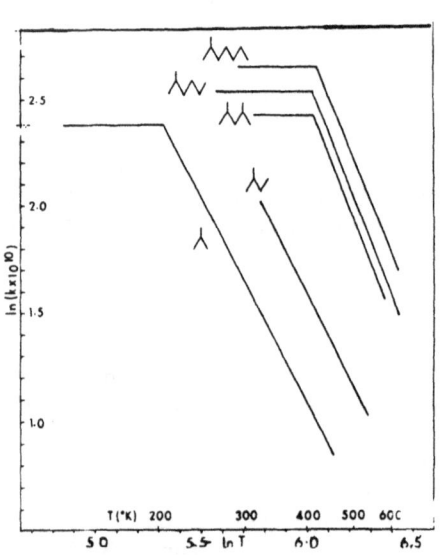

FIGURE 12. Comparison between
experimental measurements
(Meot-Ner 9,10) and calculations (19) for hydride
transfer reaction: sec-$C_3H_7^+$ +
$HC(CH_3)_3$ = C_3H_8 + $(CH_3)_3C^+$.
Calculation assumed existence
of internal electronic barrier
as in Fig. 1. ΔE_0 was treated
as parameter. Best fit
obtained with ΔE_0 shown in
Figure. Theoretical model
provides only a fair fit to
experimental data. Neither
sharp break nor linearity
at high T is reproduced by
calculation.

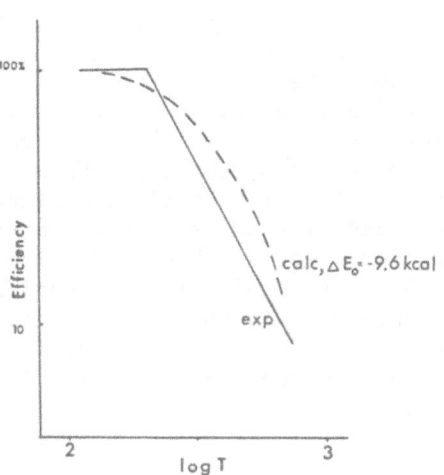

    In a search aimed to obtain better agreement between theory
and experiment, we were led to a change of the reaction model. As
shown in Fig. 13, a tight complex need not arise only because of
the presence of an electronic energy internal barrier (diagram
on left of Fig. 13). Tightness can develop also on a smooth part
of the electronic energy curve. As the two reactants approach,
at a given distance, their external rotations may become re-
stricted because of steric hindrance (see Fig. 13, right). This
effect is to be expected particularly for large reactants and a

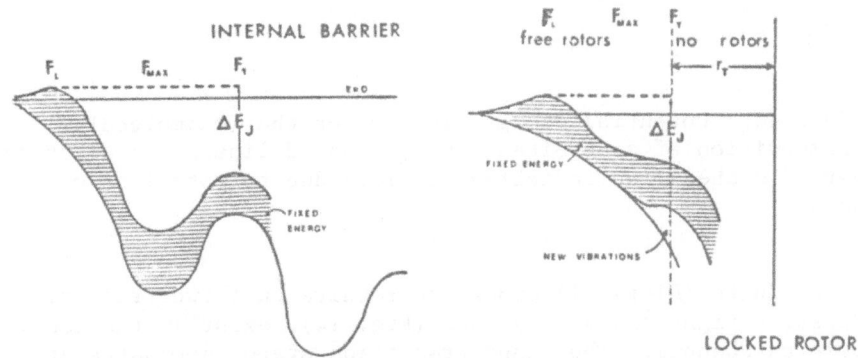

FIGURE 13. Diagram on left represents reaction coordinate when
internal electronic barrier is present. Reaction inefficiency and
negative temperature dependence can occur also in absence of
electronic barrier, as shown in diagram on the right. Locking of
external rotations of reactants on a relatively flat region of
the potential energy of the reaction coordinate can produce a
tight complex. F_L, F_T and F_{MAX} stand for the reaction fluxes at
the loose, tight and maximum flux region.

reaction in which the reaction site is not readily accessible.
For these conditions not only tightness develops due to the loss
of rotations but also a small bump is created on the reaction
coordinate because of the simultaneously appearing zero point
vibrational energies of the new vibrations which replace the lost
rotations. The "early" tight complex (Fig. 13 right) may be called
the locked rotor complex. We may expect a locked rotor complex to
develop at an intermediate distance i.e. in a region where the
long range forces are not any more completely dominant. The likli-
hood for such a tightness developing will be high for reactions in
which the fall off of the reaction coordinate with decreasing
distance r_{AB}, is small. This may be expected for molecules without a large permanent dipole, for complexes in which there is no
strong hydrogen bond, for ions in which the charge is strongly
delocalized,etc.

Support for the locked rotor model can be found from studies
of the unimolecular decomposition of ions. For such systems the
reaction coordinate may be expected to be as shown in Fig. 14
(solid line). Studies of translational energy distributions of

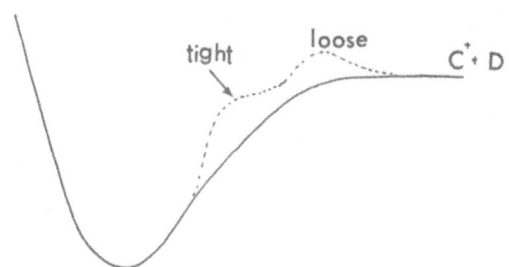

FIGURE 14. Potential energy diagram for the (unimolecular)
decomposition of an excited ion AB*. Solid line gives electronic
energy, dashed line indicates barriers due to rotations of A
and B.

the products (Klots 21) appear to require that two transition
states, a tight and a loose one (Fig. 14), exist in the decomposition channel. The tight transition state dominates at
high energies while the loose one, dominates at low energies. A
mathematical formalism was developed to deal with these reactions
(Chesnavitch and Bowers 22). This formalism is suitable also
for the present calculations involving the locked rotor model.
It should be noted that the equation (3) used earlier was obtained
by assuming a steady state for AB* which could either backdecompose (k_b) or go in the product channel (k_p). This approach is
not suitable for the locked rotor model where the two complexes
are not separated by a deep well and there is not a well characterized AB* complex. A more general treatment based on probability branching analysis (Miller 23) leads to an equation for

the rate constant for the bimolecular reaction given in eqn. (11).
The rate constant is expressed in terms of reaction fluxes F,
where the subscripts, T refer to the tight complex and L to the
loose complex and max to the maximum flux occurring somewhere
between L and T. N is the density of states for a given E and J.

$$k \propto \frac{F_T F_L}{F_T + F_L - \dfrac{F_T F_L}{F_{max}}} \cdot \frac{1}{N} \qquad (11)$$

Equation (11) becomes equivalent to eqn. (3) when a deep well is
present between T and L. For these conditions $F_{max} \gg F_T$ or F_L
so that eqn (11) reduces to (12).

$$k \propto \frac{F_T F_L}{F_T + F_L} \cdot \frac{1}{N} \qquad (12)$$

For the locked rotor case F_{max} will be close to either F_T
or F_L (depending on the values of E and J). Bowers and
Chesnavitch (22) proposed that one should set F_{max} equal to F_L or
F_T whichever is the larger. With this assumption eqn. (11) is
reduced to the much simpler form (13). F_{min} switches from F_T
to F_L whichever is the smaller one. The assumption expressed in
eqn. 13 is called the switching model (22).

$$k \propto F_{min} \cdot \frac{1}{N} \qquad (13)$$

We have calculated (19) the rate constants for the hydride
abstraction reaction (10) using the switching model. The results
are shown in Fig. 15. Results for different ΔE_o are shown. These
theoretical results also do not give a complete fit to the
experimental temperature dependence. However the initial fall
off is more abrupt and fits the sharp break of the experimental
result better than the results in Fig. 12 without the switching
model. Examining Fig. 12 one can see that a $-\Delta E_o \approx 7$ kcal/mol
would give quite a respectable fit to the experimental result.

FIGURE 15. Results from calculations (19) for the hydride abstraction: sec-$C_3H_7^+$ + $HC(CH_3)_3$ = C_3H_8 + $C(CH_3)_3^+$ using the locked rotor model (Fig. 13 right) and switching equation 13. Theoretical results give somewhat better fit to experimental measurements than electronic barrier model (Fig. 12).

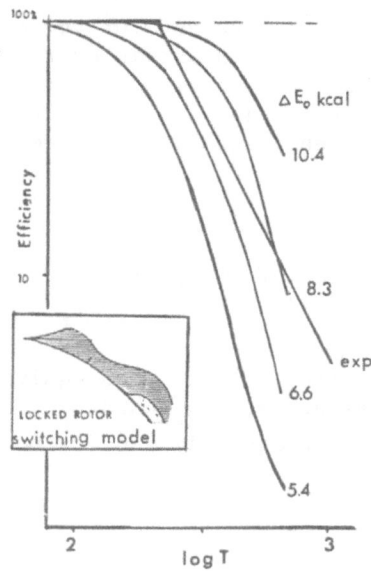

SUMMARY OF CONDITIONS WHICH MAY LEAD TO REACTION INEFFICIENCY AND NEGATIVE TEMPERATURE DEPENDENCE

a Reactions with internal (electronic) barriers

One may expect internal barriers in nearly all ion molecule reactions where an atom or a group of atoms is transferred. The size of the barrier will of course vary with reaction type. Thus for proton transfer the barrier is generally quite small and these reactions generally proceed at collision rates. For the nucleophilic displacement reactions where an alkyl group is transferred with inversion of configuration, larger barriers are to be expected, the same is true for the alkylation reactions by chloronium ions which were discussed earlier. Since ΔE_o is the actual quantity entering the equations, of importance is not only the size of the internal varrier but also the depth of the potential well for the AB* formation. A shallow well coupled with a moderate barrier will lead to a small $-\Delta E_o$ and thus to a large reaction inefficiency. While there is little accurate information on internal barriers there is quite plentiful information on the depth of potential wells for the adducts AB^+. These data are obtained from ion molecule equilibria measurements of adduct forming reactions (Kebarle 24). Shallow wells are observed with negative ions A^- and molecules B when the molecule is aprotic i.e. does not have hydrogens with large net atomic positive charge. Thus the complexes $A^-.B$ occurring in the nucleophilic displacement reactions have shallow wells. Shallow wells

are observed also when the molecule has no permanent dipole or
when the ion is relatively large and its charge delocalized. The
alkyl chloronium ions are ions with delocalized charge. Thus
while Cl is formally positive, the positive charge is distributed
mostly over the two alkyl groups. The adducts formed in the
reactions (7) and (8) have also shallow potential wells (see Fig.
10). Shallow potential wells are expected also in the hydride
transfer reactions. The adducts in these reactions should be
basically similar to those occurring in reactions (7) and (8).
Thus the adduct in reaction (8) may be the same as that leading
to the symmetric hydride transfer (14).

$$CH_3CH_2^+ + CH_3CH_3 = CH_3CH_3 + CH_2CH_3^+ \tag{14}$$

b   Reaction inefficiency due to rotational locking

       Reaction inefficiency and negative temperature dependence
may be caused also by rotational locking. This effect will be
enhanced by the presence of weak medium range attractive forces
between the ion and the molecule. Generally such weak forces
will be present when there is a shallow potential well for the
adduct AB^+. An additional condition is that the reactants should
be large and the reaction site located so, that it leads to
steric rotational hindrance between the reactants. The hydride
transfer reactions probably fall in this class.

       A special case of the locking mechanism are the sterically
hindered pyridine proton transfer reactions mentioned in the
section "Experimentally observed reactions with low collision
efficiency". Brauman (18) found that these reactions proceeded
far below collision efficiency. Very recently, Meot-Ner (ASMS
meeting 1982 Hawaii) has reported temperature dependence measurements. The reactions were found to have negative temperature
dependence. While there is no space here to discuss these interesting results, it should be mentioned that Meot-Ner's own
analysis (25) favours the operation of a locked rotor mechanism.

NEGATIVE TEMPERATURE DEPENDENCE IN BIMOLECULAR REACTIONS WITH
INSIGNIFICANT INTERNAL BARRIERS BUT WITH SMALL EXOTHERMICITY

       Bimolecular reactions with insignificant internal barriers
but of low exothermicity can be expected to show negative temperature dependence (26). The potential energy diagram of the
reaction coordinate for such a reaction is shown in Fig. 16.
The steady state assumption for AB* leads to the expression 15
which is identical to eqn. 3. However in the present case both

$$k = \frac{k_c k_p}{k_b + k_p} \tag{15}$$

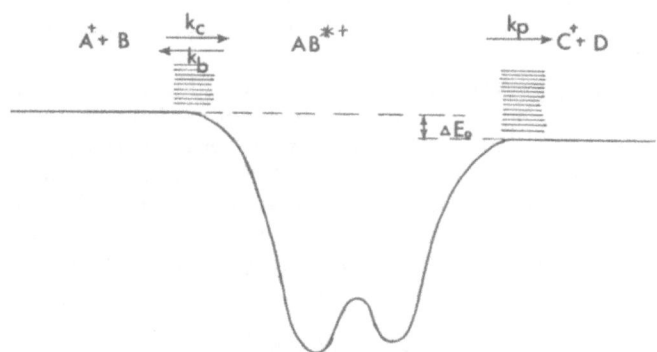

FIGURE 16. Reaction coordinate for reaction with small exothermicity and small internal barrier which does not produce a
bottle neck. Low reaction efficiency and negative temperature
dependence will be observed if the density of states for the
product loose complex is lower than that for the collision loose
complex. This will be the case when ΔS^o for the reaction is
negative.

channels available to AB* involve loose complexes and the exothermicity of the reaction corresponds to ΔE_o. For cases where
the density of states of the two complexes is approximately the
same, passage to products $C^+ + D$ will be favored at low temperatures, while at very high temperatures k_b and k_p will become
equal. This means that k will decrease to half of its low temperature value. Much more pronounced negative temperature dependence may be expected for cases where the density of states
over the product complex is significantly lower than that of the
back reaction complex. This will be the case when the entropy
change for the reaction is negative, since the loose product
complex decomposing to $C^+ + D$ must represent the properties of
$C^+ + D$ as much as the back reaction complex decomposing to $A^+ + B$
must contain those of $A^+ + B$. The lower density of states for
the product complex can lead at high temperatures to a k_p which
is significantly lower than k_b and thus to a significant negative
temperature dependence.

     Results obtained by Meot-Ner (27), confirm the above pre-
diction. For example, the proton transfer reaction 16 was found

$$(CH_3)_3NH^+ + HO(CH_2)_3NH_2 = (CH_3)_3N + \overset{\displaystyle \overset{..H}{HO \cdots NH_2^+}}{CH_2CH_2CH_2} \tag{16}$$

$\Delta H^o = -3.15$ kcal/mol; $\Delta S^o = -12$ cal/degree mol; $\Delta G^o_{483} = 2.2$ kcal/
mol   k(483) = 1.2 x 10^{-10} molecules^{-1} cm^3 sec^{-1}

to have a small exothermicity, but because of the cyclized product a relatively large $-\Delta S^{o}$. The reaction rate was much below
collision efficiency at 483K and showed negative temperature
dependence (27).

SUMMARY

        Internal (electronic barriers (see Fig. 1) are probably
present in the reaction coordinate of most exothermic ion molecule reactions. When the (largest) internal barrier for a given
reaction is small, such that $-\Delta E_{o}$ is large, the reaction will
proceed at collision efficiency. Reactions with moderate $-\Delta E_{o}$
(5-20 kcal/mol) generally proceed with collision efficiency
below or at room temperature but show negative temperature dependence at higher temperatures. Reactions with lower $-\Delta E_{o}$
reach collision efficiency only considerably below room temperature.
        Estimates of ΔE_{o} made by comparing RRKM calculated rate
constants with experimentally determined ones should be made at
several temperatures if one wishes to obtain a reliable test of
the theoretical model.

        Reactions with positive ΔE_{o} have positive temperature de-
pendence and proceed at rates much below collision efficiency at
all temperatures.

        Experimental demonstration of the presence of internal
barriers can be obtained in suitable reaction series $A^{+} + B \rightarrow$
$C^{+} + D$ in which the chemical nature of one of the reactants (B)
is changed in such a manner as to gradually increase the height
of the internal barrier. A change over from negative to positive
temperature dependence occurs when ΔE_{o} changes sign.

        Low reaction efficiency and negative temperature depen-
dence need not be caused only by internal barriers. Any loss of
rotations along the reaction coordinate could produce the same
effect. Such behavior can be expected when steric constraints
to rotation are present i.e. large reactants or bulky and inter-
fering substituent groups.

        Reactions with low exothermicity and a large negative
entropy change can also have low efficiency and negative temperature dependence.

REFERENCES

1. Eyring, H., Hirschfelder, J.O., and Taylor, H. S.: 1936, J. Chem. Phys. 4, pp. 479.
2. Giomousis, G., and Stevenson, D. P.: 1958, J. Chem. Phys. 29, pp. 294.
3. Dugan, J. V., and Magee, J. L.: 1967, J. Chem. Phys. 47, pp. 3103.
4. Dugan, J. V., and Palmer, R. W.: 1971, Chem. Phys. Lett. 13, pp. 144.
5. Su, T., Bowers, M. T.: 1979, in "Gas Phase Ion Chemistry" (Bowers, M. T. Ed.) Vol. 1, Academic Press New York Chapter 3, pp. 83; Chesnavitch, W. J., and Bowers, M. T.: 1979, in "Gas Phase Ion Chemistry" (Bowers, M. T. Ed) Vol. 1, Academic Press, New York, Chapter 4 pp. 119; Chesnavitch, W. J., Su, T., and Bowers, M. T.: J. Chem. Phys. pp. 2641.
6. Barker, R. A., and Ridge, D. P.: 1976, J. Chem. Phys. 64, pp. 4411; Celli, F., Weddle, G., and Ridge, D. P.: 1980, J. Chem. Phys. 73, pp. 801.
7. Bates, D. R.: 1981, Chem. Phys. Lett. 82, pp. 396.
8. Ausloos, P., and Lias, S. G.: 1970, J. Am. Chem. Soc. 92, pp. 5037.
9. Meot-Ner, M., and Field, F. H.: 1976, J. Chem. Phys. 64, pp. 277-281.
10. Meot-Ner, M.: 1979 in "Gas Phase Ion Chemistry" (Bowers, M. T. Ed.) Vol. 1, Academic Press, New York, Chapger 6, pp. 197-221.
11. Sen Sharma, D. K., Kebarle, P.: 1982, J. Am. Chem. Soc. 104 pp. 19-24.
12. Sen Sharma, D. K., Kebarle, P.: 1981, Can. J. Chem. 55, pp. 1592-1601.
13. Bohme, D. K., and Young, L. B.: 1970, J. Am. Chem. Soc. 92 pp. 7357.
14. Tanaka, K., Mackay, G. I., Payzant, J. D., and Bohme, D.K.: 1976, Can. J. Chem. 54, pp. 1643.
15. Olmstead, W. N., and Brauman, J. I.: 1979, J. Am. Chem. Soc. 101, pp. 3715-3723.
16. Bartmess, J. E., Hays, R. L., and Caldwell, G.: J. Am. Chem. Soc. 103, pp. 1338-1344.
17. Farneth, W. E., and Brauman, J. I.: 1976, J. Am. Chem. Soc. 98, pp. 7891-7897.
18. Jasinski, J.M., and Brauman, J. I.: 1980, J. Am. Chem. Soc. 102, pp. 2906-2913.
19. Magnera, T. F., and Kebarle, P.: to be published.
20. Hiraoka, K., and Kebarle, P.: 1975, J. Chem. Phys., 63, pp. 3947-397; 1980, Can. J. Chem. 58, pp. 2262-2270.
21. Klots, C.E., Mintz, D., and Baer, T.: 1977, J. Chem. Phys. 66, pp. 5100.
22. Chesnavitch, W.J., and Bowers, M.T.: 1976, J. Am. Chem. Soc. 98, pp. 7301.

23. Miller, W. H.: 1976, J. Chem. Phys. 65, pp. 2216.
24. Kebarle, P.: 1977, Ann. Rev. Phys. Chem. 28, pp. 445-476.
25. Meot-Ner, M.: 1982 private communication.
26. Kebarle, P.: 1975, in "Interactions between Ions and Molecules" (P. Ausloos Ed.) Plenum Publishing Corp. New York, pp. 459-487.
27. Meot-Ner, M., Hamlet, P., Hunter, E. P., and Field, F.H.: 1980, J. Am. Chem. Soc. 102, pp. 6393.

SPECTROSCOPIC STRUCTURE AND RADIATIONLESS DECAY OF

OPEN-SHELL ORGANIC CATIONS

J.P. Maier, D. Klapstein, S. Leutwyler,
L. Misev and F. Thommen

Physikalisch-Chemisches Institut der Universität Basel,
Klingelbergstrasse 80, CH-4056 Basel, Switzerland

INTRODUCTION

Spectral characterisation and quantitative description of the relaxation processes of ions are essential to the understanding of their chemistry in the gas phase. Once determined, the spectral features may be used to characterise the physical conditions of the ions' environment. Notable examples include evaluation of temperatures and densities in interstellar space, in comets and in plasmas. Concurrent to this is the continual development of the methods for such studies and for state selective identification of ionic species. Furthermore, the information derived from a particular approach is often crucial to the evolution of new spectroscopic techniques and experiments. In the investigations of the chemistry of ions, spectroscopic techniques can provide the means to induce and monitor the course of reactions (1).

The object of this article is to outline the progress made in the spectral studies of open-shell organic cations and in the quantitative determination of their fate in the gas phase under collision-free conditions (2). Three complementary techniques are used to probe their ground and lowest excited electronic doublet states. These are based on emission and laser excitation spectroscopies of rotationally cooled cations and on a photoelectron-photon coincidence technique. The heart of the measurements is the detection of the radiative decay which links the electronic states.

These methods can be applied to a variety of open-shell organic cations as is indicated by the selection given in table 1.

M. A. Almoster Ferreira (ed.), Ionic Processes in the Gas Phase, 159–178.

Table 1.   Summary of the main types of open-shell organic cations
           which decay radiatively from their lowest excited
           ($^2\tilde{B}$, $^2\tilde{A}$) to their ground ($^2\tilde{X}$) states. (Complete com-
           pilation and references can be found in review
           articles (3)).

        X–C≡N$^+$

        X–C≡C–H$^+$ X= Cl, Br, I

        X–C≡C–X$^+$

        X$(C≡C)_2$H$^+$ X= Cl, Br, I, CN, CH$_3$, C$_2$H$_5$

        X$(C≡C)_2$X$^+$ X= Cl, Br, I, CN, F, CF$_3$, C$_2$H$_5$

   CH$_3$–C≡C–X$^+$ X= Cl, Br, CN

   CH$_3$$(C≡C)_2X^+$ X= Cl, Br, CN, CH$_3$

        H$(C≡C)_n$H$^+$ n= 2, 3, 4

   Conjugated Polyenes$^+$

   Fluorobenzenes$^+$, Chlorobenzenes$^+$

   Chlorofluorobenzenes$^+$, Bromofluorobenzenes$^+$,

   Fluorophenols$^+$

The initial experiments using a crossed electron-sample beam
apparatus established that radiative decay from the lowest ex-
cited to the ground cationic states took place (i.e. $^2\tilde{B}$, $^2\tilde{A}$ → $^2\tilde{X}$)
(3). The principle of the approach is depicted on the left side
of figure 1. The main part of the information obtained from the
analyses of the emission spectra are the vibrational frequencies
in the ground cationic state. On the other hand vibrational
characterisation of the excited state lends itself to the laser
excitation technique, following the preparation of the cations
such as in the manner indicated in figure 1 (right) (4). The
laser excitation technique is, of course, also a sensitive probe
of the internal energy distribution of the initial state (5).

EMISSION AND LASER EXCITATION SPECTROSCOPY

     In both the emission and laser excitation spectra the limita-
tion in inferring the vibrational spectroscopic data is set by the
rotational and vibrational sequence broadening of the bands. Thus,
using sources yielding ions with a room temperature rotational
distribution results in uncertainties of ±5 - 10 cm^{-1} for the vibrational frequencies (2). The way to obviate this restriction is

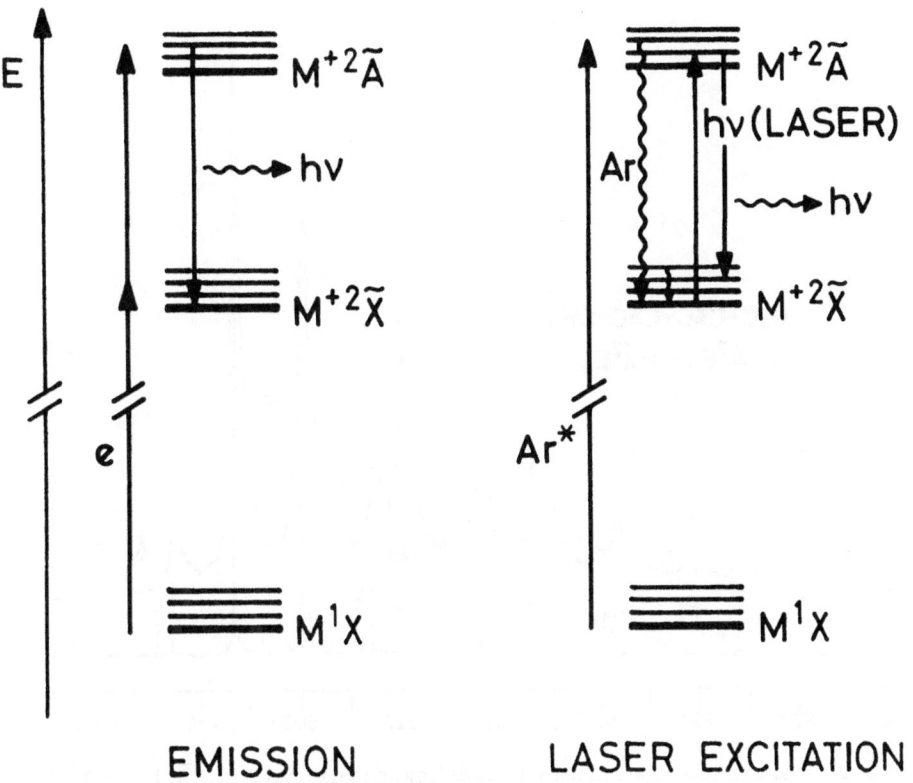

EMISSION   LASER EXCITATION

Figure 1. Schematic representation of the emission and laser excitation spectroscopic techniques.

to study rotationally cooled ions (4,6,7). This is achieved in our experiments and leads to rotational temperatures of less than 30 K for the cations probed by emission spectroscopy and of around 100 K for laser excitation spectroscopy. The narrowing of the bands in the respective spectra, and the improvement in the quantity of the spectra, is such that the vibrational frequencies can be deduced to within $1 - 2$ cm^{-1}, or better, and new spectral features become apparent. This is illustrated in figure 2 where two emission spectra of the $\tilde{A}^2E_u \to \tilde{X}^2E_g$ transition of 2,4-hexadiyne cation are shown (8). The top trace was recorded with ions characteristic of a room temperature rotational distribution whereas the lower one is representative of supercooled ions.

The cations are produced rotationally cold by electron impact excitation in a seeded helium supersonic free jet. The molecules in the jet expansion are cooled to low rotational temperatures and are

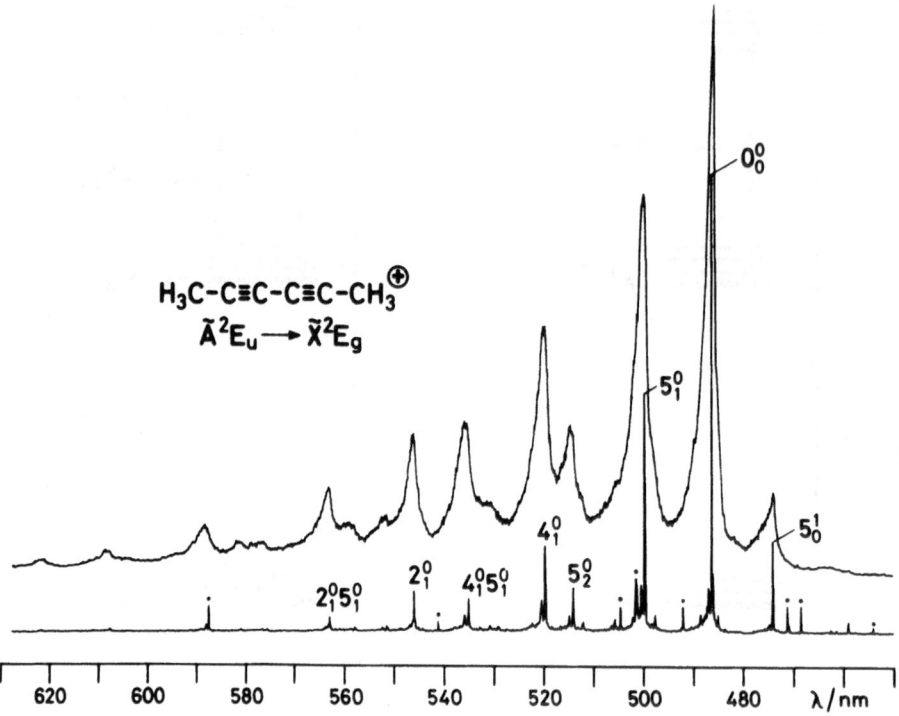

Figure 2.   Emission spectrum of 2,4-hexadiyne cation, $\tilde{A}^2E_u \rightarrow \tilde{X}^2E_g$
band system, recorded using an effusive source (top,
0.16 nm fwhm), or a seeded helium supersonic free jet
(bottom, 0.025 nm) and excited by an electron beam.

subsequently ionised by a focused electron beam impacting prior
to the Mach disk (8). The various electronic states of the cation
reached are characterised by a similar rotational temperature.
This is illustrated in figure 1 for the doublet states of concern.
The rotational temperature usually attained, <30 K, is judged on
the basis of the expansion conditions and direct temperature
evaluations from the rotational structure of di- and triatomic
cations. The molecules in the free jet are also vibrationally
cooled, ≈50 K is expected (9), but in the subsequent ionisation
process various vibrational states are populated according to the
selection rules and Franck-Condon factors. At these low temperatures, the overwhelming majority of these transitions are from the
lowest vibrational level of the neutral species and thus the totally
symmetric levels of the cation state can be reached. The resulting
emission spectrum, $^2\tilde{A} \rightarrow {}^2\tilde{X}$, is consequently similar in overall
appearance to the spectrum obtained at room temperature (effusive
source), except that the vibrational bands are narrow and well
separated (cf. fig. 2).

To sharpen up the vibronic bands in the laser excitation
spectra the ions are also rotationally cooled, though not as extensively as in the free jet experiment. The cooling is achieved
by collisions with the rare gas carrier in which the metastable
species are present for the ionisation process. The cations in
their ground electronic states, after the indicated relaxation

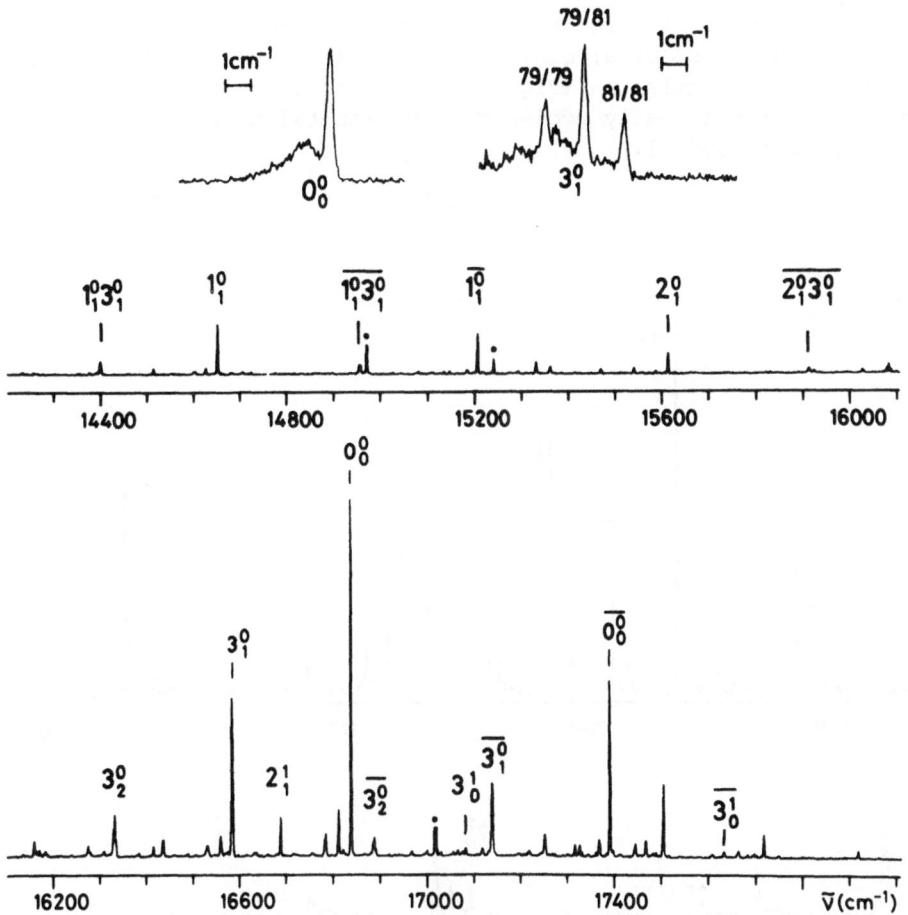

Figure 3.   Emission spectrum (0.03 nm fwhm) of dibromodiacetylene
cation, $\tilde{A}^2\Pi_{\Omega,u} \to \tilde{X}^2\Pi_{\Omega,g}$ ($\Omega = 3/2, 1/2$) band systems,
excited by $\simeq 200$ eV electrons impacting on a seeded
helium supersonic free jet. The helium lines are marked
with a dot. At the top higher resolution recordings
(≈ 0.008 nm) of the 0^0_0 and 3^0_1 bands are shown.

processes (fig. 1), are characterised by the temperature of the
rare gas bath which has been cooled by liquid nitrogen (4). The
$^2\widetilde{A} \leftarrow {}^2\widetilde{X}$ electronic transition of the cation is then laser pumped
and the wavelength undispersed fluorescence is detected to yield
the excitation spectrum (10).

The application of both techniques for the spectral studies
of open-shell cations is clearly advantageous since the information
forthcoming is complementary. Furthermore, often both sets of data
are necessary for the interpretation or confirmation of certain
features in the spectra; e.g. sequence transitions and hot bands.
As example, the emission (fig. 3) and laser excitation (fig. 4)
spectra of rotationally cooled dibromodiacetylene cation are
briefly discussed (11).

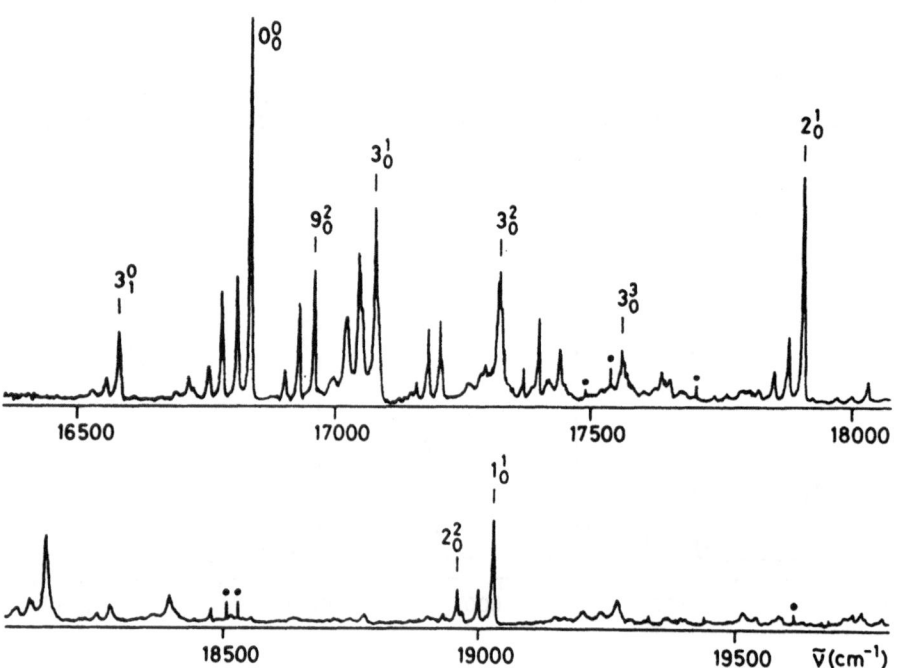

Figure 4.   Laser excitation spectrum of dibromodiacetylene cation,
$\widetilde{A}^2\Pi_{3/2,u} \leftarrow \widetilde{X}^2\Pi_{3/2,g}$, recorded at liquid nitrogen temperature in the gas phase with 0.02 nm bandwidth.
Atomic lines are marked with a dot.

Dibromodiacetylene cation.

The vibrational assignments of the stronger bands are indicated in the two spectra (figs. 3 and 4). The emission spectrum shows the excitation of the three Σ_g^+ fundamentals, their overtones and combinations mainly for the ground cationic state, $\tilde{X}^2\Pi_g$ The laser excitation spectrum shows similar features but for the excited state, $\tilde{A}^2\Pi_u$, and the excitation of low frequency degenerate modes in double quanta is also evident. The vibrational frequencies can be inferred to within 1 cm^{-1}; in fact higher accuracy can be attained if the splitting due to the natural abundance of bromine isotopes is taken into account. Due to the reduced rotational band widths the isotope bands are resolved and this is of importance in aiding spectral assignments. In the inset of figure 3 the expanded plots of the 0_0^0 and 3_1^0 bands of the $\tilde{A}^2\Pi_{3/2,u} \rightarrow \tilde{X}^2\Pi_{3/2,g}$ transition are shown. The 0_0^0 band shows no isotope splitting and the red-shaded rotational profile of the sub-bands is seen. The 3_1^0 band, however, shows the 1:2:1 abundance of the $^{79}Br\text{+}C\equiv C\text{+}_2{}^{79}Br^+$, $^{79}Br\text{+}C\equiv C\text{+}_2{}^{81}Br^+$ and $^{81}Br\text{+}C\equiv C\text{+}_2{}^{81}Br^+$ species. These data allow the force constants for the ground and excited states of the ion to be evaluated. It should also be pointed out that the vibrational frequencies obtained by these methods for the cations are more often than not more precise than the data for their molecular species. If the rotational structure of their ir transitions is not resolved, or compensated for, the inhomogeneous broadening can lead to uncertainties of several wavenumbers which, in turn, can lead to indeterminate force constant determinations.

Due to the different formation mechanism of the cations (cf. fig. 1), the emission spectrum shows the two spin-orbit separated band systems, i.e. $^2\Pi_{3/2,u} \rightarrow {}^2\Pi_{3/2,g}$ and $^2\Pi_{1/2,u} \rightarrow {}^2\Pi_{1/2,g}$ (the bands belonging to the latter are designated with the bar above the excited vibration), whereas the laser excitation spectrum only shows the $^2\Pi_{3/2,u} \leftarrow {}^2\Pi_{3/2,g}$ system. The $\Omega = 1/2$ component is collisionally relaxed since the zeroth level lies ≈ 900 cm^{-1} above that of the $\Omega = 3/2$ in the $\tilde{X}^2\Pi_g$ state (12). Clearly, in complex emission spectra this can help in the location of the origin bands. This is the case for example in chlorosubstituted acetylene and diacetylene cations where the $\Omega = 3/2$ and $\Omega = 1/2$ emission systems overlap profusely. Assignment of the sequence and hot bands (cf. figs. 3 and 4) also follows from the relative intensity changes when comparing spectra of cooled and room temperature cations. In contrast to the emission spectra, the only sequence transitions in the excitation spectra result from the residual population of the excited vibrational levels of the ground cationic state.

J. P. MAIER ET AL.

Σ_g^+ - VIBRATIONS

		$\nu_1 : \nu_S(C\equiv C)$	$\nu_2 : \nu_S(C-C)$	$\nu_3 : \nu_S(C-Br)$
.... $\pi_u^4\,\pi_g^4\,\pi_u^4\,\pi_g^4$	$X^1\Sigma_g^+$	2205	1127	250
.... $\pi_u^4\,\pi_g^4\,\pi_u^4\,\pi_g^3$	$\tilde{X}^2\Pi_g$	2186	1225	252
.... $\pi_u^4\,\pi_g^4\,\pi_u^3\,\pi_g^4$	$\tilde{A}^2\Pi_u$	2186	1071	241

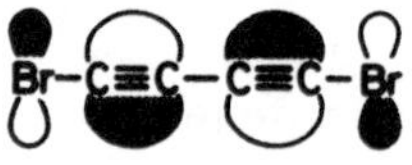

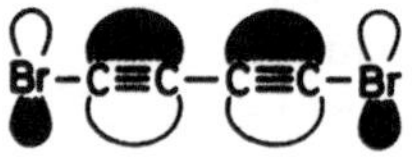

Figure 5: Comparison of the determined vibrational frequencies of dibromodiacetylene molecule and cation, of the description of their electronic structure and of the nodal characteristics of the molecular orbitals from which the electron has been removed.

Apart from the vibrational characterisation, the spectral information also leads to better energetic data for such cations than is possible from photoelectron measurements. The difference between the spin-orbit splitting in the $\tilde{X}^2\Pi_g$ and $\tilde{A}^2\Pi_u$ states is obtained from the respective emission bands (cf. fig. 3). The absolute splittings can be deduced for the $\tilde{A}^2\Pi_u$ state (450 ± 80 cm^{-1}) from the photoelectron spectrum since the adiabatic transitions are dominant (12,13). Combination of the data from the two techniques then leads to 890 ± 80 cm^{-1} for the spin-orbit splitting in the $\tilde{X}^2\Pi_g$ state (11). In this way the adiabatic ionisation energies leading to the $\tilde{X}^2\Pi_g$ state can be inferred and this is found to be ≈ 400 cm^{-1} lower than formerly estimated.

In figure 5 the measured fundamental frequencies for the three Σ_g^+ modes are given as well as the molecular values (14). This figure is meant to illustrate the links to the usual description of the electronic structure of such species. The ionic states are adequately described by the single electronic configurations indicated, which differ from the molecular ground state, $X^1\Sigma_g^+$, configuration by the removal of one electron from the highest or penultimate occupied molecular orbital (12). The symmetry characteristics of these are depicted on the left of the figure, and are nicely reflected in the frequency changes of the three Σ_g^+ fundamentals. The latter are essentially the $\nu(C\equiv C)$, $\nu(C-C)$ and $\nu(C-Br)$

symmetric stretching modes ν_1 to ν_3, respectively. The ν_1 frequencies are smaller in both the $\tilde{X}^2\Pi_g$ and $\tilde{A}^2\Pi_u$ states compared to the molecular value since the molecular orbitals have bonding characteristics in the C≡C region. On the other hand the ν_2 frequency is larger in the $\tilde{X}^2\Pi_g$, and smaller in the $\tilde{A}^2\Pi_u$ state, compared to the molecular $X^1\Sigma_g^+$ state and reflects the C-C bonding characteristics (cf. fig. 5). For the ν_3 mode the changes are smaller because of the localisation of the electron density on the bromines as indicated by the size of the spin-orbit splittings. Thus on passing from the $\tilde{X}^2\Pi_g$ to the $\tilde{A}^2\Pi_u$ state, the main geometry change is that the C-C and C-Br bond lengths increase (linear configuration is assumed), in accord with the red-shading of the bands (cf. inset fig. 3).

Application Possibilities

The means to obtain accurate vibrational frequencies for cations leads to various potential applications. For example, cations are assumed to play a crucial role in the interstellar medium (15). In the ion-molecule schemes proposed, open-shell cations also occur (eg. acetylene or diacetylene cations) and thus their identification is desirable. Since ir astronomy is becoming an important method in addition to microwave spectroscopy (16), the knowledge of the precise ionic frequencies is necessary. These are now known, or are being determined, for many of the cations listed in table 1, and their ir detection may thus eventually be possible. In other areas of ion spectroscopy (17) involving the absorption of ir photons, eg. multiphoton processes, absorption experiments in icr cells, or in double resonance approaches, the availability of accurate frequencies is an advantage for their further development. In the case of a few smaller open-shell cations, almost all the fundamental frequencies can be inferred from the emission and excitation spectra. These could be useful in model theories, such as those for mass spectra.

Another application is the use of the spectral features to provide information on the environment; an example is the deduction of the rotational, or vibrational, temperature. This is illustrated with respect to the $\tilde{A}^2A_1 \rightarrow \tilde{X}^2B_1$ emission bands of H_2O^+ which have been identified in comets (18). Following a rotational analysis of spectra recorded in the laboratory (19), the expected intensity distribution of the rotational transitions at 50 K, typical of comet tails, was predicted (18). This is reproduced in figure 6 where also displayed is a spectrum of this band of H_2O^+ recorded with the supersonic free jet apparatus (20). In the latter experiment the rotational temperature can be varied by the expansion conditions. One can see that at rotational temperatures comparable to those pertaining in comet tails, the intensity distribution

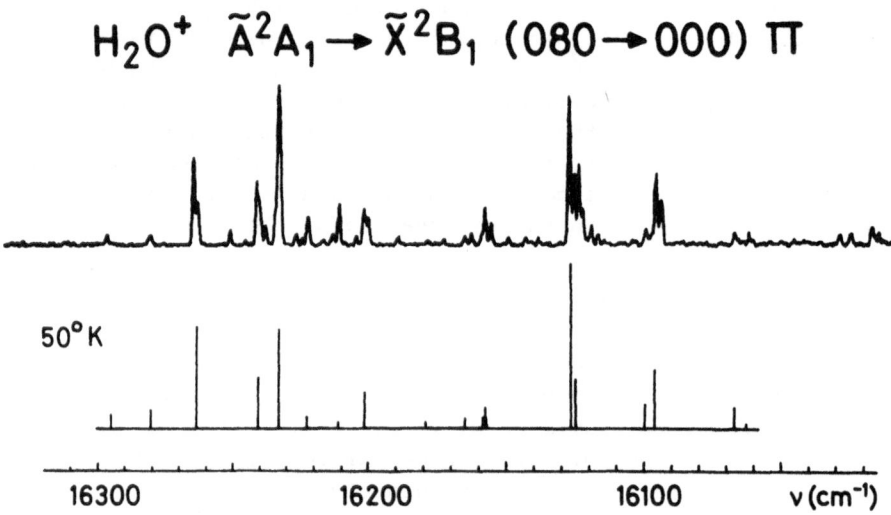

Figure 6: Rotational bands of the $\tilde{A}^2A_1 \rightarrow \tilde{X}^2B_1$ (080-000) transition
           of H_2O^+, recorded (0.044 nm band pass) using electron
           impact of a seeded helium supersonic free jet (top).
           The bottom stick plot shows the calculated intensities,
           according to ref. 18.

differs somewhat from the predicted one. However, the main point is
that if one can obtain rotationally resolved emission spectra from
comets, one can now simply simulate the temperature conditions in
the laboratory in the collision-free environment of the free jet.

High Resolution Studies

     In cases where the rotational structure of the band systems
can be resolved, the geometries of the ions in their ground and
excited electronic states can in principle be obtained. Apart from
the open-shell triatomic cations whose emission spectra have been
rotationally analysed (i.e. N_2O^+, CO_2^+, CS_2^+, H_2O^+ and H_2S^+) (21),
among the organic species this has been accomplished only for
diacetylene cation (22). Rotational structure can be resolved in
the emission spectra of the haloacetylene cations (3); the method

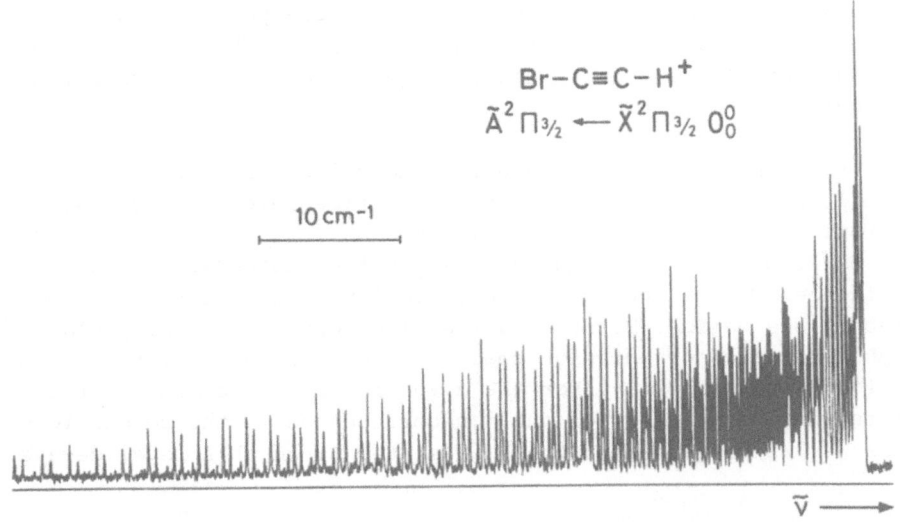

Figure 7:   Laser excitation spectrum of the 0_0^0 band of the
            $\tilde{A}^2\Pi_{3/2} \leftarrow \tilde{X}^2\Pi_{3/2}$ transition of bromoacetylene cation.
            The optical band-pass was 0.001 nm.

of choice, however, for the best resolution and quality of the
spectra is by laser excitation. This is illustrated by the laser
excitation spectrum of the 0_0^0 band of the $\tilde{A}^2\Pi_{3/2} \leftarrow \tilde{X}^2\Pi_{3/2}$ transition of bromoacetylene cation shown in figure 7, recorded with an
optical resolution of 0.001 nm (23). As can be seen, the excellent
signal to noise ratio enables the rotational transitions to be
clearly discernible. The spectrum is complicated by the overlap of
the systems due to the ^{81}Br and ^{79}Br isotopes. In contrast to the
emission spectrum (24), only the $\Omega = 3/2$ component is detected
because the population of the $\Omega = 1/2$ component in the $\tilde{X}^2\Pi$ state
is very low.

RADIATIONLESS DECAY - QUANTITATIVE STUDIES

     Once the spectroscopic features of the cations are established,
complementary data concerning the fate of state selected cations
are desirable. For the cases where fragmentation channels are
accessible, various techniques may be applied for this purpose,
e.g. photodissociation spectroscopies (26) or ion-photoelectron
coincidence (27) methods. When the ions decay radiatively (cf.
table 1), the photon emission can be used as a probe of the intramolecular relaxation behaviour. To follow this quantitatively, not
only the lifetimes of selected levels (i.e. vibrational or even

rotational) have to be known but also the corresponding fluorescence quantum yields. The latter can be obtained absolutely since
individual photons and photoelectrons can be counted and detected
in coincidence (27).

In this approach energy selected photoelectrons define the
internal energy of the ion and provide a time reference following
photoionisation with a monochromatic source (e.g. He(Iα) radiation)
(28). Wavelength undispersed photons are then detected in delayed
coincidence. This is depicted schematically in figure 8 for dibromodiacetylene cation. Dispersing the electrons or photons yields
the photoelectron or emission spectrum (recorded at room temperature
cf. fig. 3) respectively. For the detection of coincidences, the
selected internal energy of the ion corresponds to the zeroth level
of the $\Omega= 3/2$ component of the $\tilde{A}^2\Pi_u$ state. Since this level decays

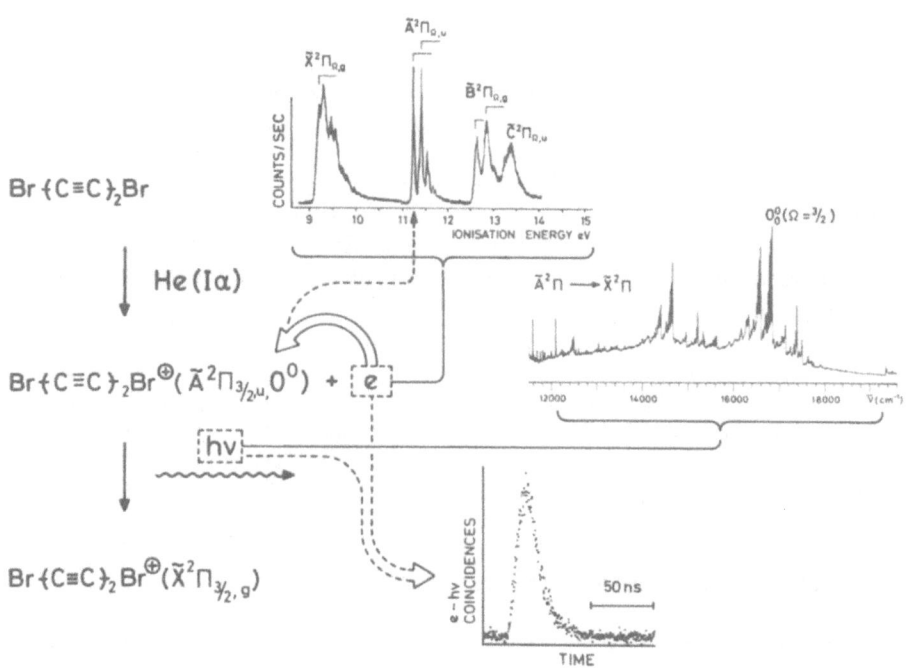

Figure 8:   Schematic representation of the principles of the photo-
electron-photon coincidence measurements using state
selected $(\tilde{A}^2\Pi_{3/2,u}\ 0^0)$ dibromodiacetylene cations as an
example. Shown are the photoelectron (top) and emission
(middle) spectra obtained by dispersing the electrons
and photons respectively, and the curve by detecting
these particles in coincidence (bottom).

radiatively, eventually a coincidence curve superimposed on a background of random coincidences is observed (cf. fig. 8); i.e. the
time difference between the creation of the cation and the emission
of the photon is measured. True coincidences occur only if the detected photon and photoelectron stem from the chosen level of the
same cation. It can be shown that the rates of detection of true
coincidences, N_T, and true photoelectrons, N_e, are related to the
fluorescence quantum yield, $\phi_F(v')$, of the selected level or energy
designated by v', by $N_T/N_e = f_{h\nu}(\lambda) \; \phi_F(v')$ (29). An apparatus calibrated for the probability of detection of photons, $f_{h\nu}(\lambda)$, can
therefore be used to determine $\phi_F(v')$ of the open-shell organic
cations decaying radiatively. In practice the calibration is accomplished by coincidence measurements on di- and triatomic cations
with unity fluorescence quantum yields (e.g. $N_2^+ \tilde{B}^2\Sigma_u^+$, $CO_2^+ \tilde{A}^2\Pi_u$).
The lifetimes $\tau(v')$ can at the same time be inferred from the
decay part of the coincidence curve (fig. 8).

Such a coincidence apparatus has been used to systematically
determine the $\phi_F(v')$ and $\tau(v')$ data of most of the types of organic
cations listed in table 1. The radiative, $k_r(v')$, and non-radiative,
$k_{nr}(v')$, rate constants for the decay channels depleting the level
v' are then directly available (27). In the succeeding discussion
some examples illustrate the quantitative information that can be
acquired on the relaxation behaviour of open-shell cations under
collision-free conditions.

Non-unity fluorescence quantum yields of triatomic cations indicate that predissociation pathways are also followed. This is the
observation for vibrationally excited levels of N_2O^+ in the $\tilde{A}^2\Sigma^+$
state and for all the levels of COS^+ in the $\tilde{A}^2\Pi$ state (29). The
predissociation (via a $^4\Sigma^-$ state) leads to NO^+ and S^+ fragment ions,
respectively, and competes with the $^2\tilde{A} \rightarrow {}^2\tilde{X}$ radiative decay. The
$\phi_F(v')$ and $\tau(v')$ values allow the determination of the predissociation rate constants for the various levels (30).

For the majority of organic cations studied the fluorescence
quantum yields are non-unity even though fragmentation channels
are not accessible. This corresponds to the statistical limit
decay (31) where the dense manifold of isoenergetic vibronic levels
of the cationic ground state leads to irreversible decay on the
time scale, and under the conditions, of such experiments. The
result is that the internal conversion, $^2\tilde{A} \rightsquigarrow {}^2\tilde{X}$, competes with
the radiative decay, $^2\tilde{A} \rightarrow {}^2\tilde{X}$. This situation also appears to hold
for some cations where the non-radiative decay leads ultimately to
fragment ions, e.g. cis-1,2-difluoroethylene and alkyl substituted
diacetylene cations (27). The non-unity fluorescence quantum yields,
monoexponential decay and the increase (exponential or linear) in

Table 2. Radiative and Non-Radiative Decay Data for Electronically Excited Open-Shell Organic Cations determined by Photoelectron-Photon Coincidence Measurements (with uncertainty of 5-10%). Full compilation and references are given in a recent review (27).

Cation	Transition	$\Delta E(0^0_0)$ (cm^{-1})	$\phi_F(0^0)$	$\tau(0^0)$ (ns)	$k_r(0^0)$ (s^{-1})	$k_{nr}(0^0)$ (s^{-1})
$Cl-C\equiv C-C\equiv N^+$	$\tilde{A}^2\Sigma^+ \to \tilde{X}^2\Pi$	18990	0.73	376	1.9×10^6	7.2×10^5
$Br-C\equiv C-C\equiv N^+$	$\tilde{A}^2\Pi \to \tilde{X}^2\Pi$	18601	0.30	17	1.8×10^7	4.1×10^6
$H-C\equiv C-C\equiv H^+$	$\tilde{A}^2\Pi_u \to \tilde{X}^2\Pi_g$	19762	0.72	72	1.0×10^7	3.9×10^6
$D-C\equiv C-C\equiv D^+$	"	19762	0.80	79	1.0×10^7	2.5×10^6
$H-C\equiv C-C\equiv Cl^+$	$\tilde{A}^2\Pi \to \tilde{X}^2\Pi$	19710	0.79	41	1.9×10^7	5.1×10^6
$H-C\equiv C-C\equiv Br^+$	"	17940	0.59	23	2.6×10^7	1.8×10^7
$Cl-C\equiv C-C\equiv Cl^+$	$\tilde{A}^2\Pi_u \to \tilde{X}^2\Pi_g$	19090	0.47	21	2.2×10^7	2.5×10^7
$Br-C\equiv C-C\equiv Br^+$	"	16585	0.19	10	1.9×10^7	8.1×10^7
$N\equiv C-C\equiv N^+$	$\tilde{A}^2\Pi_g \to \tilde{X}^2\Pi_u$	16780	0.12	13	9.2×10^6	6.8×10^7
cis-1,2-difluoroethylene$^+$	$\tilde{A}^2A_1 \to \tilde{X}^2B_1$	28880	0.07	320	2.2×10^5	2.9×10^6
$CH_3-C\equiv C-Cl^+$	$\tilde{A}^2E \to \tilde{X}^2E$	28150	0.33	18	1.8×10^7	3.7×10^7
$CH_3-C\equiv C-Br^+$	"	21780	0.25	15	1.7×10^7	5.0×10^7
$CH_3-C\equiv C-C\equiv N^+$	$\tilde{A}^2A_1 \to \tilde{X}^2E$	18390	0.26	210	1.2×10^6	3.5×10^6
$CH_3-C\equiv C-C\equiv H^+$	$\tilde{A}^2E \to \tilde{X}^2E$	20375	0.94	50	1.9×10^7	1.2×10^6
$CH_3-C\equiv C-C\equiv Cl^+$	"	19840	0.55	22	2.5×10^7	2.0×10^7
$CH_3-C\equiv C-C\equiv Br^+$	"	18170	0.23	10	2.3×10^7	7.7×10^7

Table 2, continued

				$\tilde{A}^2E \rightarrow \tilde{X}^2E$	
				$\tilde{A}^2E_u \rightarrow \tilde{X}^2E_g$	
				"	
				"	
				$\tilde{B}(\pi^{-1}) \rightarrow$	
				$\tilde{A}(\pi^{-1}),\ \tilde{X}(\pi^{-1})$	
$CH_3-C\equiv C-C\equiv N^+$	17720	0.10	7	1.4×10^7	1.3×10^8
$CH_3-C\equiv C-CH_3^+$	20562	0.74	25	3.0×10^7	1.0×10^7
$CD_3-C\equiv C-CD_3^+$	20560	0.92	32	2.9×10^7	2.5×10^6
$CF_3-C\equiv C-CF_3^+$	19510	0.88	46	1.9×10^7	2.6×10^6
1,2,3-trifluorobenzene$^+$	22288	0.74	48	1.5×10^7	5.4×10^6
1,2,4-trifluorobenzene$^+$	24277	0.14	10	1.4×10^7	8.6×10^7
1,3,5-trifluorobenzene$^+$	21867	1.00	59	1.7×10^7	$\leq 1 \times 10^6$
d_3-1,3,5-trifluorobenzene$^+$	21867	1.00	58	1.7×10^7	$\leq 1 \times 10^6$
1,3,5-trichlorobenzene$^+$	15410	0.41	22	1.9×10^7	2.7×10^7
1,3,5-trifluorotrichlorobenzene$^+$	16950	0.66	35	1.9×10^7	9.7×10^6
1,2,3,4-tetrafluorobenzene$^+$	23277	0.90	50	1.8×10^7	2.0×10^6
1,2,3,5-tetrafluorobenzene$^+$	23320	0.99	50	2.0×10^7	$\leq 1 \times 10^6$
1,2,4,5-tetrafluorobenzene$^+$	24437	0.61	32	1.9×10^7	1.2×10^7
d_2-1,2,4,5-tetrafluorobenzene$^+$	24437	0.74	38	1.9×10^7	6.8×10^6
Pentafluorobenzene$^+$	23137	0.98	48	2.0×10^7	$\leq 1 \times 10^6$
Hexafluorobenzene$^+$	21645	1.00	49	2.0×10^7	$\leq 1 \times 10^6$

the non-radiative rate with increasing internal energy can all be rationalized within the framework of the statistical decay model (31).

In table 2 are compiled the determined $\phi_F(0^0)$, $\tau(0^0)$, $k_r(0^0)$ and $k_{nr}(0^0)$ data for some of the organic cations showing the above mentioned features. The general observation is that for dipole allowed transitions, designated as $\tilde{A}(\pi^{-1}) \rightarrow \tilde{X}(\pi^{-1})$ and lying in the visible part of the spectrum, the $k_r(0^0)$ values are in the $1-3\times10^7$ s^{-1} range. On the other hand, the $k_r(0^0)$ rates for $\tilde{A}(\sigma^{-1}) \rightarrow \tilde{X}(\pi^{-1})$ transitions are an order of magnitude or two smaller. The energy gap between the zeroth levels of the ground and excited state, $\Delta E(0_0^0)$, is also given in the table; in structurally related species, e.g. X$+$C$\equiv$C$+_2$X$^+$ X= H, Cl, Br, a trend in the exponential decrease of $k_{nr}(0^0)$ as function of $\Delta E(0_0^0)$ is apparent. This is again a prediction of the statistical limit scheme. Since the $\phi_F(0^0)$ and $\tau(0^0)$ data are known, the oscillator

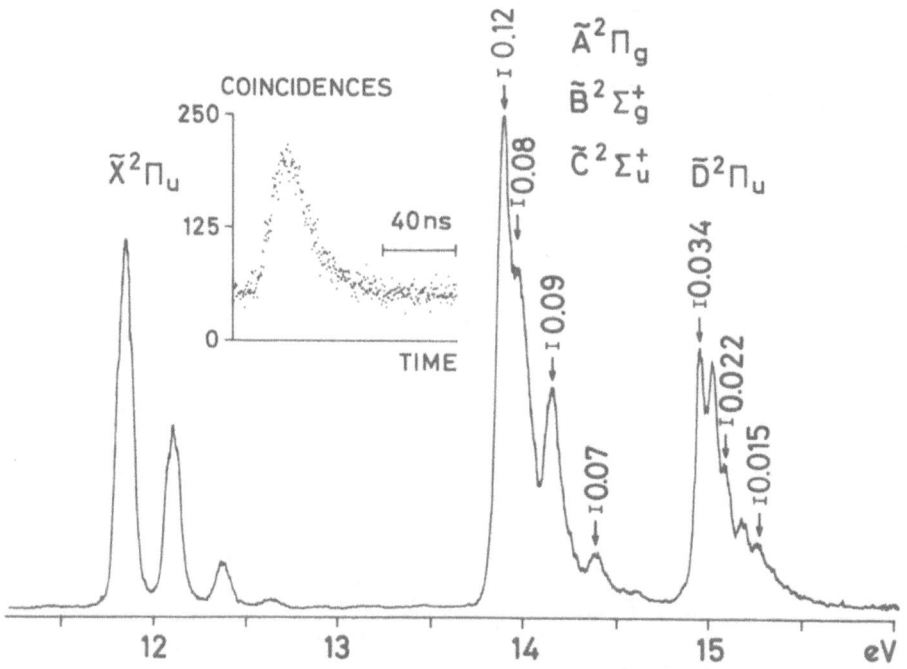

Figure 9: Fluorescence quantum yields for dicyanoacetylene cation determined at the locations indicated by the arrows above the He(Iα) photoelectron spectrum. The inset shows a coincidence curve for the lowest level of the $\tilde{A}^2\Pi_g$ state accumulated in 32 h; N_e= 310 Hz and N_T= 0.09 Hz.

strengths at the origin for the respective transitions of such
open-shell cations can be evaluated.

The coincidence measurements also provide information on the
decay of higher excited electronic states for some of these
cations. The detection of true coincidences from several excited
electronic states has been established for halogenated benzene,
dihaloacetylene and dicyanoacetylene cations (27). The latter is
chosen to illustrate the phenomenon.

Figure 9 shows the photoelectron spectrum of dicyanoacetylene
and the locations where coincidences were obtained (32). At these
energies, the $\phi_F(v')$ (given above the arrows) and $\tau(v')$ values
were determined and in addition the wavelength region of the
emitted photons could be ascertained by conducting the measurements with optical filters. The latter show that the wavelength

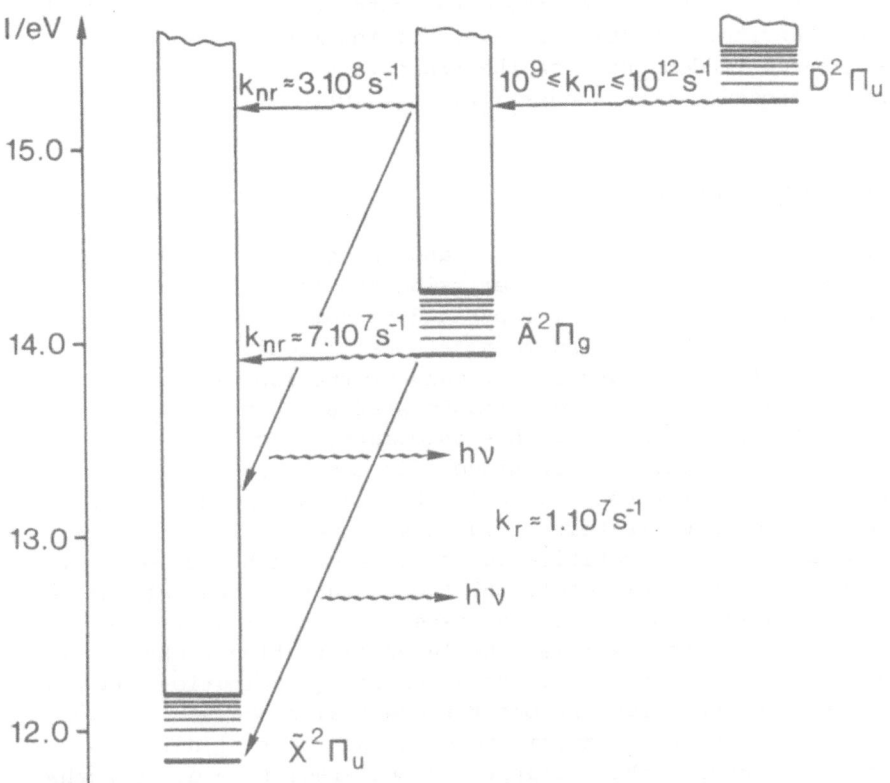

Figure 10:   Summary of the inferred relaxation behaviour of
             "isolated" dicyanoacetylene cation prepared initially
             either in the lowest level of the $\tilde{A}^2\Pi_g$ or $\tilde{D}^2\Pi_u$ state.

of the radiation emitted upon initial generation of the cations in the $\tilde{D}^2\Pi_u$ state is in the same region as the $\tilde{A}^2\Pi_g \rightarrow \tilde{X}^2\Pi_u$ band system.

The inferred scheme of the relaxation behaviour of dicyano-acetylene cation in the lowest level of either the $\tilde{A}^2\Pi_g$ or $\tilde{D}^2\Pi_u$ state is shown in figure 10. The lowest excited state, $\tilde{A}^2\Pi_g$, decays either radiatively or by an internal conversion to the ground state, $\tilde{X}^2\Pi_u$, with the given rate constants. As far as the higher excited states are concerned, the evaluated $k_{nr}(v')$ data show a monotonous trend, i.e. there is no discontinuity in passing from one excited electronic state to another. This and the wavelength of the fluorescence imply that a rapid intramolecular (non-radia-tive) transition takes place to vibrationally excited levels of the $\tilde{A}^2\Pi_g$ state. The upper rate of this process from the $\tilde{D}^2\Pi_u$ state is $10^{12}s^{-1}$ due to the well resolved peaks in the photoelectron spectrum; the lower value is a consequence of the detection limits of the coincidence experiment. The vibrationally excited $\tilde{A}^2\Pi_g$ state then decays at the indicated rates either radiatively to highly excited vibrational levels of the $\tilde{X}^2\Pi_u$ state, causing a red Franck-Condon shift of the transition to $\lambda > 550$ nm, or by an internal conversion process (cf. fig. 10).

CONCLUDING REMARKS

Spectral characterisation and investigations of open-shell organic cations by means of emission and laser excitation spectroscopies has been illustrated by some examples. These tech-niques can now be applied to rotationally cooled cations so that the vibrational frequencies of many of the fundamentals in the ground and lowest excited cationic states can be obtained to within ± 1 cm^{-1} or better. This information should be useful in the development of related techniques for the vibrational studies of cations and as reference data in theoretical approaches. Since ions play a role in extra-terrestrial environments, the frequencies may be of use in the identification of some of such species by ir astronomy. Also, the rotational temperature of the cations can be tuned in a supersonic free jet experiment to simulate the conditions in comet tails for example. The laser excitation technique is suitable for probing the internal energy distribution and state selective preparation of ions for chemical processes and ion-molecule reactions. Complementary to the spectral data are the rate constants of the radiative and non-radiative decay pathways of state selected cations. This enables a quantitative picture of their fate to be established, for comparison with theories of radiationless transitions and for future studies of state specific processes of such excited open-shell cations.

Acknowledgement

The studies summarized in this article as a result of the research projects carried out in Basel have been financed throughout the years by the "Schweizerischer Nationalfonds zur Förderung der wissenschaftlichen Forschung".

REFERENCES

1. See for example "Kinetics of Ion Molecule Reactions", ed. P. Ausloos (Plenum Press, New York, 1979); "Gas Phase Ion Chemistry" ed. M.T. Bowers, Vol. I and II (Academic Press, New York, 1979).

2. For previous reviews, see J.P. Maier, Acc. Chem. Res. $\underline{15}$, 18 (1982); J.P. Maier, Ang. Chem. Int. Ed. Engl. $\underline{20}$, 638 (1981); J.P. Maier in "Kinetics of Ion-Molecule Reactions", ed. P. Ausloos (Plenum Press, New York, 1979), 437.

3. J.P. Maier, Chimia $\underline{34}$, 219 (1980); J.P. Maier, O. Marthaler, L. Misev and F. Thommen, in "Molecular Ions", ed. J. Berkowitz (Plenum Press, New York, 1982).

4. T.A. Miller and V.E. Bondybey, J. Chim. Phys. Phys.-Chim. Biol. $\underline{77}$, 695 (1980) and references therein.

5. R.N. Zare and P.J. Dagdigian, Science $\underline{185}$, 739 (1974).

6. A. Carrington and R.P. Tuckett, Chem. Phys. Letters $\underline{74}$, 19 (1980); R.P. Tuckett, Chem. Phys. $\underline{58}$, 151 (1981).

7. T.A. Miller, B.R. Zegarski, T.J. Sears and V.E. Bondybey, J. Phys. Chem. $\underline{84}$, 3154 (1980).

8. D. Klapstein, S. Leutwyler and J.P. Maier, Chem. Phys. Letters $\underline{84}$, 534 (1981).

9. D.H. Levy, Ann. Rev. Phys. Chem. $\underline{31}$, 197 (1980).

10. J.P. Maier and L. Misev, Chem. Phys. $\underline{51}$, 311 (1980).

11. D. Klapstein, J.P. Maier and L. Misev, J. Chem. Phys., 1983 in press.

12. E. Heilbronner, V. Hornung, J.P. Maier and E. Kloster-Jensen, J. Am. Chem. Soc. $\underline{96}$, 4252 (1974).

13. M. Allan, E. Kloster-Jensen, J.P. Maier and O. Marthaler, J. Electron Spectrosc. $\underline{14}$, 359 (1978).

14. P. Klaboe, E. Kloster-Jensen, E. Bjarnov, D.H. Christensen and D.F. Nielsen, Spectrochim. Acta $\underline{31A}$, 931 (1975).

15. See for example D. Smith and N.G. Adams, Int. Rev. of Phys. Chemistry $\underline{1}$, 271 (1981).

16. H. Kroto, Int. Rev. of Phys. Chemistry $\underline{1}$, 309 (1981).

17. See for example, "Ions and Light", Vol. III of "Gas Phase Ion Chemistry" ed. M.T. Bowers (Academic Press, New York, 1983); "Physical Chemistry Advances in the Study of Molecular Ions", eds. T.A. Miller and V.E. Bondybey (North-Holland,

Amsterdam, 1983).

18. P.A. Wehinger, S. Wyckoff, G.H. Herbig, G. Herzberg and
     H. Lew, Astrophys. J. $\underline{190}$, L43 (1974).

19. H. Lew, Can. J. Phys. $\underline{54}$, 2028 (1976).

20. D. Klapstein, S. Leutwyler and J.P. Maier, Chem. Phys.,
     1983 in press.

21. G. Herzberg, Quart. Rev. Chem. Soc. $\underline{25}$, 201 (1971); S. Leach
     in "The Spectroscopy of the Excited State" ed. B. Di Bertolo
     (Plenum Press, New York, 1976), 369.

22. J.H. Callomon, Can. J. Phys. $\underline{34}$, 1046 (1956).

23. J.P. Maier and L. Misev, unpublished results.

24. M. Allan, E. Kloster-Jensen and J.P. Maier, J.C.S. Faraday
     II, $\underline{73}$, 1406 (1977).

25. R.C. Dunbar in "Gas Phase Ion Chemistry" Vol. II, ed. M.T.
     Bowers (Academic Press, New York, 1979).

26. T. Baer in "Gas Phase Ion Chemistry, Vol. I, ed. M.T. Bowers
     (Academic Press, New York, 1979).

27. See J.P. Maier and F. Thommen, in "Ions and Light", Vol. III
     of "Gas Phase Ion Chemistry", ed. M.T. Bowers (Academic
     Press, New York, 1983) for a recent review of these studies.

28. M. Bloch and D.W. Turner, Chem. Phys. Letters $\underline{30}$, 344 (1975).

29. J.P. Maier and F. Thommen, Chem. Phys. $\underline{51}$, 319 (1980).

30. D. Klapstein and J.P. Maier, Chem. Phys. Letters $\underline{83}$, 590
     (1981).

31. K.F. Freed, Top. Appl. Phys. $\underline{15}$, 23 (1976); P. Avouris,
     W.M. Gelbart and M.A. El-Sayed, Chem. Rev. $\underline{77}$, 793 (1977).

32. J.P. Maier, L. Misev and F. Thommen, J. Phys. Chem. $\underline{86}$,
     514 (1982).

GAS-PHASE ION PHOTODISSOCIATION

Robert C. Dunbar

Chemistry Department
Case Western Reserve University
Cleveland, Ohio 44106

ABSTRACT

Ion photodissociation is discussed from a number of points of view, including: (1) Ion spectroscopy; (2) Ion structures; (3) Dynamics and energetics of fragmentation; (4) Mechanisms of ion multiphoton chemistry; and (5) Collisional and radiative relaxation of excited ions. Some survey of older results is undertaken, along with consideration of recent developments mainly involving multiphoton dissociation and related chemistry involving excited ions.

INTRODUCTION

Gas-phase photodissociation of ions has proven to be informative in a variety of ways. Among the points of view which have emerged as especially fruitful we can include:

-Optical and infrared spectroscopy of ions;
-Structures of ions;
-Dynamics, rates and energetics of ion fragmentations;
-Mechanisms of ion multiphoton chemistry;
-Collisional and radiative relaxation processes in excited ions.

Some aspects of all of these different points of view will appear in this chapter; reflecting some recent

M. A. Almoster Ferreira (ed.), Ionic Processes in the Gas Phase, 179–203.
© *1984 by D. Reidel Publishing Company.*

interests of our own laboratory, there will be more to
say about the last two of these five areas, whereas
the corresponding report from the last NATO Institute
in this area (Dunbar, 1979a) expanded more fully on
the first three. However, this has been a field of
rapidly developing interest and activity, so an
overview of some already well known developments will
still not be ancient history, and we will try to give
some overall perspective on the field. Some recent
reviews might also be of interest, describing areas of
drift tube dissociation, (Moseley, 1980), ion beam
dissociation (Moseley and Durup, 1980, 1981), ICR
dissociation (Dunbar, 1979a, 1980, 1981, 1982a), ion
spectroscopy (Saykally and Woods, 1981) and infrared
dissociation (Woodin et al., 1979).

THE EXPERIMENTS

We will describe only work involving ion beams and
ICR ion traps, without giving the consideration they
deserve to other fruitful and successful techniques
such as drift tubes, tandem quadrupoles, and rf
quadrupole ion traps. Ion beam experiments are simple
in principle: A beam of mass-selected ions is formed
and passed through an interaction region in which
light intersects the ion path, either along or
perpendicular to the ion flight path. Fragment ions
are sorted by mass and detected. Fragment detection
at different points along the path permits a
straightforward measurement of fragmentation rate of
light-induced metastable ions, while analysis of peak
broadening arising from fragment velocity spread gives
a determination of kinetic energy release. Polarized
light also gives angular dependence information when
the light beam is perpendicular to the ion beam, if
there is sufficient kinetic energy release to give a
noticeable angular asymmetry in the fragment ion
velocity distribution.

Greatest interest in ion beam photodissociation
has centered around the extraordinary optical
resolution possible in Doppler-tuned fast beam studies
(Moseley and Durup, 1980, 1981). In an ion beam
accelerated to kilovolt energies, the phenomenon of
kinematic compression reduces the Doppler broadening
of the optical absorptions by a large factor, and
virtually Doppler-free spectroscopy at optical
resolutions of tens of MHz has been carried out in
this way on a number of small ions. The corresponding
kinematic expansion which takes place on decelerating

the fast beam to low energies also means that kinetic
energy release occurring before deceleration is
greatly magnified: This effect, which has long been
used to study metastable ion decompositions (Cooks et
al., 1973), is readily applied to measuring photodissociation energy releases in the meV range.

Photodissociation in the ICR ion trap exploits the
ability of the trap to contain ions for tens of
seconds or minutes, and then mass analyze the residual
ion population. In a standard mode of experiment,
ions are produced by an electron beam pulse through
the neutral gas at 10^{-8} or 10^{-7} torr. Following
trapping and optical irradiation for a time which
might be about 3 s, the photodissappearance of the
parent ion is ascertained by the ICR detect pulse or
by a Fourier-transform excitation-detection sequence,
and then the cell is purged for a new cycle. Fig. 1
shows an example of a photodissociation spectrum
obtained by repeating this process at a series of
wavelengths: The ion illustrated here, methylcyclo-
pentadiene parent ion, is in fact interesting, because
the photodissociation spectrum is identical with that
of 1,3-cyclohexadiene parent ion, and no doubt this
reflects complete rearrangement of the former ion to

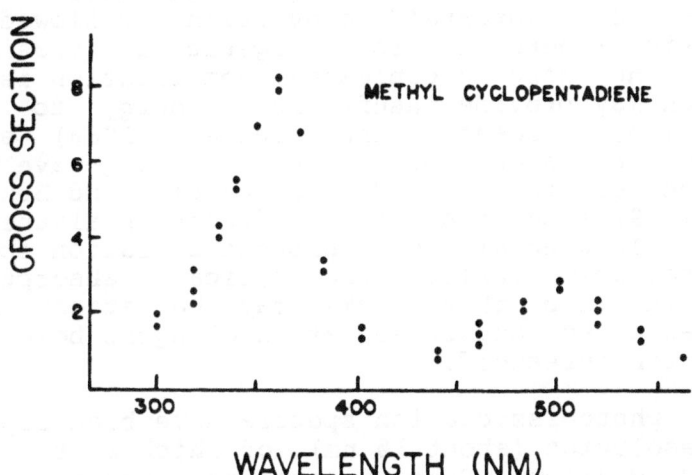

Fig. 1. ICR photodissociation spectrum showing the
disappearance of the parent ion of methylcyclopenta-
diene as a function of wavelength (Hays, 1980).

the latter, which is far more stable (Hays, 1980).

Using the ICR double resonance capabilities a
variety of more elaborate experiments can be arranged
to sort out details of fragmentation products and
their subsequent ion-molecule reactions and photo-
chemistry. The spatial anisotropy of the ICR cell can
also be used to determine kinetic energy releases and
angular dependences of photodissociation processes
(Dunbar, 1979a).

ION SPECTROSCOPY

Photodissociation spectroscopy is the term which
has been used to denote observations of dissociation
rate as a function of wavelength. Since photon
absorption is implied by observing dissociation, these
spectra have been extensively used as a source of
information about the gas-phase absorption spectroscopy of ions. Of course, since photon absorption
does not necessarily result in dissociation, this
connection is not always one-to-one. One possible
complication of this sort is fluorescence, and in fact
a number of ions are now known for which fluorescence
competes with dissociation to some extent (Maier,
1980, 1982; Dujardin, 1980). However, probably no
polyatomic ions are known which fluoresce to the
exclusion of internal conversion followed by
dissociation (where it is energetically feasible).
Another, and more troublesome, complication is that
the photon may provide insufficient energy to allow
dissociation. Reents and Freiser (1980) showed
convincing evidence for such a long-wavelength
truncation of the $C_6H_5NOH^+$ spectrum, and Benz and
Dunbar (1979) suggested similar effects in alkane ion
spectra. In general, then, photodissociation spectra
should reliably reflect the optical absorptions,
bearing in mind that peaks may be attenuated by
fluorescence, and suppressed at wavelengths below the
dissociation threshold.

Many photodissociation spectra have been reported
at low resolution (about 10 nm), of which a few will
be discussed below. Spectra taken at laser resolution
(better than 1 nm) are especially interesting because
of the resolution of vibrational fine structure, but
these have been rarer, partly because of the
difficulty of the experiments, and partly because the
vibrational structure often appears to be smeared by a
combination of lifetime broadening and spectral

congestion. Among ions whose laser spectra have shown notable vibrational structure we might mention the well known cases of CH_3I^+ (McGilvery and Morrison, 1977), hexatriene ion (Dunbar and Teng, 1978) and CO_3^- (Moseley et al., 1976), and the more recently reported $C_3H_5Cl^+$ (Orth and Dunbar, 1982) and $Fe(CO)_3$ (Rynard and Brauman, 1980).

Some recent experiments (Honovich and Dunbar, 1981) have suggested the possibility of what might be called "proton-labelling" spectroscopy, in which a proton is attached to a gas-phase molecule, to obtain a photodissociation spectrum of the protonated species. Assuming that the spectroscopic properties of the protonated molecule are interpretable in a way similar to neutral molecules, the useful capabilities of optical-spectroscopic identification of samples developed long ago for neutral molecules might be applied to the extremely small samples which can be handled in mass spectrometers. These experiments are based on the observation of Freiser and Beauchamp (1977) that in a variety of unsaturated organic molecules the photodissociation spectrum of the protonated molecule closely resembles the optical spectrum of the neutral, but with a shift to longer wavelength. They related this observation to the difference in ground- and excited-state proton affinities via a Forster-cycle argument.

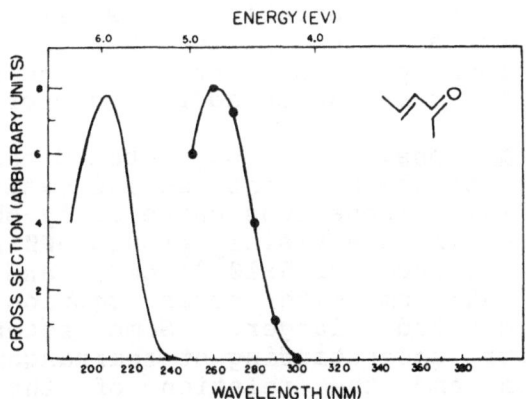

Fig. 2. Photodissociation spectrum of the protonated ion of 3-penten-2-one (——●——) compared with the neutral absorption spectrum (———), showing a red shift of 1.17 eV (Honovich and Dunbar, 1981).

The effect of substituents on the peak positions
in the neutral spectra of acrolein derivatives is very
well characterized and understood. In Fig. 2 is shown
an illustration from our results on the protonated
acroleins, exhibiting the typical red shift upon
protonation. A series of protonated acrolein
derivatives showed substituent effects on the peak
positions closely similar to the neutral-molecule
patterns. The extent of red shift depends on the
nature of the pi system, and is about 1.15 eV for
simple acroleins, 1.35 eV for triene analogs, and 1.6
eV for aromatic derivatives. Within each of these
classes of compound, the peak positions varied with
substituent in the same way as with the neutrals. It
seems reasonable to hope that the well worked out
spectroscopy of many neutral chromophoric groups will
be directly applicable to interpreting photodissociation spectra of proton-labelled molecules in this
way.

ION STRUCTURES

The widest interest and application of the
capabilities of photodissociation spectroscopy have
centered on the linking of spectra with ion
structures. A frequent point of view takes the
spectra as reliable fingerprints of the structures of
different isomeric ions; more demanding and ambitious
efforts attempt to assign specific structures to ions,
based on systematic understanding of their spectral
features. Two illustrations here will have to serve
to suggest the sorts of problem that can be attacked,
and the kinds of ion structure and rearrangement
considerations that can be addressed profitably.

Diene Radical Ions. The conjugated diene ion chromophore is one of the most intense and easily recognized
(Dunbar, 1976). These ions universally show a pair of
peaks: one in the visible around 500 nm with cross
section of the order of 5×10^{-18} cm^2; and one in the
UV around 350 nm with cross section typically an
order of magnitude larger. Some recent work has
expanded our understanding of rearrangements of some
of these ions and the relation of the spectra to
structural details.

The spectra of isomeric hexadiene ions (Benz et
al., 1981) provide a good example of the mapping of
ion structure/rearrangement chemistry by photodissociation. The conjugated isomers 1,3- and

2,4-hexadiene give ions whose spectra clearly identify them as conjugated ions according to the pattern noted above. (These two isomers do give distinguishable spectra, however, and do not interconvert.) The 1,4 isomer gives an ion spectrum identical with that of 2,4-hexadiene ion, and clearly the 1,4 to 2,4 rearrangement following electron impact ionization is rapid compared with the 1-second time scale of the experiment. Such a 1-3 hydrogen shift to bring a double bond into conjugation seems to be universal: Among many other instances known, 1,4-octadiene and 1,4-pentadiene ions similarly rearrange into conjugation (Dunbar and Fitzgerald, 1982). On the other hand, the 1,5-hexadiene isomer, and similarly the 1,5-, 2,6- and 1,7-octadiene isomers, yield spectra clearly characteristic of the isolated double bond chromophore in radical ions (Dunbar, 1976), with strong dissociation in the far UV, and certainly do not rearrange into conjugation to an appreciable extent.

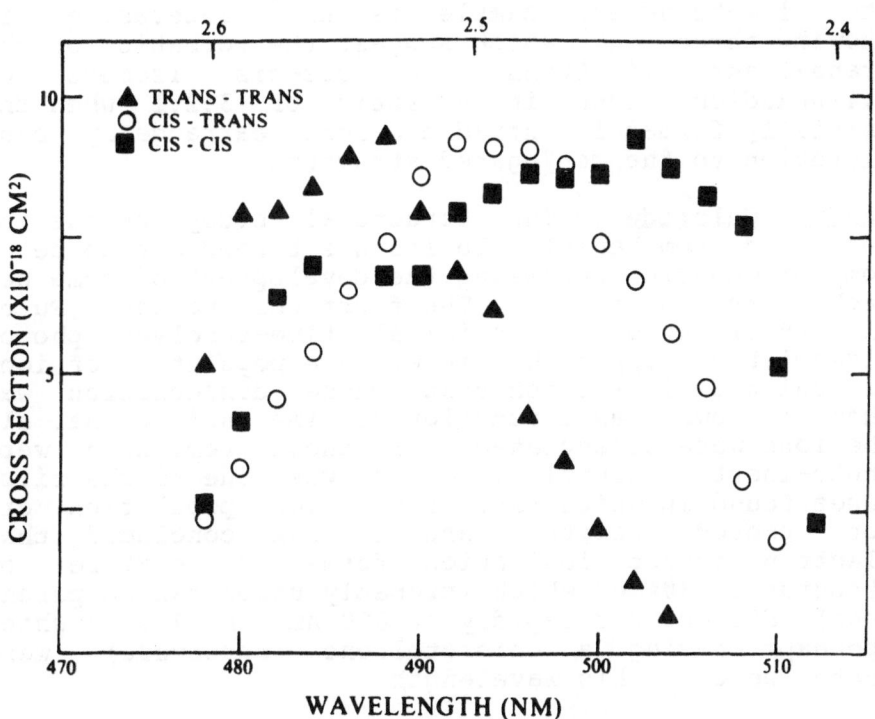

Fig. 3. Laser photodissociation spectra of the three cis-trans isomers of 2,4-hexadiene parent ion.

More subtle is the question of cis-trans isomerization in these ions. The three cis-trans isomers of 2,4-hexadiene ion are just barely distinguishable from their low-resolution spectra, but laser photodissociation at about 1 nm resolution resolves the three structures comfortably (Fig. 3). Observing clearly distinct peaks for these three isomers rules out their interconversion by rotation about the double bonds (which is far easier than in the neutral diene). Molecular orbital calculations prove to be useful in predicting the peak shifts due to cis-trans isomerism in these ions: The ordering of the photodissociation peaks for the three cis-trans isomers and the spacing between them of about 7 nm, is reproduced exactly by a Koopmans' Theorem calculation using MINDO/3.

Separate cis-trans isomers of 1,3-hexadiene have not been studied, but the visible-light spectrum of the ions from a mixed-isomer sample shows two peaks, which may well corrrespond to the two non-inter-converting isomers. The visible-region spectrum of the 1,4-hexadiene sample is also interesing, in showing three peaks which suggest the formation of the trans-trans, cis-trans and cis-cis isomers of 2,4-hexadiene ions in a ratio of 3:2:1 when the initially-formed 1,4-hexadiene ions rearrange by bond migration to the conjugated structure.

Benzyl Chloride. The structural study of the ion generated from benzyl chloride has turned out to be a complex question focussing the development of some new techniques and ideas. The first observations (Fu et al., 1976) used the original time-resolved photo-dissociation approach, in which a population of ions was built up in the ion trap whose dissociation was then followed as a function of time until either all the ions were dissociated, or those remaining were photo-inert. Benzyl chloride was one of the first cases found in which part of the ion population did not photodissociate, and it was concluded that electron impact ionization forms a mixture of structures, 30% of which (probably unrearranged parent ions) dissociated rapidly at 600 nm, and 70% of which (perhaps having a chlorotoluene structure) were photo-inert at this wavelength.

This attractive picture was overturned by Morgenthaler and Eyler's observation (1981) that in their pulsed-laser instrument, the ions behaved as a homogeneous population which dissociated cleanly at 600 nm. The contradiction between these two apparently valid experiments was resolved with a new set of experiments (Honovich and Dunbar, 1982a) using pulsed ion production followed by dissociation with gated lasers at varying times after ion formation. It was found that the two ion structures are indeed present, but are not both formed by the initial electron impact. The chemistry is as shown in Eq. 1:

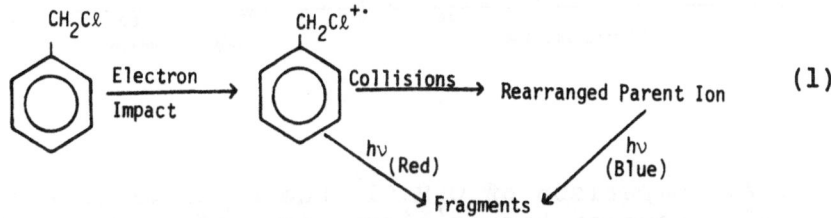

$$(1)$$

Electron impact forms the red-absorbing ions observed in both previous studies. At high pressures and long times, a collisional process converts these into ions of different structure which dissociate in the blue but not in the red, requiring about 50 collisions for the conversion to happen. Fig. 4 shows clearly the increase in blue photodissociation and decrease in red photodissociation with time, as the isomerization proceeds. Morgenthaler and Eyler, working at short times and modest pressures, naturally saw no collisionally converted ions.

ION PHOTOCHEMISTRY

As suggested pictorially in Fig. 5, a great many different processes can be important in gas-phase ion photochemistry. Photodissociation offers a means of probing many of these. In particular, the long time scale of trapped-ion experiments gives a unique chance to observe directly some very slow processes like infrared-fluorescence ion deactivation, while ion beam studies are particularly apt for study of kinetic energy release, fragmentation rate and other fragmentation characteristics. We will explore a few recent ideas for putting some of these possibilities to work.

TWO-PHOTON DISSOCIATION AND RELAXATION

A fruitful discovery was Freiser and Beauchamp's observation (1975) of two-photon dissociation, which they correctly interpreted in terms of the sequential

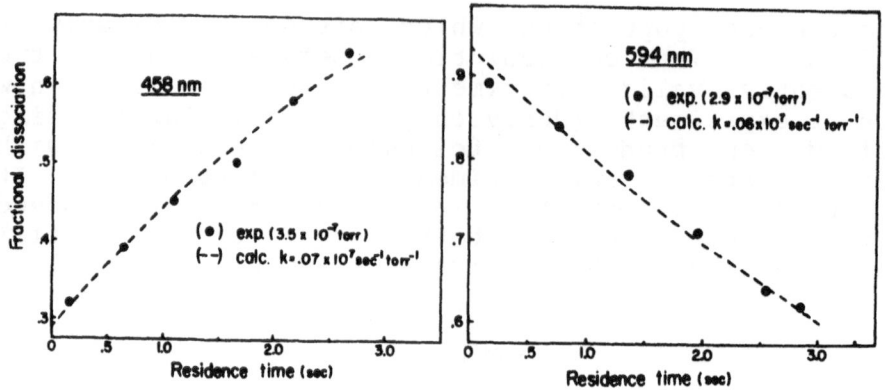

Fig. 4. Comparison of $C_7H_7Cl^+$ photodissociation at red and blue wavelengths as a function of ion residence time. Following ion formation, ions are stored and undergo collisions for the indicated residence time, and are then sampled by a brief (300 ms) pulse of laser irradiation. The increase in blue dissociation and decrease in red dissociation match the expectation for a simple collisional rearrangement process (solid lines).

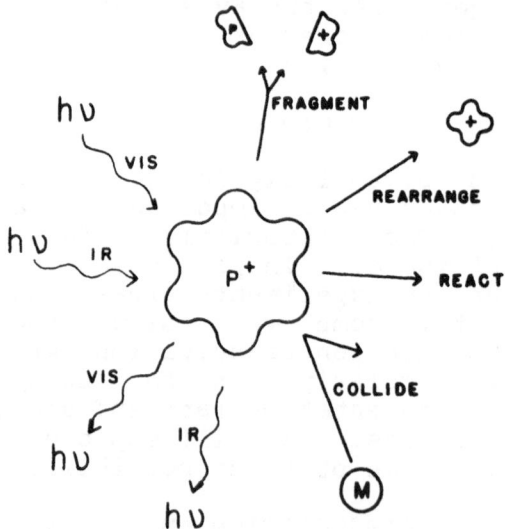

Fig. 5. Some important processes in ion photo-chemistry.

mechanism

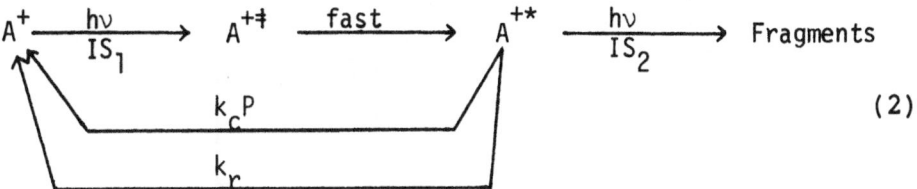

$$\tag{2}$$

where A^+ is an electronically excited ion; A^{+*} is a vibrationally excited ground-state ion; I is the light intensity; S_1 and S_2 are optical absorption cross sections; P is the neutral pressure; k_c is the bimolecular collision quenching rate constant; and k_r is the infrared radiative relaxation rate constant.

PULSED-LASER TWO-PHOTON EXPERIMENTS

When the light source is not continuous, but delivers light in spaced pulses, an interesting interplay develops between the occasional photo-excitation of the ions and their steady relaxation by collisions and by infrared fluorescence. Our group (Lev and Dunbar, 1981) explored the effect of changing from continuous to repetitively pulsed radiation in the two-photon dissociation of iodobenzene ion. The effect is simply understood in a qualitative way: If the laser pulse repetition rate is fast compared with all the ion relaxation rates, then the light source excites ions just as if it were actually continuous, and the time-average light intensity is the important variable. But if the pulse rate becomes slow compared with the fastest relaxation, then the ions can relax between each pulse, and the experiment is equivalent to a series of relaxation-free single pulse experiments for which the pulse energy and number of pulses, but not the pulse shape or repetition rate, are the important variables.

Recently results of higher precision have been obtained for bromobenzene ion, using a cw laser chopped at rates varying from 1 to 12 s^{-1}, and the radiative and collisional parts of the relaxation have been clearly separated (Dunbar, 1982b). Using a chopper, the average light intensity is conveniently kept constant, while the repetition rate (and pulse length) are varied, so that the dissociation should be pulse-rate-independent at high pulse rates and low

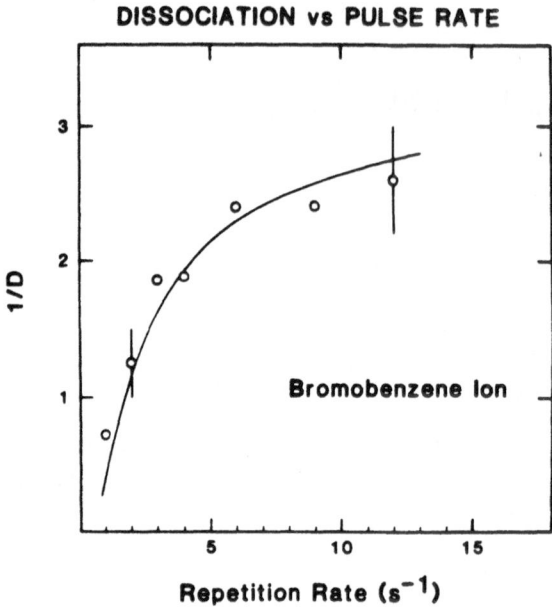

DISSOCIATION vs PULSE RATE

Fig. 6. Photodissociation of bromobenzene ion as a function of pulse rate at $2x10^{-7}$ torr. The solid line is the theoretical curve assuming a radiative relaxation rate of 1.5 s^{-1}.

relaxation rates. Fig. 6 shows the strong dependence of dissociation on pulse rate at rather high pressure, compared with the theoretical curve calculated using the collisional relaxation rate already determined from the quenching studies. At high pulse rates (4 s^{-1} and above) the dissociation is strongly pressure-dependent in the $2-30x10^{-8}$ torr region. However, at 1 s^{-1} there is little pressure dependence, and it is clear that at this low repetition rate, relaxation by infrared radiation is largely completed between each pulse. At 2 s^{-1} and 3 s^{-1} the pressure dependence is intermediate, indicating that at these pulse rates radiative relaxation is partial, allowing some collisional relaxation in addition to the radiative relaxation. Using a radiative relaxation rate of around 2 s^{-1} the basic kinetic scheme Eq. 2 gives a good account of the pressure and pulse rate dependences in this experiment, and we believe this firmly establishes the radiative rate k_r in the range $1-3 \text{ s}^{-1}$.

REARRANGEMENT DISSOCIATIONS

The characteristic kinetic behavior of two-photon dissociation gives an unexpected means of viewing some rearrangement-dissociation chemistry. For example, the dissociation of p-iodotoluene, <u>1</u>, yields $C_7H_7^+$.

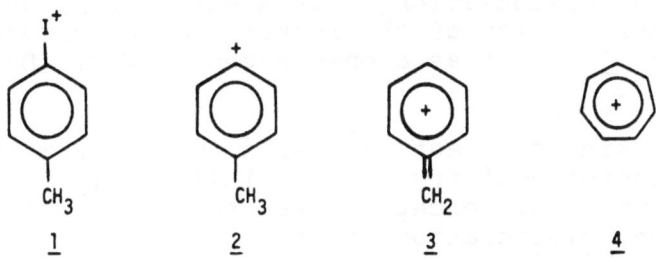

The simple bond-cleavage product, tolyl ion <u>2</u>, has its one-photon threshold near 530 nm (by analogy with the accurately known iodobenzene thermochemistry). The

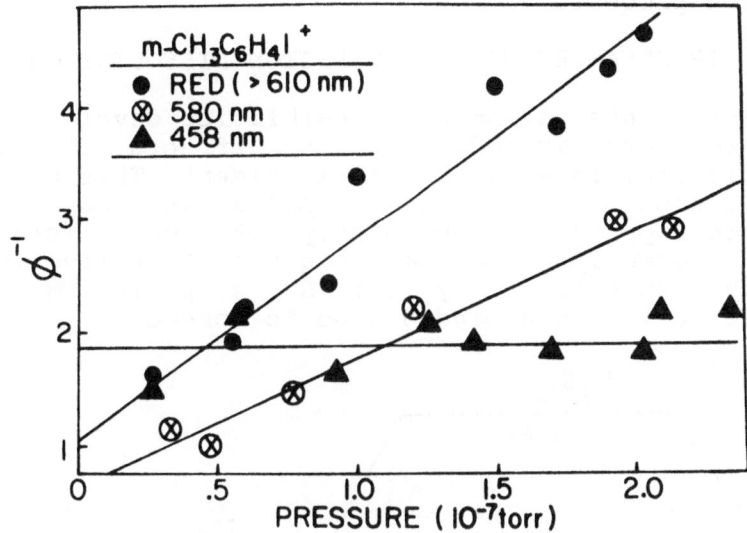

Fig. 7. Pressure dependence of the iodotoluene dissociation at wavelengths of 458 and 580 nm.

rearrangement dissociation to yield benzyl (<u>3</u>) or tropylium (<u>4</u>) product ions is much easier, and is accessible by one-photon dissociation at all visible wavelengths. As shown in Fig. 7, photodissociation at 458 nm is a one-photon process, as expected (signalled by the lack of pressure dependence). But the strong pressure dependence at 580 nm indicates a clear two-photon dissociation at this wavelength, and rules out the possibility of the rearrangement dissociation to yield <u>3</u> or <u>4</u> as a one-photon product (Dunbar et al., 1982).

A striking contrast is seen with bromotoluene, whose thermochemistry is similar to that of iodotoluene ion, except that the threshold for one-photon dissociation to tolyl ion <u>2</u> is around 440 nm. For this ion, pressure independent one-photon dissociation is seen not only in the blue, but also at wavelengths longer than 600 nm. So dissociation at long wavelengths must proceed through a rearrangement path to yield more stable products, presumably <u>3</u> or <u>4</u>. It is not known, and will not be easy to find out, whether the intact parent ion rearranges after absorption of the second photon and then dissociates, or whether there is a concerted rearrangement/dissociation process.

IR ENHANCEMENT OF TWO-PHOTON DISSOCIATION

Some interesting new possibiities develop when an infrared laser is added to the effects of visible irradiation, in a two-laser experiment. This has given us access to some new aspects of the infrared spectroscopy and photochemistry of ionic molecules. The infrared laser can be thought of as delivering the energy represented by the S_1 step in Eq. 2, as suggested in the expanded kinetic scheme

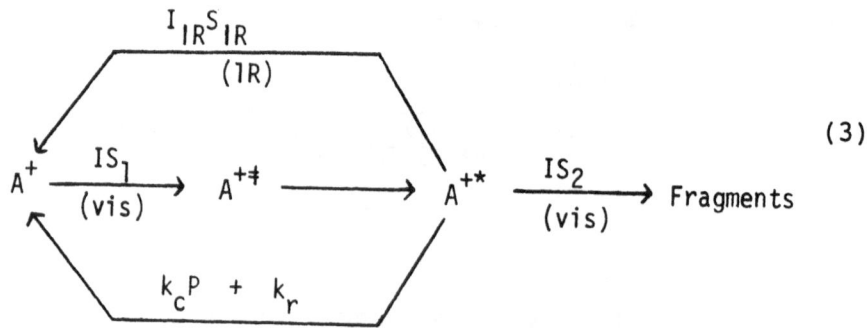

(3)

This effect is clearly observed as a strong increase in dissociation (typically by up to a factor of three) when a CO_2 laser of several Watts power is added to the visible irradiation (Dunbar et al., 1980).

Some recent experiments using gated CO_2 and visible lasers have proven that in the cases of iodobenzene and bromobenzene ions, at least, the IR radiation serves to give initial excitation to the ions, rather than to push excited ions over the brink of dissociation (Honovich and Dunbar, 1982b): if the ions are first irradiated with the IR laser and then immediately with the visible laser, the IR enhancement is practically undiminished; while if the visible-laser irradiation precedes the IR irradiation, no enhancement is seen; nor did IR irradiation alone have any effect. (Wight and Beauchamp (1981) have reported a very small IR enhancement effect in cyanobenzene ion which seems to require simultaneous irradiation by both lasers, and very likely is a different sort of effect than the large effect discussed here).

For iodobenzene ion, which has been studied the most extensively, photodissociation data were taken as a function of pressure and of the intensities of both lasers. These data were compared with simulations based on numerical integration of the kinetic scheme of Eq. (3), with the aim of establishing as much about the mechanism of IR enhancement as possible (Lev and Dunbar, 1982b). Extensive exploration of the parameter values led to the conclusion that several features of the kinetics are essential to a successful model:
-Most or all of the ions are susceptible to both infrared and visible up-pumping processes at the wavelengths used.
-The infrared photoexcitation has a bottleneck somewhere below 4000 cm^{-1}, and is also too slow to compete with relaxation processes near the dissociation threshold at 20,000 cm^{-1}.
-Infrared radiative relaxation is significant at pressures of a few times 10^{-8} torr, but collisional relaxation dominates above about 10^{-7} torr. The radiative relaxation time constant is around 500 ms.
-The visible photoexcitation cross section at 610 nm decreases by a factor of 2 to 5 with increasing internal energy from zero to 20,000 cm^{-1}.

INFRARED ENHANCEMENT SPECTROSCOPY

The IR enhancement effect depends on the absorption of IR photons by near-thermal ions, based on the conclusions of the previous section, so the wavelength dependence of the effect gives an indirect route to finding the IR absorption peaks of gas-phase ions. This is extremely difficult information to come by in any more direct way.

Fig. 8 shows an example of the spectra we have been able to obtain (Honovich and Dunbar, 1982c): the spectrum does indeed show wavelengths of high and low IR enhancement, and the main obstacle to measuring a complete IR spectrum is the sparse wavelength coverage of the CO_2 laser. Such IR-enhancement spectra of several halobenzene and halotoluene ions all show sharply wavelength-dependent effects. In most cases there is strong IR absorption in the ion at wavelengths longer than any of the ring vibrations in the coresponding neutral molecules. This seems to reflect some red shift of the ring vibrational frequencies upon ionization, not surprising since it is a bonding ring pi electron which is removed.

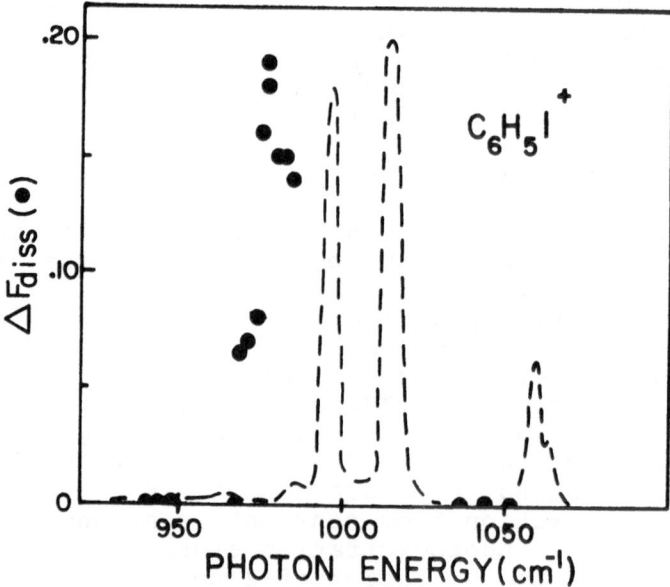

Fig. 8. IR enhancement spectrum of iodobenzene ion. The visible wavelength was 610 nm.

The IR-enhanced two-photon approach described here gives a complementary alternative to the direct IR multiphoton dissociation spectra reported by Beauchamp's group for a number of ions (Woodin et al., 1979). Direct IR dissociation is most favorable for ions with low dissociation threshold, while the IR enhancement approach is favorable for ions having a high dissociation threshold.

TWO-LASER RELAXATION STUDIES

Since it is easy to separate the IR and visible irradiation steps in these two-laser experiments, some very revealing relaxation studies are possible. Following irradiation with the CO_2 laser to pump ions above the one-photon threshold, the subsequent visible irradiation can be delayed, giving the ions time to relax back below the one-photon threshold. As a result, the IR enhancement effect will diminish to zero as the delay is lengthened, with a time constant reflecting the sum of ion relaxation processes. Fig. 9 shows data from an experiment in which iodobenzene

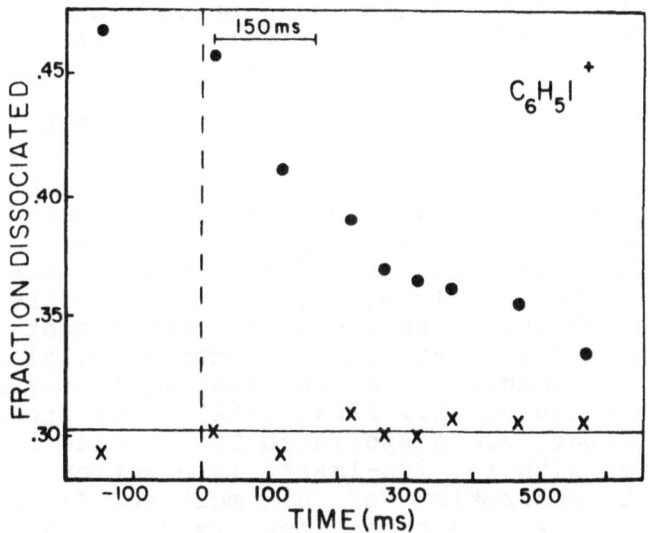

Fig. 9. Two-laser photodissociation of iodobenzene ion as a function of the delay between the CO_2 laser irradiation and the visible irradiation. The (X) values are visible dissociation in the absence of IR.

ion was first irradiated for one second with the CO_2
laser, was allowed a time T to relax, and was finally
interrogated by a 150 ms pulse of visible irradiation
(Honovich and Dunbar, 1982b). A curve is also shown
giving the theoretical expectation for a relaxation
time constant of 500 ms. At a pressure of $2x10^{-8}$ torr
the relaxation does not vary much with pressure, and
is undoubtedly due mostly to infrared fluorescence.
Relaxation becomes faster at higher pressures, and the
collisional relaxation rate k_c can be extracted
separately from the radiative component k_r. The rates
for iodobenzene measured in this way are $k_r = 2$ s^{-1},
and $k_c = 6x10^{-10}$ cc-molec^{-1}-s^{-1}.

This is a collisional rate slightly faster than
half the orbiting collision rate, and is thus slightly
faster than found in the quenching experiments
described above. The relaxation rates for IR-excited
ions might be expected to be somewhat faster than the
corresponding rates in visible two-photon processes,
since it is likely that the average energy of the
IR-excited ions is less than that of the
visible-photon-excited ions. This seems to be true
also in the bromobenzene ion, where the IR-enhancement
experiment described here gives a radiative relaxation
rate of 4.5 s^{-1}, while the pulsed-laser visible
results discussed above definitely indicate a k_r rate
less than 3 s^{-1} for the visible two-photon process.

PHOTODISSOCIATION IN A FAST ION BEAM

Infrared photodissociation or enhanced photodissociation experiments like those described above yield
information about the lower part of the IR excitation
ladder, since the important IR absorption events are
normally those involving ions far below the dissociation threshold. A complementary experiment dealing
with the top of the excitation ladder is possible with
ion beam instruments. In the experiments carried out
at SRI (Coggiola et al., 1980, 1982), ions produced by
electron impact are dissociated by infrared radiation
in the beam, with an ion-laser interaction time so
short that absorption of at most one IR photon is
possible. Thus only those ions are dissociated which
already have internal energy nearly equal to the
dissociation energy. It may be said that this
photodissociation corresponds to the last photon in a
multiphoton IR photodissociation experiment.

One interesting aspect of the results is the ap-

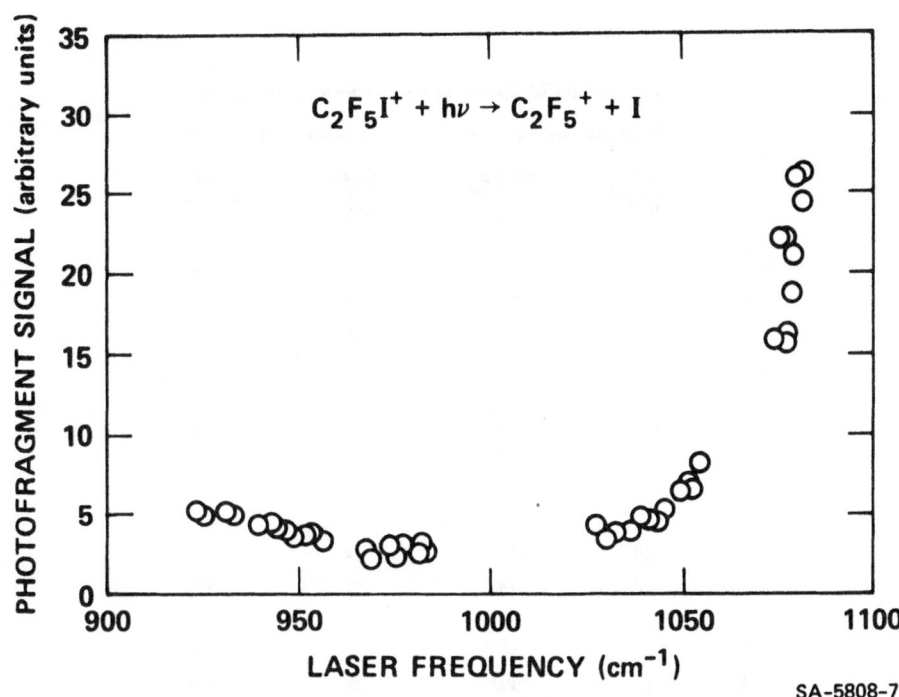

Fig 10. Fast-ion-beam one-photon IR spectrum of highly vibrationally excited $C_2F_5I^+$.

pearance of the photodissociation spectra as a function of IR wavelength. As is illustrated in Fig. 10 for a typical polyatomic ion, the spectra typically show well defined peaks in the spectra, and it is certainly true that the presence of several eV of internal energy in the ion does not destroy the IR spectral peaks of the cool ions. This is easily understood in a nearly-harmonic picture of polyatomics, because in such large molecules the individual vibrational modes are not highly excited. The v=0 to v=1 transitions of the IR active modes are still dominant, although possibly broadened and shifted by anharmonic perturbations.

One molecule for which a comparison has been made between the one-photon hot-ion spectra of the present experiments and a cool-ion multiphoton IR dissociation spectrum is $C_3F_6^+$. The multiphoton IR spectrum shows its peak at 1043 cm^{-1} (Bomse et al., 1978), while the one-photon beam experiments show the peak red-shifted to 1030 cm^{-1}, no doubt due to anharmonic coupling.

The ion-beam experiment yields a wealth of

TABLE I

OBSERVED AND RRKM QUANTITIES FOR IR ONE-PHOTON FRAGMENTATION

	METASTABLE FRAGMENTATION		IR-INDUCED FRAGMENTATION	
	ENERGY RELEASE (MEV)	AVERAGE LIFETIME (μSEC)	ENERGY RELEASE (MEV)	AVERAGE LIFETIME (μSEC)
CF_3I^+				
OBS.	—	—	4.4	$\ll 1$
CALC.	—	—	24	$\ll 1$
$C_2F_5I^+$				
OBS.	2.8	26	10	2.2
CALC.	2.5	25	12	2.8
$C_3F_7I^+$				
OBS.	10	>100	14	5.5
CALC.	10	170	21	5.5
$C_6H_5I^+$				
OBS.	29	>100	28	N.A.
CALC.	39	250	50	30

information about the photofragmentation process, including kinetic energy release and fragmentation rates. Table I shows such data for a few polyatomic ions which have been carefully analyzed. Corresponding metastable fragmentations (laser-independent) are also tabulated. The calculated values are from RRKM calculations using the assumptions of statistical (quasiequilibrium) theory, which imply complete randomization of internal energy prior to dissociation. For the larger ions, the RRKM model evidently gives a very satisfactory description of the fragmentation process. However, for CF_3I^+ the kinetic energy release is much too small, and it is concluded that this dissociation occurs by a non-statistical mechanism. The slightly counterintuitive observation that $C_3F_7I^+$ releases more kinetic energy than $C_2F_5I^+$ is an interesting feature of the beam experiment, arising because many of the $C_3F_7I^+$ ions arrive at the photon interaction region with substantially more internal energy than the dissociation threshold, while such above-threshold ions are more nearly depleted with $C_2F_5I^+$.

COMPARISON OF ION BEAM AND ION TRAP DISSOCIATION

Recent photodissociation studies of n-butylbenzene ion make an interesting contrast between beam and trap methods. Beynon's group (Griffiths et al.; Mukhtar et al.) have reported a series of studies of this dissociation in an ion beam. The dissociation products are $C_7H_7^+$ (the direct-cleavage product) and $C_7H_8^+$ (a rearrangement product), according to the scheme

$$C_{10}H_{14}^+ \xrightarrow{\ h\nu\ } \begin{array}{c} \xrightarrow{-C_3H_6} C_7H_8^+ \qquad \Delta H = 1.4 \text{ eV} \\ (92) \\ \xrightarrow{-C_3H_7} C_7H_7^+ \qquad \Delta H = 2.6 \text{ eV} \\ (91) \end{array}$$

(134)

The $C_{10}H_{16}^+$ ions emerging from the ion source have a spread of internal energies from 0 to 1.4 eV, and many of the more excited ions have suffficient energy to yield the more energetic m/e 91 product. Accordingly, a mixture of photo-produced m/e 91 and m/e 92 is observed, with the relative amount of m/e 91 increasing at higher photon energy. The ratio of products is in accord with the RRKM calculations of the group.

In our Fourier transform ICR spectrometer (Chen and Dunbar, 1982), on the other hand, no primary photoproduct m/e 91 is observed at visible wavelengths, although the m/e 91 ion is formed as a secondary product according to the scheme

$$C_{10}H_{14}^+ \xrightarrow[-C_3H_6]{h\nu} C_7H_8^+ \xrightarrow[-H]{h\nu} C_7H_7^+$$

(134) (92) (91)

The relative extent of formation of m/e 92 and m/e 91 as a function of light intensity is in quantitative accord with this scheme. The exclusive formation of m/e 92 as the primary product is just as expected for thermal parent ions. The contrast with the beam results illustrates the important point that the ions in the ICR trap are highly thermalized (by collisional and radiative processes) relative to the ions in typical beam experiments, and the photodissociation processes are typically those of ions with little excess internal energy.

ACKNOWLEDGMENTS

Many collaborators contributed to the work described here, to whom a large collective expression of gratitude is extended. Also with gratitude, the author acknowledges the support of the National Science Foundation and of the donors of the Petroleum Research Fund, administered by the American Chemical Society.

MEMORY

This chapter has been written with the constant memory of Henry Rosenstock, whose warmth and penetrating understanding enriched the days at Vimeiro. As a friend and as a scientist he always set for us all the highest standard.

REFERENCES

Benz, R.C., and Dunbar, R.C., (1979). J. Am. Chem. Soc. 101, 6363.

Benz, R.C., Claspy, P.C., and Dunbar, R.C., (1981). J. Am. Chem.Soc. 103, 1799.

Bomse, D.S., Woodin, R.L. and Beauchamp, J.L. (1978). In "Adv. Laser Chem.," A. Zewail, Ed., Springer.

Chen, J.H., and Dunbar, R.C. (1982). To be published.

Coggiola, M.J., Cosby, P.C., and Peterson, J.R. (1980). J. Chem. Phys. 72,6507.

Coggiola, M.J., Cosby, P.C., Helm, H., Peterson, J.R., and Dunbar, R.C. (1982). J. Chem. Phys., Submitted for publication.

Cooks, R.G., Beynon, J.H., Caprioli, R.M., and Lester, G.R (1973). "Metastable Ions," Elsevier.

Dujardin, G., Leach, S, and Taieb, G. (1980). Chem. Phys. 46,407.

Dunbar, R.C. (1976). Anal. Chem. 48, 723.

Dunbar, R.C. (1979a). In "Ion Photodissociation" (P. Ausloos, Ed.) Plenum Press.

Dunbar, R.C. (1980). In "Physical Methods of Modern Chemical Analysis" (T. Kuwana, Ed.) Vol. 2, Academic Press.

Dunbar, R.C. (1981). In "Mass Spectrometry" (R.A.W. Johnstone, Ed.) Vol. 6, The Royal Society, p. 100.

Dunbar, R.C. (1982a). In Proceedings of the Conference on Ion Cyclotron Resonance in Mainz, Germany, 1981. In press.

Dunbar, R.C. (1982b). J. Phys. Chem., Submitted for publication.

Dunbar, R.C., and Fitzgerald, G.B. (1982). To be published.

Dunbar, R.C., and Teng, H.H. (1978). J. Am. Chem. Soc. 100, 2279.

Dunbar, R.C., Hays, J.D., Honovich, J.P., and Lev, N.B. (1980). J. Am. Chem. Soc. 102, 3950.

Dunbar, R.C., Honovich, J.P., and Segall, J. (1982). To be published.

Freiser, B.S. and Beauchamp, J.L. (1975). Chem. Phys. Lett. 35,35.

Freiser, B.S. and Beauchamp, J.L. (1977). J. Am. Chem. Soc. 99,3214.

Fu, E.W., Dymerski, P.P. and Dunbar, R.C. (1976). J. Am. Chem. Soc. 98, 337.

Griffiths, I.W., Mukhtar, E.S., Harris, F.M. and Beynon, J.H. (1981). Int. J. Mass. Spect. Ion Phys. 38, 333.

Griffiths, I.W., Mukhtar, E.S., R.E. March, Harris, F.M. and Beynon, J.H. (1981). Int. J. Mass. Spect. Ion Phys. 39, 125.

Griffiths, I.W., Harris, F.M., Mukhtar, E.S., Beynon, J.H. (1981). Int. J. Mass. Spect. Ion Phys. 41, 83.

Hays, J.D. (1980). PhD Thesis, Case Western Reserve Univ.

Honovich, J.P., and Dunbar, R.C. (1981)., J. Phys. Chem. 85, 1558.

Honovich, J.P. and Dunbar, R.C. (1982a). Int. J. Mass Spectrom. Ion Phys. 42, 333.

Honovich, J.P., and Dunbar, R.C. (1982a). J. Chem. Phys.,Submitted for publication.

Honovich, J.P., and Dunbar, R.C. (1982b). J. Am. Chem. Soc., In press.

Honovich, J.P., and Dunbar, R.C. (1982c). J. Phys. Chem., In press.

Kim, M.S., and Dunbar, R.C. (1979). Chem. Phys. Lett. 60, 247.

Lev, N.B. and Dunbar, R.C. (1981). Chem. Phys. Lett. 84, 483.

Lev, N.B. and Dunbar, R.C. (1982a). J. Phys. Chem., Submitted for publication.

Lev, N.B. and Dunbar, R.C. (1982b). J. Am. Chem. Soc., Submitted for publication.

Maier, J.P. (1980). Chimia 34, 219.

Maier, J.P. (1982). Accts. Chem. Res. 15, 18.

McGilvery, D.C., and Morrison, J.D. (1977). J. Chem. Phys. 67, 368.

Morgenthaler, L.N. and Eyler, J.R. (1981). Int. J. Mass. Spectrom. Ion Phys. 37, 153.

Moseley, J.T., Cosby, P.C., and Peterson, J.R. (1976). J. Chem. Phys. 65,2512.

Mukhtar, E.S., Griffiths, I.W., March,R.E., Harris, F.M., and Beynon, J.H. (1981a). Int. J. Mass. Spectrom. Ion Phys. 41, 61.

Mukhtar, E.S., Griffiths, I.W., Harris, F.M., and Beynon, J.H. (1981b). Int. J. Mass. Spectrom. Ion Phys. 37, 159.

Orth, R.G., and Dunbar, R.C.   (1982).   J.   Am.   Chem.
Soc., In press.

Rynard, C.M and Brauman, J.I.   (1980).   Inorg.   Chem.
19,3544.

Reents, W.D. Jr., and Freiser, B.S.   (1980), J.   Am.
Chem. Soc. 102, 271.

Wight, C.A., and Beauchamp, J.L. (1981).   Chem.   Phys.
Lett. 77, 30.

Woodin, R.L., Bomse, D.S., and Beauchamp, J.L. (1979).
In "Chemical and Biochemical Applications of Lasers"
(C.B. Moore, Ed.) Vol. IV, Academic Press.

DISSOCIATION DYNAMICS OF ENERGY SELECTED IONS STUDIED BY ONE AND
MULTIPHOTON IONIZATION

Tomas Baer

Department of Chemistry
University of North Carolina
Chapel Hill, NC 27514

INTRODUCTION

The study of excited ions with well defined energies has
provided us with a wealth of information concerning ion lifetimes,
excess energy partitioning among dissociation products, and the
role of internal energy in ion molecule reactions. The technique
which has made this advance possible is photoelectron photoion
coincidence (PEPICO) in which ions are formed by UV
photoionization. Because the ionization step is a bound to
continuum transition, ions of all internal energies between 0 and
$h\nu$ -IP can be formed, while the electron carries off the excess
energy. In order to select an ion of a given energy, it is
therefore necessary to measure it in coincidence with an electron
of an appropriate energy.

In recent years the technique of multiphoton ionization (MPI)
has been developed. As with most new areas, the initial work has
focused on the phenomenon of MPI itself. In addition a great deal
of research has centered on the spectroscopy of neutral states
which are often accessible via 2 photon transitions, but
inaccessible via the traditional 1 photon absorption. It is now
rapidly becoming apparent that MPI under appropriate conditions
may allow us to state select the ions. This approach may thus
evolve into a complementary, perhaps competitive, technique with
PEPICO for the study of energy selected ion dynamics.

This paper will review some of the highlights in gas phase
ion chemistry studied by PEPICO and MPI, and discuss the
conditions necessary and the problems encountered with state
selection by MPI.

M. A. Almoster Ferreira (ed.), Ionic Processes in the Gas Phase, 205–225.
© *1984 by D. Reidel Publishing Company.*

Experimental Approach

A. PEPICO

Two types of PEPICO experiments are being performed in the various laboratories engaged in this research. The two approaches are distinguished by the type of light source employed. With the fixed wavelength He(I) source at 21.2eV, ions are collected in coincidence with electrons energy selected by the use of dispersive electron energy analyzers(1). The ion energy is then varied by changing the electron energy according to the relation:

$$E_{ion} = h - E_{el} + E_{thermal} \qquad (1)$$

The last term in Equation (1) accounts for the initial thermal energy of the molecule. This is of course a distribution of energies which in large molecules such as benzene can extend over 0.2eV. The other PEPICO approach uses a variable wavelength light source so that the electron energy in Equation (1) remains fixed at zero energy, and the ion energy is selected with the photon monochromator (2).

The experimental details have been discussed in previous publications (3) and so will not be further dealt with here. However it is instructive to point out some of the differences between the two PEPICO experiments.
1. Electron Energy Resolution

Because of the electric field required to study the ions, the electron energy resolution of the fixed wavelength experiment is generally not better than 0.1eV. By contrast the threshold PEPICO resolution as measured by its full width half maximum is typically .03eV and often better. On the other hand there exists a long tail of energetic electrons which pass through the threshold electron filter (steradiancy analyzer). If these are not eliminated by a dispersive analyzer (2a) or in the case of pulsed synchrotron radiation by their time of flight (2b), the coincidence data may need to be corrected for these unwanted electrons.
2. Accessible Ion States

The fixed wavelength, He(I) PEPICO experiment makes accessible those ion states which can be reached with good Franck-Condon factors from the ground neutral state. On the other hand, the threshold PEPICO approach makes possible the production of ions in a much larger variety of states because intermediate autoionizing states are often excited. Figure 1 shows two examples of the differences encountered. Because the threshold technique can excite high vibrational levels, uncertainties may exist when exciting for instance the first excited state of 2,4

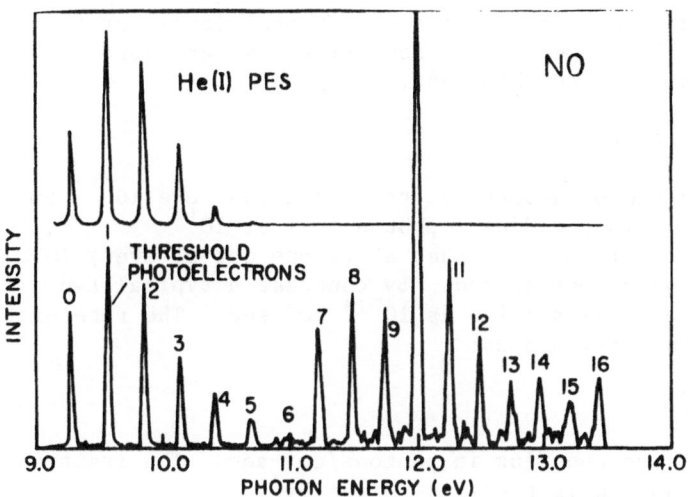

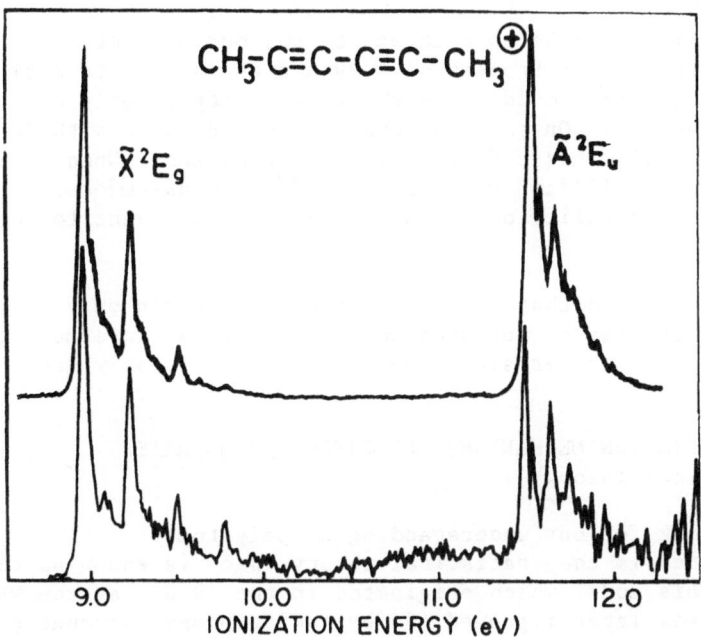

Figure 1. Comparison of He(I) PES (upper spectra and threshold photoelectron spectra (TPES) of NO and 2, 4 hexadyine. Spectra taken in order from top to bottom from references: 4, 5, 6, and 7.

hexadiyne in that a mixture of high vibrational levels of the
ground state may be formed at the same time. We are just
beginning to uncover such phenomena.

B. Multiphoton ionization

Typical one photon absorption cross sections are 10^{-17} cm^2.
With a resonance lamp yielding a photon flux of 10^{12}
photons/cm -sec, it is evident that about one out of every 10^5
molecules will ionize per second. By contrast a typical two
photon absorption cross section is 10^{-54} cm^4 sec. The rate of two
photon absorptions is given by:

$$\frac{dN}{dt} = \sigma\,I^2\,N$$

in which I is the photon flux in photons/cm^2-sec. The fraction of
molecules which absorb is therefore given by:

$$\frac{dN}{N} = \sigma\,I^2\,dt$$

The resonance lamp would be able to excite one out of every 10^{30}
molecules by a two photon transition. Even a CW laser with a flux
of 10^{22} photons/cm^2-sec would raise the probability to only one
out of 10^{10} molecules. On the other hand a pulsed laser with 10^{16}
photons per 5 ns pulse has a flux of 10^{24} photons/sec. When it is
focused to a spot of 10^{-5}cm^2 the flux is 10^{29} photons/cm^2-sec.
This raises the probability of two photon absorption event to one
out of 10^4.

It should be noted that the above discussion deals only with
true two photon processes, not ones which pass via an intermediate
resonant state. The latter situation gives rise to vastly greater
probabilities.

DISSOCIATION MECHANISMS AND DISSOCIATION RATES.
A. The Statistical Theory

The framework for our understanding of polyatomic
dissociation rates is the statistical theory which is known as the
RRKM or QET. This model which originated in the 1920's as the RRK
theory (8) and was later improved (9) by taking proper account of
the quantum nature of vibration is virtually the only tractable
theory we have today. It is based on the assumption of random
energy flow of vibrational energy in an isolated molecule. The
simplest way to visualize the model is in terms of the phase
space. Consider a molecular ion with an energy E and a minimum
energy for dissociation of E_o (Figure 2). When the ion is in the
configuration of the molecular ion there is a total energy, E,
which can be partitioned among all the vibrational degrees of
freedom of the ion. That is within a small $/\backslash$E at a total energy E

there will be some large number of combinations of vibrational quantum numbers (micro states) such that for each combination $E \approx \geq$ $h\nu_i(v_i+\frac{1}{2})$. We represent each such micro state by a square in phase space. The horizontal bottle in Figure 2 is divided into a number

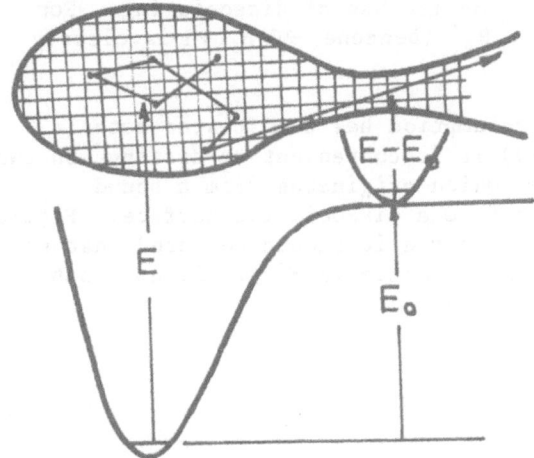

Figure 2
Potential energy surface showing a random walk in phase space leading to dissociation of a polyatomic ion or molecule.

of these squares. In the transition state configuration the number of possible microstates is reduced because the available energy, $E-E_o$, for distribution among the vibrational modes is greatly reduced. The whole bottle along with its bottle neck is the phase space of the dissociating ion. At each energy, E, a new phase space must be drawn.

We can visualize the dissociation as a random walk in this phase space. The statistical assumption is that all steps between squares are equally probable. Hence the probability for dissociation is simply related to the area of the bottle neck and inversely related to the area of the bottle. Mathematically we write the rate, k(E) as

$$k(E) = \frac{\sigma \int_o^{E-E_o} \rho^{\neq}(E)\,dE}{h\,\rho(E)} \qquad (2)$$

in which $\rho(E)$ in the denominator is the density of states (number of states per unit energy interval) when the ion is in the molecular ion configuration, and the integral over the transition state density of states, $\rho^{\neq}(E)$, represents the number of ways the ion can pass through the transition state. The coefficient, σ, is the number of equivalent ways the ion has of dissociating. For instance σ for the reaction $C_6H_6^+$ (benzene)$\rightarrow C_6H_5^+ + H$ is clearly equal to six.

The random energy flow assumption has been tested both in neutral and ionic systems.(10) It is convenient to distinguish two classes of dissociations, one which originates from a bound electronic surface, the other from a dissociative surface. Figure 3 shows two such surfaces. Of course it should be noted that we show only one coordinate of the molecule in Figure 3, and that

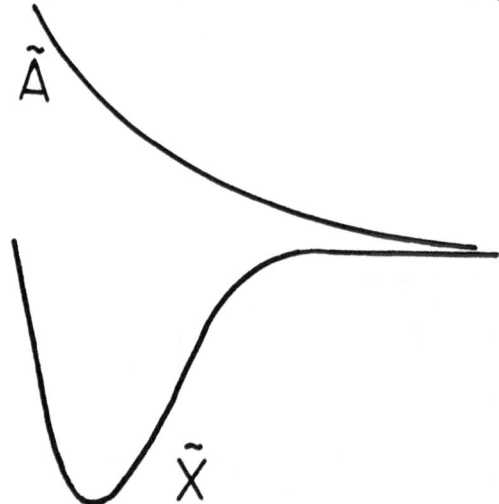

Figure 3.   Bound (X) and a dissociative (A) potential energy curves.

the other coordinates are bound. The statistical theory is expected to describe only the dissociation from the bound state with any reasonable accuracy because the dissociation from the dissociative state will take place on the 10^{-12}–10^{-13} sec time scale. Pure repulsive states such as B in Figure 3, are rare in polyatomics. Some notable examples are in $C_2F_6^+$,(11) $C_2H_2F_2^+$ (12), CF_4^+,(13) and CF_3Cl^+ (13) to be discussed in the following section.

The case of Br loss from bromobenzene serves as a good test of the statistical assumption for the case of a bound state dissociation.(14)   The C-H vibrations are not directly involved in the C-Br bond breaking step. However, they should, according to the statistical hypothesis, participate fully in the energy flow.

If we replace the hydrogen atoms by deteurium atoms, the energetics of the reaction remain the same, but the density of states will vary because the C-D vibrational frequencies are a factor of $\sqrt{2}$ less than the C-H vibrational frequencies. The difference in the dissociation rates should be a maximum at the dissociation limit where the numerator in Equation (2) is unity and the ratio $k_H/k_D = \rho_D(E_o)/\rho_H(E_o)$. As the energy increases, the ratio eventually goes to unity because numerator and denominator in Equation (2) will approach their classical value thereby becoming independent of the frequency.

An example of the experimental data which consist of time of flight distributions of $C_6H_5^+$ ions are shown in Figure 4. The asymmetry of the peaks results from dissociation events which are occuring while the ion is being accelerated. The more asymmetric the peak, the slower is the decay rate. From such data we can construct the experimental $k(E)$ vs E results of Figure 5. The RRKM/QET calculations are the solid lines which were adjusted to give a good fit for the $C_6H_5Br^+$ dissociation data. The theoretical curve for the $C_6D_5Br^+$ dissociation was constructed using the same E_o and the same vibrational frequencies as in $C_6H_5Br^+$ except that all CH frequencies were lowered by the factor $\sqrt{2}$. Clearly the excellent fit indicates that all CH frequencies fully participate in the energy flow as assumed by the statistical theory.

This dissociation reaction also illustrates the well known kinetic shift, which is the difference between the observed dissociation onset of about 12.1eV and the assumed (or known) thermochemical dissociation limit of 11.73eV. The reason for the shift is that the onset is observed only when the reaction proceeds with a rate sufficiently fast to cause the ions to dissociate in the time scale of the experiment which is typically of the order of micro seconds. The thermochemical activation energy, E_o is 2.76eV in the case of the bromobenzene ion dissociation.

B) Examples
1) $CF_3Cl^+ \rightarrow CF_3^+ + Cl$

Powis and Danby (13, 15) investigated the dissociation of state selected CF_3Cl^+ ions and found some fine examples of non statistical behavior. The photoelectron spectrum (16) and the various dissociation limits of this ion are shown in Figure 6. The dissociation rates at all energies were too rapid to be measured by PEPICO. However the kinetic energy released upon dissociation was determined. The only dissociation products possible from the ground electronic state are CF_3^+ and Cl. The kinetic energy release distributions (KERDs) shown in Figure 7, are very well described by the statistical theory (solid lines).

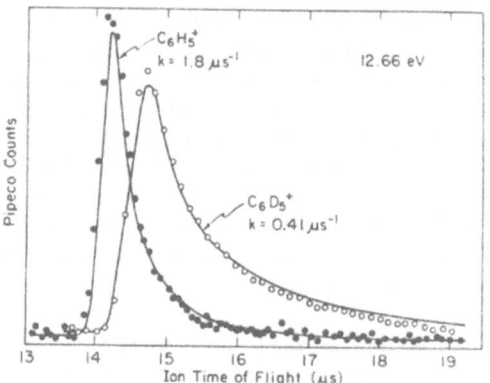

Figure 4.   Bromobenzene dissociation product TOF distributions.
Solid lines are calculated assuming the indicated decay
rates.   Taken from reference 14.

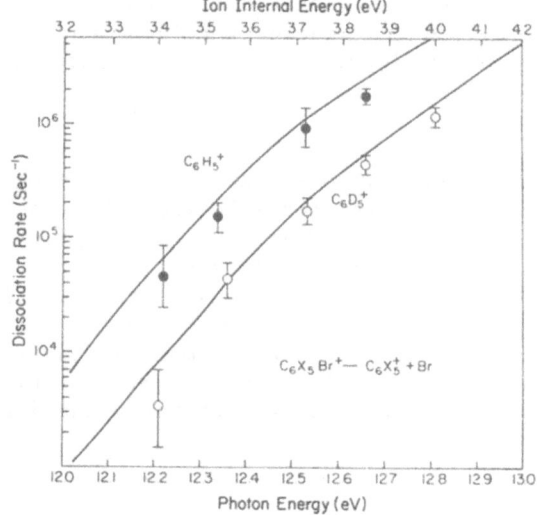

Figure 5.   Bromobenzene ion dissociation rates.   Points are
experimental.   Lines are RRKM/QET calculations.   Taken
from reference 14.

The dissociation from higher electronic states is considerably more complex. Figure 8 shows some KERDs which result from the dissociation of the A state. These KERDs are not at all statistical. In fact if we ignore the contribution at low energies in the two He(I) PEPICO results, we conclude that they are the result of a direct dissociation from a repulsive state as is also observed when using the Ne resonance lamp. The feature which is difficult to explain in these data is the availability of the low energy release dissociation path when the ionizing radiation is He(I). Powis(13) rationalized this in terms of an autoionizing state which is excited at 21.2eV and which produces vibrationally excited ground state ions having the same total energy as the electronically excited A state ions. The vibrationally excited X state ions would dissociate statistically with a KERD such as is shown in Figure 7, while the A state dissociation proceeds along a repulsive surface. It is further argued that photoionization with the Ne source does not result in the formation of vibrationally excited X state ions.

These results demonstrate beautifully that energy selection is not the same as state selection. More such examples will surely be found.

2. $C_4H_8O_2^+$ (dioxane) $\rightarrow C_3H_6O^+ + CH_2O$

A rather unusual and interesting reaction is that of the dioxane ion which dissociates to three products within about 0.2eV.(17) These ions and their OK appearance energies are $C_2H_4O^+$ (10.50eV); $C_2H_5O^+$ (10.57eV); and $C_3H_6O^+$ (10.67eV). The latter is particularly interesting because $C_3H_6O^+$ has the same chemical composition as acetone whose ion energy is well known.(18) If we assume this ion to have the acetone structure, we can calculate an appearance energy by the relations:

$$AE(OK) = \underline{\Delta H^o_{fo}} \text{ (products)} - \underline{\Delta H^o_{fo}} \text{ (reactant)}$$

This gives an expected appearance energy of 9.51eV compared to the experimentally observed onset of 10.67eV. The discrepancy of 1.16eV in the expected and observed onset cannot be explained in terms of a kinetic shift because a thermochemical onset of 9.51eV and an ionization energy of 9.19eV (see Figure 9) would result in an $E_o = 0.32eV$, and a rate which even at threshold would be extremely fast. The measured rates (Figure 10) are slow and are consistent with an onset for $C_3H_6O^+$ of 10.65eV as determined by the RRKM/QET calculation. As an independent check the average kinetic energy release was noted to go to zero at 10.65eV onset. We conclude therefore that at the dissociation onset a structure other than acetone is produced.

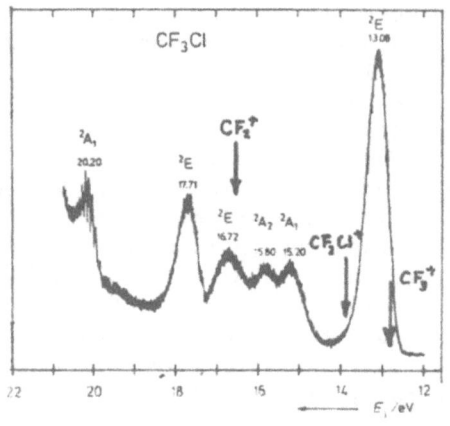

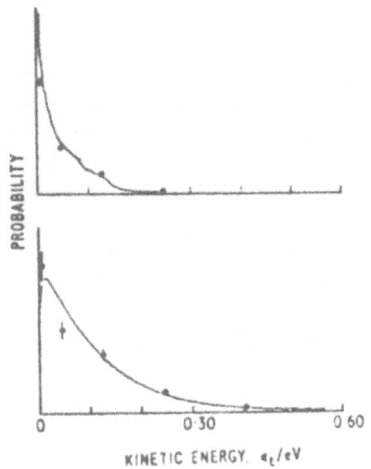

Figure 6.   He(I) PES of CF_3Cl from reference 16.
Figure 7.   KERDs for CF_3^+ formation from X state of CF_3Cl^+ at
            excess energies of 0.0eV and 0.4eV. From reference 15.

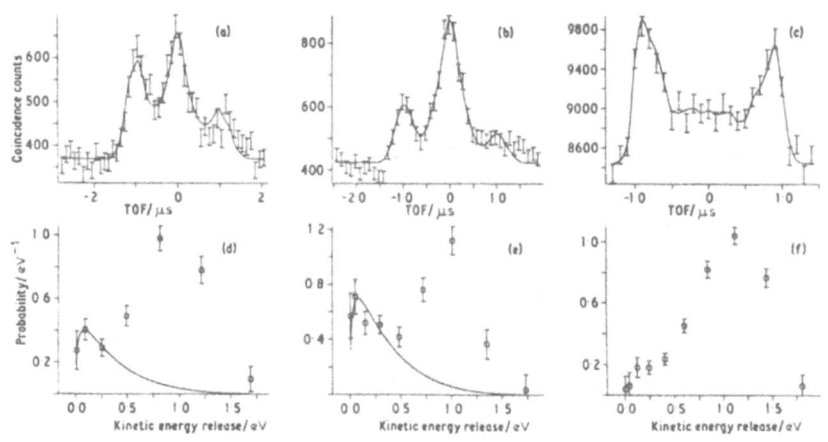

Figure 8.   Examples of the results for A $CF_3Cl^+ \rightarrow CF_3^+ + Cl$. Coinci-
            dence TOF peaks were recorded at nominal ionization
            energies of (a) 15.1eV (b) 15.4eV (both using HeI radi-
            ation) and (c) 15.4eV (using neon radiation); (d), (e)
            and (f) are the kinetic energy release distributions
            corresponding to (a), (b) and (c) respectively. The
            smooth curves drawn in (d) and (e) are distributions
            calculated from statistical-dynamic theory, arbitrarily
            scaled for comparison with the low energy portions of
            the experimental distributions. Taken from reference 13.

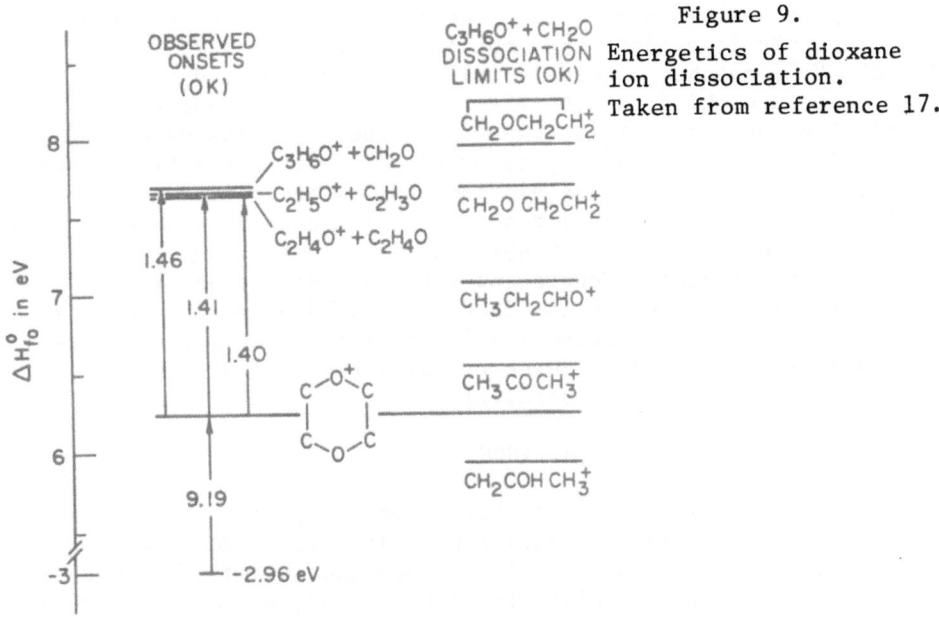

Figure 9.
Energetics of dioxane
ion dissociation.
Taken from reference 17.

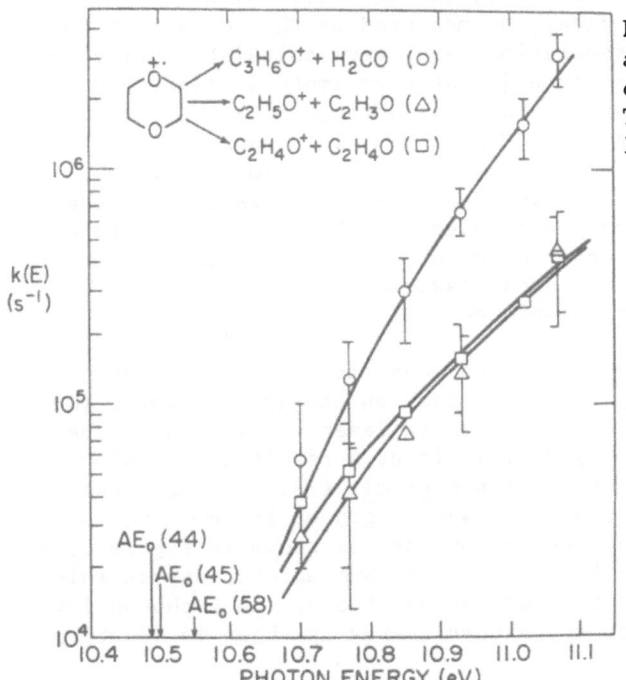

Figure 10.
Experimental (points)
and RRKM/QET (lines)
dissociation rates.
Taken from reference
17.

Figure 9 shows some of the structures for $C_3H_6O^+$ and their energies. It is evident that the calculated structure,(19) $CH_2OCH_2CH_2^+$, matches extremely well with the observed onset. It is a structure which is naturally derived from dioxane by two simple bond breaks. It is significant that the rate data of Figure 10 can be fit only by assuming a "tight" transition state, that is a structure such as dioxane itself rather than a linear, or "loose" structure. This reaction is remarkable because it is a very slow dissociation which does not produce the lowest energy dissociation product. Consistent with this very specific dissociation mechanism is the fact that the dioxane ion retains its structure even 2eV above its ionization potential. It does not isomerize to the much lower energy structures of ethyl acetate(20) or n-butanoic acid.(21)

3. $C_6H_5NH_2^+$ (aniline) $\rightarrow C_5H_6^+ + HCN$

This dissociation has now been studied by one, (22) three, (23) and five (22) photon ionization. It will serve as a transition between the discussion of one photon and multiphoton ionization phenomena. Unlike the dioxane ion, which formed a very unstable product ion, the dissociation of the aniline ion forms a stable ion and neutral fragment which cannot be produced via simple bond ruptures. A rearrangement must take place. The energetics and rates of decay, as measured by PEPICO, are shown in Figure 11. RRKM/QET dissociation rates were calculated using the activation energy, E_o, as an adjustable parameter. Tight and loose transition state structures gave a range of E_o's which are shown by the hatched region in Figure 11. This range of E_o's is clearly consistent with the formation of the cyclopentadiene product ion. Previous electron impact measurements (24) of the $C_5H_6^+$ onsets had mistakenly concluded that the linear structures were formed because the observed onset was at about 12.5eV, which just corresponds to the measured dissociation rate of 10^4 sec^{-1}. That is, we have another example here of the kinetic shift.

The decay rates have also been measured by UV multiphoton ionization using three photons to reach an energy of 13.5eV. (23) Figure 12 shows the TOF data at several laser wavelengths. The m/z 66 ($C_5H_6^+$) peak is very asymmetric as a result of its slow formation while accelerating. Proch et al. (23) derived a rate of about 10^6 sec^{-1} when using a laser wavelength of 266nm. The ion energy reached with three such UV photons is shown in Figure 11 to be in the vicinity of 13.5eV. There is some uncertainty in this final ion energy because the ionization step at the 2 photon level produces an electron with an unknown energy as shown in Equation 3.

$C_6H_5NH_2 + h\nu \quad \rightarrow C_6H_5NH_2^+ + e(KE)$ \hspace{2cm} (3)

$C_6H_5NH_2^+ + h\nu \quad \rightarrow C_6H_5NH_2^{+*}$ \hspace{2.5cm} (4)

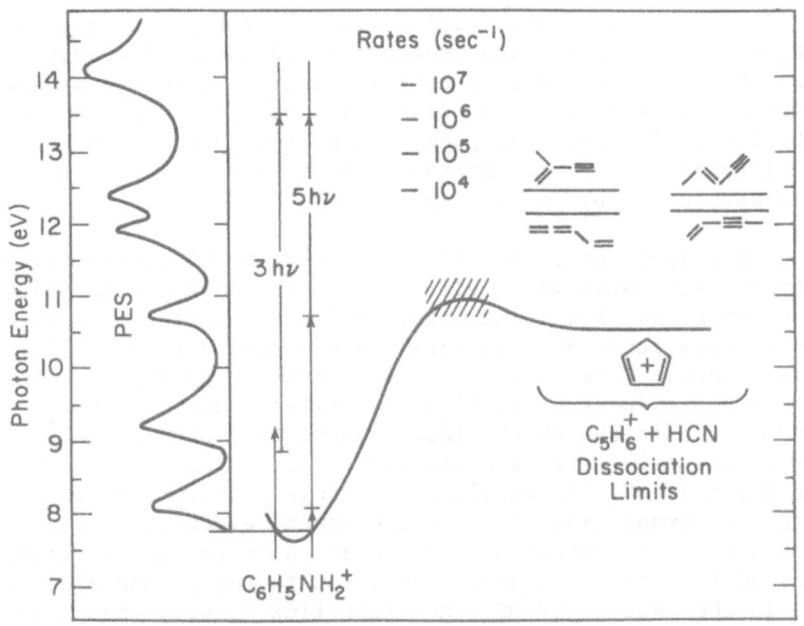

Figure 11.   Potential energy curve and dissociation rates of
             aniline ions.   Taken from reference 22.

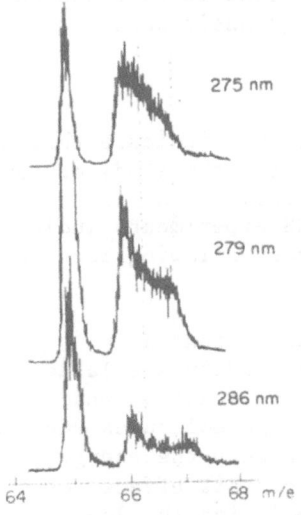

Figure 12.
UV MPI mass spectra of
aniline at several laser
wave lengths.   Taken
from reference 23.

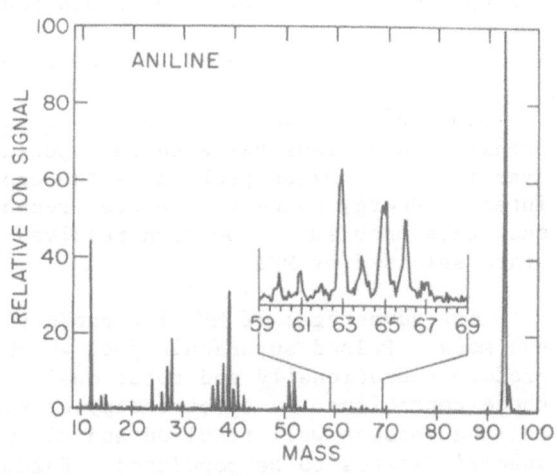

Figure 13.   Visible MPI mass spectrum
             at 462 nm of aniline.
             Taken from reference 22.

The measured rate of 10^6 sec^{-1} indicates that the ion was formed
with 13.6eV of energy. This implies that in the photoionization
step (Equation 3) the ion was produced at an internal energy of
about 8.9eV while the electron carried 0.5eV. These are of course
just average energies because the ion may have been formed with a
distribution of internal energies. However without a PES under
these MPI conditions it is not possible to know the internal
energy distribution of the aniline ions.

The mass spectrum of aniline under the UV MPI conditions show
a number of mass peaks which clearly result from the absorption of
many more photons. How is it possible for the photon absorption
to stop in some cases at 3 photons thereby producing long lived
low energy ions? This question is particularly appropriate in
view of the mass spectrum of Figure 13 taken under 5 photon
visible MPI conditions which clearly shows that no metastable ions
of 13.6eV internal energy are allowed to exist long enough for
them to dissociate. The resolution of this problem lies in the
radiationless transition rates to the ground electronic state. It
is certain that the metastable dissociation of aniline at 13.6eV
is preceded by a radiationless transition to the ground electronic
state. In the case of UV MPI the light flux is weak enough to
allow this transition to take place. In fact about 50% of all the
ions evidently decay in this manner. However under the much more
tightly focused laser beam conditions of the visible MPI, the
photon flux is about 1000 times greater and thus further photon
absorption dominates over the radiationless transitions.

STATE SELECTION BY MPI

The aniline results of Proch et al. clearly indicate that MPI
preparation of ions has a certain potential for the study of ion
dynamics. The major problem is the uncertainty in the ion
internal energy content. However recent PES experiments indicate
that this problem may be soon resolved thereby allowing ions to be
state selected by MPI.

The advantages of MPI ion production are potentially
enormous. Pulsed supersonic jets which are readily available
produce vibrationally and rotationally cold molecules. Under
these conditions the spectroscopy of even large molecules such as
aniline becomes well resolved and allows not only the energy, but
specific states to be populated. Examples of this effect for
chlorobenzene are shown in Figure 14. In addition the low
translational temperature will reduce thermal broadening of ion
TOF distributions giving much greater precision in determining
ion dynamics. Finally the pulse nature of the experiment should
allow crossed beam experiments to be performed with pulsed beams.
This will in fact be trivial because the pumping requirements of

such pulsed valves are very modest. A single diffusion pump of
moderate size should be sufficient.

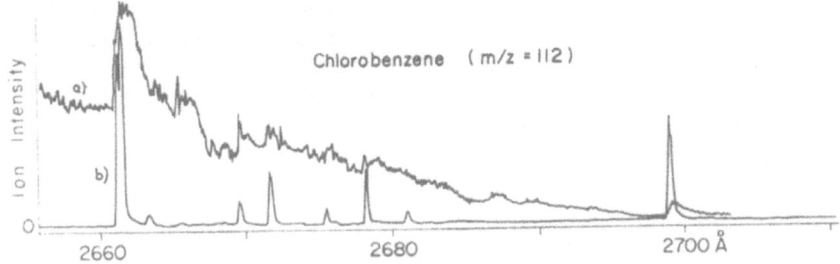

Figure 14.   Ion intensity versus laser wavelength for room
              temperature (spectrum a) and cooled (spectrum b)
              chlorobenzene.   Taken from reference 25.

     As already indicated, the major problem lies in the control
of the ion internal energy content.   It has recently been
demonstrated for the case of NO that if MPI can proceed by
populating an intermediate Rydberg state (Figures 15 and 16) whose
geometry is very similar to that of the ion, the Franck-Condon
factors connecting these two states insures that an ion is
produced in the same vibrational level as the Rydberg state.   An
example of such a situation is shown in Figure 17.   What has

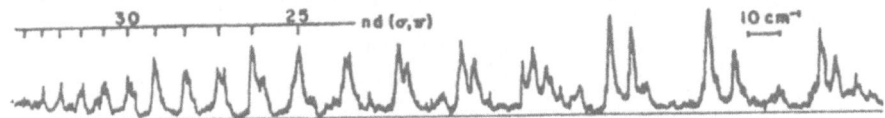

Figure 15.   NO$^+$ signal resulting from two color MPI.   The 2 photon
              pump laser was fixed to the X→C (3pπ, v=0, N=3) transi-
              tion.   Scanning the probe laser excited the nlλ, v=0,
              N=2,3,4 Rydberg states shown in the spectrum.   A 4th
              photon ionized the Rydberg states.   Taken from refer-
              ence 26.

become very evident is that in order to insure the production of a
single vibrational level, it is essential to populate pure Rydberg
states.   This can only be accomplished by rotationally resolving
the states.

     As an example of the possibility of investigating the
dissociation by state selected ions we consider the recent results
of Anderson et al. (25)   They investigated the dissociation rate
of chlorobenzene ions in selected energy states by a method
similar to that used in the aniline dissociation.   The $C_6H_5^+$ time
of flight (TOF) spectra at four different laser wavelengths are

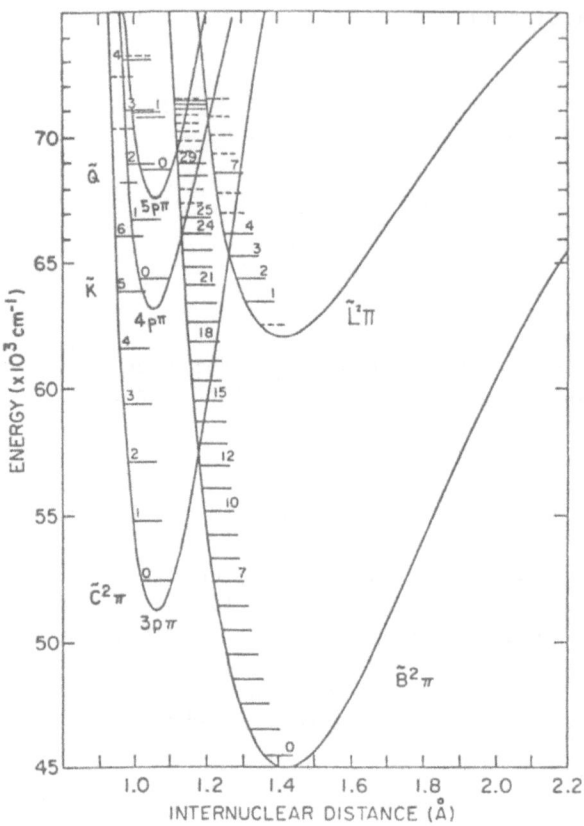

Figure 16
ome of the potential
urves for NO reached by
visible 2 photon transi-
tions. Taken from refer-
ence 26.

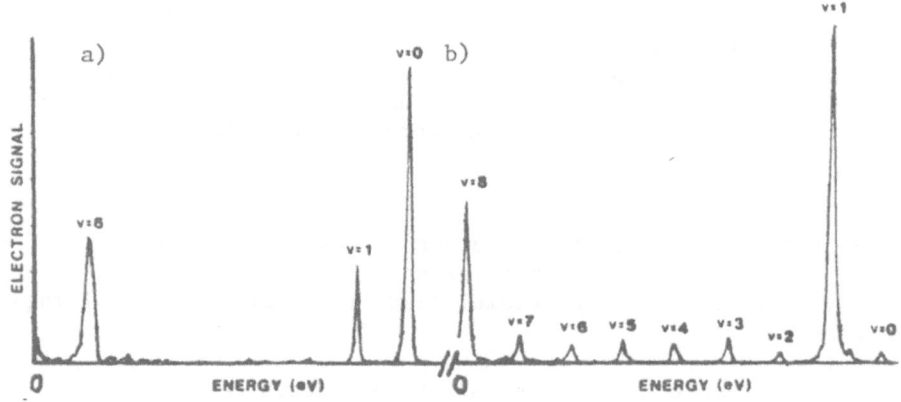

Figure 17.   Two PES of NO for 4 photon ionization (single color).
a) λ = 451.1 nm via $A^2\Sigma^+$ (v=o). b) λ = 430 $A^2\Sigma^+$ (v=1).
Taken from reference 27.

$$C_6H_5Cl^+ \rightarrow C_6H_5^+ + Cl$$

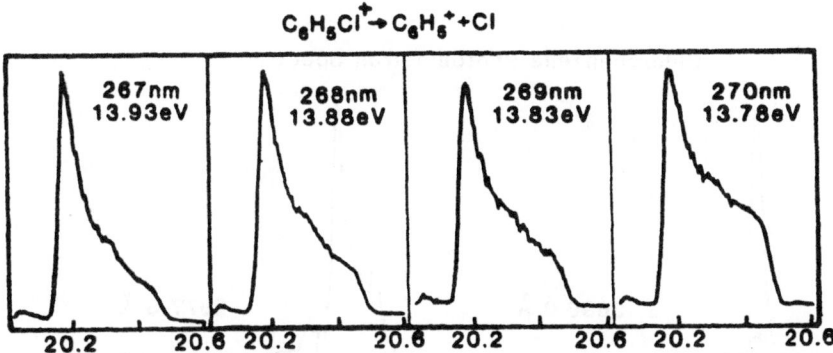

Figure 18. $C_6H_5^+$ TOF distributions at 4 laser wavelengths.   The
            asymmetric peak shapes are a result of the metastable
            $C_6H_5Cl^+$ dissociation at the indicated ion internal
            energies.   Taken from reference 25.

shown in Figure 18. The dissociation rates were derived from the
asymmetric TOF distributions. In this study, a considerable
effort was made to determine the $C_6H_5Cl^+$ ion internal energy.
With the UV laser, only two photons were needed to ionize the
molecule. The third photon excited the ion to the dissociating
state. Photoelectron spectra taken at the four wavelengths are
shown in Figure 19. At 2699A, there was just enough energy to
form the ion in the ground state. However as the photon energy
increases, it is evident that the ion is produced in a rather
broad distribution of internal energies. This causes a
considerble uncertainty in the final ion energy. Nevertheless the
ion energy distribution is known if we can assume that all
vibrational levels of the $C_6H_5Cl^+$ ion produced at the two photon
level have equal cross section for further photon absorption. The
resulting rate data are shown in Figure 20 where they are compared
with PEPICO results of Baer et al. (2c) and Rosenstock et al. (28)
The agreement is very satisfactory.

It is clear from Figure 19 that the result would be a great
deal more clear cut if a two color laser were used. The first
could be the 2699A beam which would prepare the ion in the ground
vibrational state. A second, more powerful laser, could then be
tuned to excite the $C_6H_5Cl^+$ ions to various final states. The
only restriction is that the second laser not accidentally be in
resonance at the first photon level with a neutral C_6H_5Cl
molecule. The possibilities of carrying out this experiment at
high electric fields would allow rates as high as 10^9 sec^{-1} to be
measured thereby greatly extending the currently feasible PEPICO
data.

Chlorobenzene Photoelectron Spectra

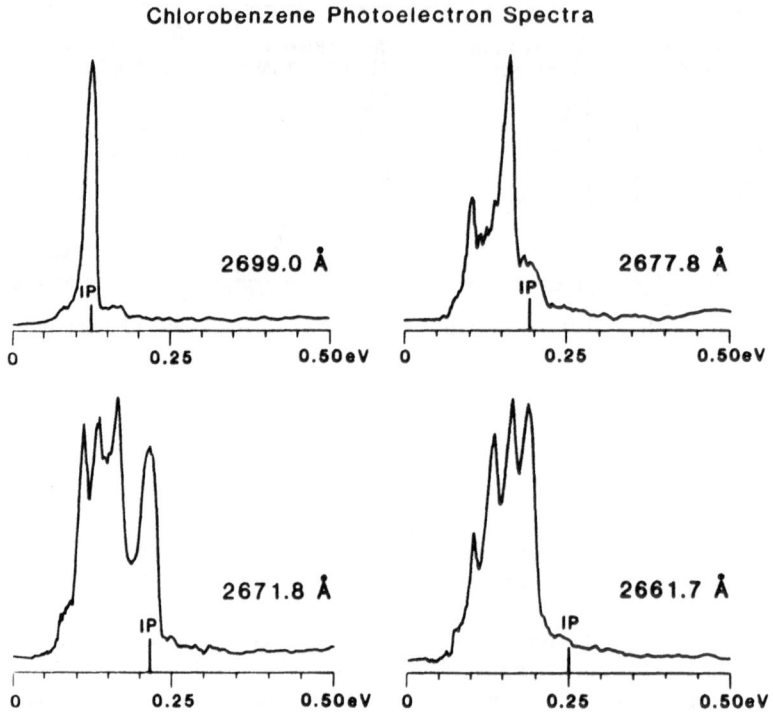

Figure 19.   PES of chlorobenzene resulting from two photon
             ionization.   A third photon excited these ions to
             the dissociative state.   Taken from reference 25.

Acknowledgments
The National Science Foundation and the Department of Energy are
gratefully acknowledged for financial support of the work in the
author's laboratory.

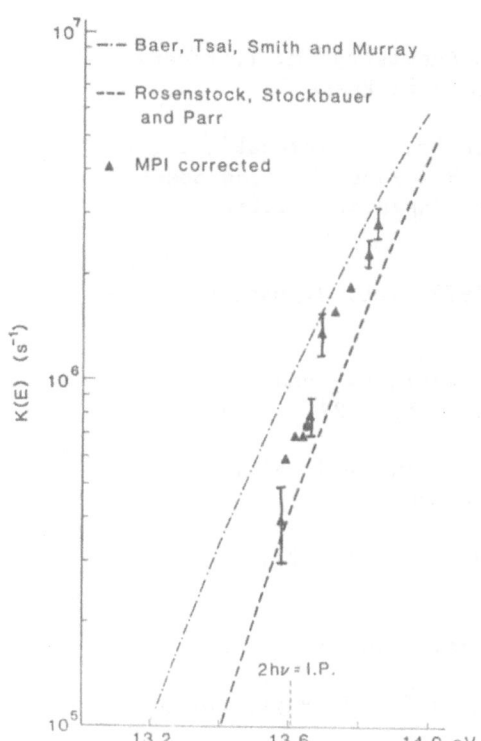

Figure 20
Dissociation rates for chlorobenzene ion dissociation.
Lines are obtained by PEPICO,
points by MPI.   Taken from
reference 25.

REFERENCES

1. a) Bombach, R., Dannacher, J., Stadlemann, J. P., Vogt,
      J.: 1981, Int. J. Mass Spectrom. Ion Phys. 40, p. 275.

   b) Eland, J. H. D.: 1979, J. Chem. Phys. 70, p. 2926.

   c) Mansell, P. I. Danby, C. J., and Powis, I.: 1981, J.
      Chem. Soc. Faraday Trans. 2, p. 1449.

2. a) Rosenstock, H. M., Stockbauer, R., and Parr, A. C.:
      1981, Int. J. Mass Spectrom. Ion Phys. 38, p. 323.

   b) Nenner, I. Guyon, P. M., Baer, T., and Govers, T. R.:
      1980, J. Chem. Phys. 72, p. 6587.

   c) Baer, T., Tsai, B. P., Smith, D., Murray, P. T.,: 1976,
      J. Chem. Phys. 64, p. 2460.

   d) Meisels, G. G., Hsieh, T., and Gilman, J. P.: 1980, J.
      Chem. Phys. 73, p. 4126.

e) Kato, T., Tanaka, K., Koyano, I.: 1982, J. Chem. Phys. 77, p. 834.

3. Baer, T.: 1979 "Gas Phase Ion Chemistry" M. T. Bowers ed. vol. 1, ch. 5, p. 153, Academic Press.

4. Kimura, K., Katsumata, S., Achiba, Y., Yamazaki, T., and Iwata, S.: 1981, "Handbook of HeI Photoelectron Spectra of Fundamental Organic Molecules" Japan Scientific Societies Press, Tokyo.

5. Murray, P. T. and Baer, T.: 1978, Int. J. Mass Spectrom Ion Phys. 30, p. 165.

6. Allan, M., Maier, J. P., Marthaler, O., and Kloster-Jensen, E.: 1978, Chem. Phys. 29, p. 331.

7. Baer, T., Willett, G. D., Smith, D., and Phillips, J. S.: 1979, J. Chem. Phys. 70, p. 4076.

8. a) Rice, O. K., and Ramsperger, H. C.: 1927, J. Am. Chem. Soc. 49, p. 1617.

   b) Kassel, L. S.: 1928, J. Phys. Chem. 32, p. 225.

9. a) Marcus, R. and Rice, O. K.: 1951, J. Phys. Colloid Chem. 55, p. 894.

   b) Rosenstock, H. M., Wallenstein, M. B., Warhaftig, A. L., and Eyring, H.: 1952, Proc. Natl. Acad. Sci. USA, 38, p. 667.

10. Forst, W.: 1973, "Theory of Unimolecular Reactions," Academic Press.

11. Simm, I. G., Danby, C. J., and Eland, J. H. D.: 1974, Int. J. Mass Spectrom. Ion Phys. 14, p. 285.

12. Stadelmann, J. P. and Vogt, J.: 1980, Int. J. Mass Spectrom. Ion Phys. 35, p. 83.

13. Powis, I.: 1980, Mol. Phys. 39, p. 311.

14. Baer, T., and Kury, R.: 1982, Chem. Phys. Lett. in press.

15. Powis, I. and Danby, C. J.: 1979, Chem. Phys. Lett. 65, p. 390.

16. Cvitas, T., Gusten, H., and Klasinc, L.: 1977, J. Chem. Phys. 67, p. 2687.

17. Fraser-Monteiro, M. L., Fraser-Monteiro, L., Butler, J. J., Baer, T., and Hass, J. R.: 1982, J. Phys. Chem. 86, p. 739.

18. Rosenstock, H. M., Draxl, K., Steiner, B. W., and Herron, J. T.: 1977, J. Phys. Chem. Ref. Data 6, number 1.

19. Bouma, W. J., MacLeod, J. K., and Radom, L.: 1979, J. Am. Chem. Soc. 101, p. 5540.

20. Fraser-Monteiro, L., Fraser-Monteiro, M. L., Butler, J. J., and Baer, T.: 1982, J. Am. Chem. Soc. 86, p. 752.

21. Butler, J. J., Fraser-Monteiro, M. L., Fraser-Monteiro, L., Baer, T., and Hass, J. R.: 1982, J. Am. Chem. Soc. 86, p. 747.

22. Baer, T. and Carney, T. E.: 1982, J. Chem. Phys. 76, p. 1304.

23. Proch, R., Rider, D. M., and Zare, R. N.: 1981, Chem. Phys. Lett. 81, p. 430.

24. Occolowitz, J. L. and White, G. L.: 1968, Aust. J. Chem. 21, p. 997.

25. Anderson, S., Rider, D., Durant, J., and Zare, R. N.: 1982, unpublished results.

26. Chupka, W. H. and Colson, S.: 1982 unpublished results.

27. Kimman, J., Kruit, P., and Van der Wiel, M. J.: 1982, Chem. Phys. Lett. 88, p. 576.

28. Rosenstock, H. M., Stockbauer, R., and Parr, A. C.: 1980, J. Chem. Phys. 73, p. 773.

SIFT-DRIFT STUDIES OF ANIONS

Charles H. DePuy
Department of Chemistry, University of Colorado,
Boulder, Colorado 80309

INTRODUCTION

Because anions are more difficult to form in the gas phase,
studies of their chemistry have lagged behind those of cations.
With the development of the ion cyclotron resonance (ICR) and
flowing afterglow (FA) techniques this situation has changed,
since in these instruments new types of anions can be prepared by
chemical reactions. Recent instrumental developments have taken
place for both techniques. In ICR spectrometry Fourier transform
methods have greatly extended the sensitivity and made it
possible to examine sequential reactions of types never before
possible. Nibbering has described some applications to the
organic chemistry of anions in an earlier paper in this volume.
Smith and Adams (1) have developed the selected ion flow tube
(SIFT), a modification of the flowing afterglow, which greatly
increases the variety of experiments possible with this
technique. For organic chemists anions are especially important
since most synthetically useful reactions proceed through anionic
intermediates. We have been studying the gas phase ion-molecule
chemistry of anions for some time using a conventional FA (2).
Recently we completed construction of a SIFT, in which it is also
possible to study reactions of ions as a function of energy in a
drift field (3). In this paper we summarize some of our recent
results using both the FA and SIFT. In addition some preliminary
Drift results are reported.

EXPERIMENTAL

A schematic of the SIFT-Drift instrument is given in Figure
I. From right to left in the diagram, it consists of an ion

M. A. Almoster Ferreira (ed.), Ionic Processes in the Gas Phase, 227–241.
© 1984 by D. Reidel Publishing Company.

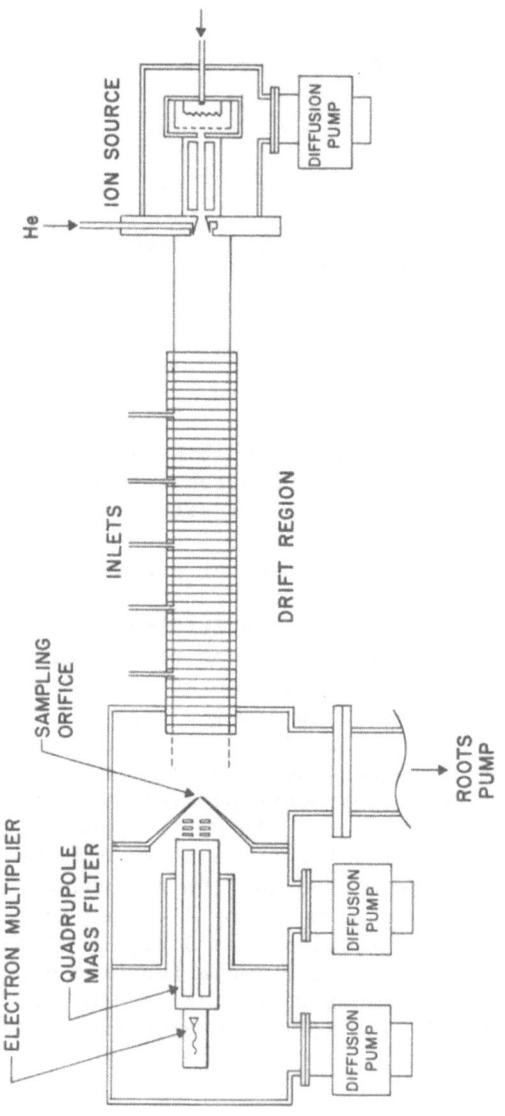

Figure I

source, a flow-drift tube in which ion-molecule reactions are carried out and a mass spectrometer for analysis and detection of the ions. Ions are formed by electron impact in a low pressure ion source and the neutrals are removed by pumping. Ions of a desired mass to charge ratio are selected by the source quadrupole mass filter and injected into the flow tube in the absence of all other ions, metastable species, electrons and photons. The helium carrier gas enters the flow tube through a narrow annulus concentric with the ion entrance orifice; this injector design allows minimum backflow of helium into the ion source region. The carrier gas is pumped through the 100 cm long x 8 cm diam. flow tube by the action of a large Roots pumping system which establishes pressures of ~0.5 torr and flow velocities of ~8000 cm/sec. In the first 30 cm of the flow tube the ions are collisionally relaxed and the diffusion and flow properties are established. The neutral reagent is added through a manifold of radial inlets which provide variable reaction distance for convenient acquisition of kinetic data. Most of the gas mixture is exhausted by the pumping system, a fraction is sampled through a small orifice, analyzed with a quadrupole mass filter and detected with an electron multiplier and pulse counting electronics. The system operation and the data acquisition are facilitated by a minicomputer.

The flow tube consists of a series of thin rings which are electrically insulated from one another. These rings are connected and grounded for the study of ion-molecule reactions at thermal energy. Alternatively, precision resistors are inserted between adjacent rings and a potential is placed across the entire assembly; in this drift mode ions are accelerated between collisions and slowed by collisions so that they quickly attain a steady state translational velocity. In this way ion-molecule reaction rate constants, branching ratios and mechanisms can be studied as a function of kinetic energy.

RESULTS

Inorganic Ions

Only a few types of reactive inorganic anions can be prepared efficiently by direct electron impact. Electron impact on ammonia gives amide ion, while hydroxide ion is formed either by electron impact on water or, preferably, on a mixture of N_2O and CH_4 (eq. 1,2). We have been investigating ways to use these

$$NH_3 \xrightarrow{e^-} H_2N^- + H\cdot \tag{1}$$

$$N_2O \xrightarrow{e^-} N_2 + O^{-\cdot} \xrightarrow{CH_4} HO^- + CH_3\cdot \tag{2}$$

precursor anions to synthesize other types of anions in a flowing afterglow (4).

Of course the simplest method is by proton abstraction (e.g., eq. 3). This can be an excellent method if the acid RH is

$$H_2N^- \ + \ CH_2=CH-CH_3 \ \longrightarrow \ CH_2=CH-CH_2^- \ + \ NH_3 \qquad (3)$$

available and volatile. Many times, however, the acid is either unknown, hazardous or otherwise not useful. In that event, other methods of synthesis of the anion are necessary. Amide ion reacts readily with a number of neutrals by elimination of water or other small molecules (5). For example it reacts with CO_2 to form cyanate ion; we formulate the reaction as occurring by the path shown in eq. 4

$$H_2N^- \ + \ CO_2 \ \longrightarrow \ \left[H_2N{-}\overset{\overset{O}{\|}}{C}{-}O^- \right] \ \longrightarrow \ \left[H\bar{N}{-}\overset{\overset{O}{\|}}{C}{-}OH \right] \ \longrightarrow$$
$$\left[HN=C=O \ + \ HO^- \right] \ \longrightarrow \ {}^-N=C=O \ + \ H_2O \qquad (4)$$

In this process amide adds to CO_2, but the resulting adduct is not observed because there is no solvent to remove the reaction exothermicity. Proton transfer from nitrogen to oxygen is followed by loss of hydroxide ion to form a relatively long-lived complex. Proton transfer within this complex, followed by dissociation, leads to the formation of cyanate ion. A key step in this process is the initial intramolecular proton transfer; prototropic rearrangements within anions are much less frequent than within cations.

Analogous reactions occur with other neutrals (6), for example CS_2 and SO_2, (eq. 5 and 6). In the first case the

$$H_2N^- \ + \ CS_2 \ \longrightarrow \ {}^-N=C=S \ + \ H_2S \qquad (5)$$
$$H_2N^- \ + \ SO_2 \ \longrightarrow \ {}^-N=S=O \ + \ H_2O \qquad (6)$$

product NCS^- ion is, of course, well known in solution. Much less is known about NSO^-. Thus these synthetic methods can be used to prepare ions in the gas phase whose solution chemistry is as yet unexplored.

Amide ion reacts at every collision with CO_2, SO_2 and CS_2. It is somewhat less reactive with N_2O, but azide ion is formed readily (eq. 7). The reaction mechanism is different in this

$$H_2N^- \ + \ N{\equiv}\overset{+}{N}{-}\overset{-}{O} \ \longrightarrow \ \left[\begin{matrix} HNH \quad O^- \\ \diagdown \quad \diagup \\ N=N \end{matrix} \right] \ \longrightarrow \ \left[\begin{matrix} HN^- \quad OH \\ \diagdown \quad \diagup \\ N=N \end{matrix} \right] \ \longrightarrow$$
$$\left[HN=N=N \ + \ HO^- \right] \ \longrightarrow \ N_3^- \ + \ H_2O \qquad (7)$$

case, with addition occurring at the end of the molecule. The
intramolecular proton transfer must occur over more atoms, yet
seems also to occur readily. We will discuss this point further
below. An examination of the reaction of amide ion with carbon
oxysulfide (SCO) presents an opportunity to learn more about the
product controlling steps in these reactions, since in theory
either loss of H_2S could occur with the formation of cyanate ion
or H_2O could be lost to form thiocyanate. In fact only products
arising from the former pathway (eq. 8) are observed in this
competition. Since proton transfer among the various atoms is

$$H_2N^- + S=C=O \longrightarrow \left[H_2N-\overset{\overset{O}{\parallel}}{C}-S^-\right] \underset{\rightleftarrows}{\overset{\rightleftarrows}{}} \begin{array}{l} \left[H\bar{N}-\overset{\overset{O}{\parallel}}{C}-SH\right] \longrightarrow \left[HN=C=O + HS^-\right] \quad (8) \\[2em] \left[H\bar{N}-\overset{\overset{OH}{\mid}}{C}=S\right] \nrightarrow \left[HN=C=S + HO^-\right] \quad (9) \end{array}$$

probably fast, we believe the result reflects the easier loss of
the less basic anion, HS^- rather than HO^-.

Products arising from addition to carbon account for only
48% of the ions in this reaction (7). The major product results
from attack on sulfur with expulsion of carbon monoxide (eq. 10).

$$H_2N^- + S=C=O \longrightarrow \left[H_2N-\bar{S}=C=O\right] \longrightarrow H_2NS^- + CO \qquad (10)$$

This type of direct displacement on sulfur has proven to be a
general one for carbon oxysulfide, and is extremely useful for
the gas phase synthesis of many sulfur anions whose conjugate
acids are too noxious for use. In many cases, too, the
corresponding thiol is not readily available (eq. 13).

$$C_6H_5^- + S=C=O \longrightarrow C_6H_5S^- + CO \qquad (11)$$

$$(CH_3)_2N^- + S=C=O \longrightarrow (CH_3)_2NS^- + CO \qquad (12)$$

$$(CH_3)_3SiCH_2^- + S=C=O \longrightarrow (CH_3)_3SiCH_2S^- + CO \qquad (13)$$

Oxygen Exchange Reactions

In view of the rapidity with which amide ion reacts with
CO_2, it seemed reasonable to suppose that hydroxide ion also
reacts. However if CO_2 and HO^- come together in the gas phase,
only a slow, termolecular addition reaction is observed (8),
because the initially formed adduct dissociates unless cooled by
collisions with the buffer gas.

$$HO^- + CO_2 \rightleftarrows \left[HO-\overset{\overset{O}{\parallel}}{C}-O^-\right]^* \overset{He}{\longrightarrow} HO-\overset{\overset{O}{\parallel}}{C}-O^- \qquad (14)$$

If, however, there is rapid enough intramolecular proton transfer within the adduct, oxygen isotopic exchange will be observed even in the absence of stabilization. To investigate this possibility we prepared $H^{18}O^-$ by low pressure electron impact on a mixture of H_2O and $H_2^{18}O$ in our SIFT. From the ions formed (eq. 15) only those of m/z 19 were injected and allowed to

$$H_2O/H_2^{18}O \xrightarrow{e^-} O^{-\cdot},\ HO^-,\ ^{18}O^{-\cdot},\ H^{18}O^- \xrightarrow{SIFT} H^{18}O^- \quad (15)$$

$$m/z \quad 16, \quad 17, \quad 18, \quad 19$$

react with CO_2. A very rapid conversion of $H^{18}O^-$ to HO^- occurred (9), as predicted by eq. 16. By comparing the rate of

$$
H^{18}O^- + O=C=O \rightleftharpoons \left[H^{18}O-\overset{\overset{\textstyle O}{\|}}{C}-O^- \right] \rightleftharpoons \left[-^{18}O-\overset{\overset{\textstyle O}{\|}}{C}-OH \right] \longrightarrow
$$

$$HO^- + {}^{18}O=C=O \tag{16}$$

exchange to the collision rate it is evident that proton transfer among the oxygens is indeed several times faster than dissociation of the adduct. SO_2 exchanges oxygen with $H^{18}O^-$ even faster than does CO_2, while again N_2O reacts more slowly. Nevertheless an easily detectable exchange occurs in the latter case.

The ease with which intramolecular proton transfer occurs in the adducts of CO_2 and SO_2 is rather surprising since 1,3-proton shifts of this type are forbidden by simple Woodward-Hoffman considerations. In the adduct with N_2O, however, the proton transfer can be considered as an allowed 1,4 6-electron process (eq. 17) and indeed an analogous process is seen when carbanions

$$
\left[\begin{array}{c} RCH_2\ \ O^- \\ \diagdown\ \diagup \\ N=N \end{array} \right] \longrightarrow \left[\begin{array}{c} R\bar{C}H\ \ OH \\ \diagdown\ \diagup \\ N=N \end{array} \right] \longrightarrow R-\bar{C}=N=N + H_2O \tag{17}
$$

are added to N_2O. However only addition and no proton transfer occurs between carbanions and CO_2. Rapid intramolecular proton

$$
RCH_2^- + CO_2 \longrightarrow \left[RCH_2-\overset{\overset{\textstyle O}{\|}}{C}-O^- \right] \longrightarrow R\bar{C}=C=O + H_2O \tag{18}
$$

transfer does not occur in the adduct of formaldehyde and $H^{18}O^-$ even though the distance which the proton must move between the

$$
H^{18}O^- + H_2C=O \rightleftharpoons \left[\begin{array}{c} O^- \\ | \\ H-C-^{18}OH \\ | \\ H \end{array} \right] \not\rightleftharpoons \left[\begin{array}{c} OH \\ | \\ H-C-^{18}O^- \\ | \\ H \end{array} \right] \longrightarrow HO^- + H_2C=^{18}O
$$

$$\tag{19}$$

oxygen atoms in the latter adduct is shorter than in the adduct between $H^{18}O^-$ and CO_2. We hope that experiments presently underway will shed more light on these proton transfer processes.

Phosphorous-containing Anions

Since phosphorous lies just below nitrogen in the periodic table, it should be possible to use the synthetic methods we have developed for nitrogen-containing anions to produce analogous phosphorous-containing anions. Again we begin with amide ion and generate phosphide by reaction with phosphine. This anion is then allowed to react with various neutrals (10) as shown in eq. 20-22. Little is known about these ions in solution, and we hope

$$H_2P^- + S=C=O \longrightarrow {}^-P=C=O + H_2S \qquad (20)$$

$$H_2P^- + S=C=S \longrightarrow {}^-P=C=S + H_2S \qquad (21)$$

$$H_2P^- + O=S=O \longrightarrow {}^-P=S=O + H_2O \qquad (22)$$

an investigation of their chemistry in the gas phase will provide insight into a new and interesting field.

In general phosphide reacts more slowly (by about a factor of 100) and less selectively than amide. For example phosphide reacts with N_2O in two ways, eq. 23 and 24. The slower rate of

$$H_2P^- + N=N=O \quad \overset{8\%}{\longrightarrow} \quad {}^-P=N=N + H_2O \qquad (23)$$

$$\overset{92\%}{\longrightarrow} \quad H_2PO^- + N_2 \qquad (24)$$

reaction of H_2P^- compared to H_2N^- is probably due to its much weaker basicity. A wide variety of phosphorous-containing anions can be synthesized. Some unusual reactions of phosphide are given in eq. 25 and 26. With a wide variety of synthetic

$$H_2P^- + O_2 \longrightarrow PO_2^- + H_2 \qquad (25)$$

$$H_2P^- + CS_2 \longrightarrow PS^- + H_2CS \qquad (26)$$

methods available, it should be possible to begin a systematic study of phosphorous containing anions in the gas phase, and to compare them to their nitrogen analogs.

S_N2 Reactions at Silicon

A fundamental question of interest in silicon chemistry is whether simple silicon halides undergo nucleophilic substitution reactions by the same concerted pathway as their carbon analogs (eq. 27 path a) or whether the ability of silicon to expand its octet leads to reaction by way of a pentacovalent intermediate (eq. 27 path b). We have recently examined this question in the

$$\text{Nu}^- + \underset{\diagdown}{\overset{\diagup}{\text{Si}}}\text{-X} \quad \overset{a}{\longrightarrow} \quad \overset{\delta-}{\text{Nu}\cdots\underset{|}{\overset{\diagup}{\text{Si}}}\cdots}\overset{\delta-}{\text{X}} \longrightarrow$$

$$\overset{b}{\longrightarrow} \quad \text{Nu}\longrightarrow\underset{|}{\overset{\diagup}{\text{Si}}}\overset{\diagup}{=}\text{X} \longrightarrow \qquad \text{Nu-}\overset{\overset{\diagup}{\overset{\diagup}{\text{Si}}}}{\underset{\diagdown}{}} + \text{X}^- \quad (27)$$

gas phase by comparing the rates of reaction of methyl iodide and
trimethylsilyl chloride with a series of anions of varying
basicity (11). The results are summarized in Figure II. As has
been observed before, as the basicity of the nucleophile
decreases its rate of reaction with methyl iodide also decreases.
This is the same general pattern as is observed in solution.
Trimethylsilyl chloride responds totally differently. All anions
which react do so at the same, very fast collision rate. When
the anion becomes less basic than Cl^-, reaction ceases. Only
isotopically labelled chloride ($^{37}Cl^-$) reacts with a measurable
rate constant that is less than the collision rate.

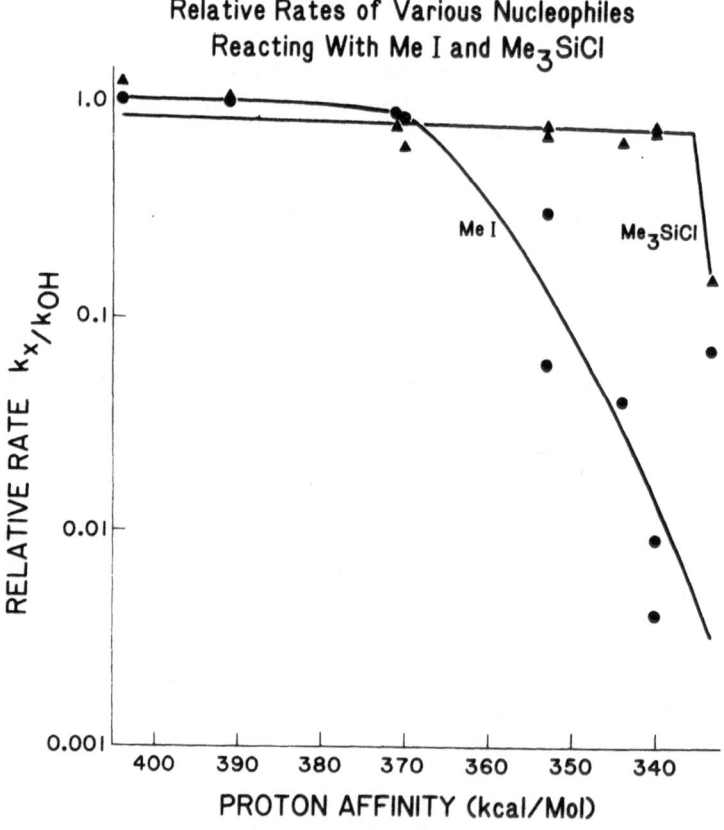

Figure II

These results are easily accommodated if the silyl chloride reacts with an anion at each encounter to form a pentacoordinate intermediate (path b). This activated intermediate will decompose in the thermodynamically favorable direction, on to products if that is favorable, back to reactants if not. In the case of $^{37}Cl^-$ the intermediate is symmetrical so that the rate should be at most half of the collision rate. In fact it is somewhat slower than that and appreciable amounts of the pentacovalent adduct are produced, giving additional support to this mechanism.

H–D Exchange

In a conventional FA we have carried out a number of studies of hydrogen–deuterium exchange between carbanions and D_2O (12). We have shown, for example, that the anion derived from 1-pentene will exchange all of its protons with D_2O. We made a number of

$$CH_3CH_2\bar{C}HCH=CH_2 \xrightarrow{D_2O} C_5H_8D^-, C_5H_7D_2^-, \ldots C_5D_9^- \qquad (28)$$

qualitative studies of these exchange reactions in order to try to understand what structural features in the reactants facilitate exchange. One easily recognizable factor is the difference in basicity between the carbanion and the exchange reagent. For example, while an allyl anion will exchange with D_2O, the less basic dienyl anion will not. It will exchange, however, with the more acidic exchange reagent CH_3OD (eq. 29).

$$\xrightarrow{D_2O} \text{No reaction}$$
$$CH_2=\overset{-}{C}H-\bar{C}H-CH=CH_2 \qquad (29)$$
$$\xrightarrow{CH_3OD} CD_2=CH-\bar{C}H-CH=CD_2$$

Such a correlation is reasonable, since in the proposed mechanism for exchange there is an endothermic proton transfer between the exchange reagent and the anion (13). As this process becomes

$$R_2CH^- + DOR \rightleftarrows \left[R_2CHD\cdot{}^-OR \right] \longrightarrow R_2CD^- + HOR \qquad (30)$$

more and more endothermic, the exchange rate should decrease.

This cannot be the only factor, however, since there are a number of anomolies. For example the anion of acetonitrile, which is 19 kcal/mole less basic than hydroxide, exchanges with D_2O while the acetylide ion, which is only 14 kcal/mole less basic, does not. Indeed the acetylide ion does not even exchange with CH_3OD even though the basicity difference in this case is only 4 kcal/mole.

$$\bar{C}H_2CN \xrightarrow{D_2O} \bar{C}D_2CN \qquad (31)$$

$$HC\equiv C^- \overset{D_2O}{\not\longrightarrow} \text{No exchange} \qquad (32)$$

Clearly if other factors which influence the exchange rate are to be identified, a more thorough kinetic investigation is necessary. This proved extremely difficult to do in a FA for at least two reasons. First most carbanions have a number of different types of hydrogens available for exchange, making it difficult to examine a single process. Secondly carbanions may undergo multiple exchanges in a single encounter with D_2O (14) as shown in eq. 33. Both these factors greatly complicate the kinetic analysis.

$$CH_2=CH-CH_2^- + D_2O \rightleftarrows \left[C_3H_4D^- \cdot HOD\right] \rightleftarrows \left[C_3H_3D_2^- \cdot H_2O\right]$$
$$\downarrow \qquad\qquad\qquad\qquad \downarrow (33)$$
$$C_3H_4D^- + HOD \qquad\qquad C_3H_3D_2^- + H_2O$$

By working in a SIFT, kinetic data can be obtained much more simply. In these experiments we inject DO^- into the flow tube in the absence of HO^- or of neutral precursors. Acids, RH, which are less acidic than water are added and the rate of conversion of DO^- to HO^- is measured (15). The pathway for exchange is given in eq. 34. Excellent kinetic data can be obtained. In

$$DO^- + HR \rightleftarrows \left[DO^- \cdot HR\right] \rightleftarrows \left[DOH \cdot R^-\right] \rightleftarrows \left[HO^- \cdot DR\right] \rightleftarrows HO^- + DR \quad (34)$$

selected cases deuterium isotope effects can be measured by injecting HO^- and allowing it to react with RD. Some selected rates are given in Table I.

Table I

$$DO^- + MH \rightarrow HO^- + MD$$

MH	$k_{obs} \times 10^{10}$	Reaction Efficiency	$\Delta|\Delta H^o_{acid}|$
H_2	0.379	0.024	9.6
NH_3	3.80	0.175	12.8
C_6H_6	6.65	0.338	~7
$H_2C=CH_2$	≤0.002	≤0.0001	>12.8
$H_2C=CH\diagdown C(CH_3)_3$	1.13	0.047	<12.8

Three factors have been identified as of primary importance in determining the rates of these exchange reactions (15). The

first of these, as mentioned before, is the relative acidity of
the anion and the exchanging neutral. The second is the
magnitude of the ion-dipole complex energy between DO^- and RH,
for it is this energy which is used to fuel the endothermic
proton transfer. Consider, for example, two molecules RH and R'H
of exactly the same acidity. Suppose one, RH, is highly
polarizable so that it is strongly attracted to DO^-, while the
other is not. Then in the first complex $[DO^- \cdot RH]$ there will be
more energy, and exchange will be likely to occur more rapidly
than in the second complex $[DO^- \cdot R'H]$. For example ammonia is
polarizable and has a dipole moment, while H_2 has no dipole
moment and is only weakly polarizable. So the $[DO^- \cdot NH_3]$ complex
will be effectively "hotter" and exchange can occur more rapidly
than in $[DO^- \cdot H_2]$.

Unfortunately there are not many pairs of neutrals, like H_2
and NH_3, which have the same or nearly the same acidity but which
differ greatly in their polarizability. However it is
instructive to compare ethylene with t-butylethylene. Ethylene

$$D_2N^- + CH_2{=}CH_2 \ \longrightarrow \ HDN^- + CH_2{=}CHD \quad k = 2.7{\times}10^{-11} \ cm^3 \ s^{-1} \quad (35)$$

$$DO^- \ + CH_2{=}CHC(CH_3)_3 \ \longrightarrow \ HO^- + CH_2{=}CDC(CH_3)_3 \ k = 1.1{\times}10^{-10} \ (36)$$

is about 5 kcal less acidic than ammonia yet exchanges only
slowly with D_2N^-. t-Butylethylene is about 10 kcal less acidic
than water, but exchanges more rapidly. Since ethylene is only
slightly polarizable there is little energy in its complex with
amide, and exchange is slow. Attachment of a t-butyl group
increases its polarizability so that in its complex with DO^-
there is more energy, and a more endothermic exchange can occur
more rapidly.

There is a third factor which can play an important role in
determining H-D exchange rates in the gas phase, namely the
dipole and induced dipole forces within the complex after
endothermic proton transfer. Again consider the cases of DO^-
exchanging with ammonia and hydrogen. DO^- and NH_3 are attracted
by forces of about 20 kcal/mole. There would be a comparable
amount of energy in a complex between DOH and NH_2^-; there is
little or no change in complexation energy associated with the
proton transfer, and the difference in energy between the two
complexes is reasonably approximated by the difference in
basicity between DO^- and NH_2^-. This situation is shown in a
potential energy diagram in Fig. III.

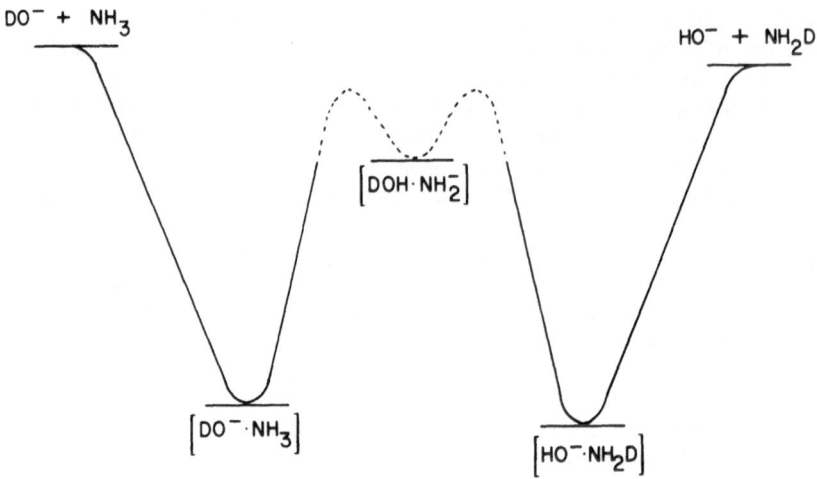

Figure III

DO$^-$ and H_2 are attracted by forces which amount to only a
few kcal/mole, not nearly enough to allow an endothermic proton
transfer of 10 kcal/mole. However as the proton is transfered
from H_2 to DO$^-$ there is a large increase in complexation energy
since H$^-$ is "solvated" by HOD, a polarizable molecule with a
dipole moment. This increase in complexation energy more than
compensates for the fact that H$^-$ is a stronger base than DO$^-$ and
leads to a potential energy surface like that given in Fig. IV.
Indeed we predicted that the ion H_3O^- would have the structure
H$^-$•H_2O rather than HO$^-$•H_2, and this has now been borne out by
experiment.

· · · · · · · · · · ·

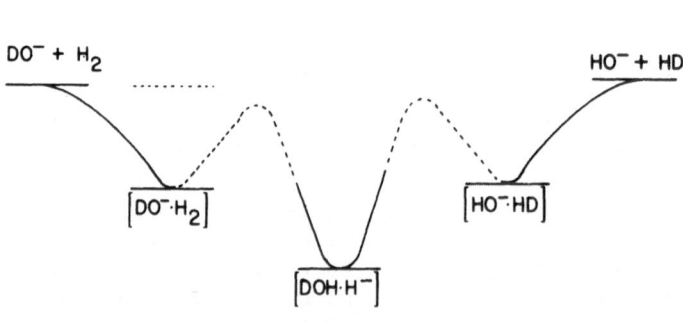

Figure IV

With these factors in mind we can now understand why the
acetonitrile anion exchanges with D_2O and the acetylide ion does
not. Consider the two key proton transfers (eq. 37,38). In the

$$\left[D_2O \cdot \bar{C}H_2CN \right] \quad \rightleftharpoons \quad \left[DO^- \cdot CH_2DCN \right] \tag{37}$$

$$\left[D_2O \cdot \bar{C} \equiv CH \right] \quad \rightleftharpoons \quad \left[DO^- \cdot DC \equiv CH \right] \tag{38}$$

first case acetonitrile is at least as good at stabilizing the complex as is D_2O; there is little or no change in dipole complex energy. In eq. (38), by contrast, because acetylene is so weakly polarizable the proton transfer is accompanied by a large decrease in complex energy making the equilibrium (38) much less favorable than (37).

Some recent SIFT results also appear to shed light on the role molecular complexity may play in extending the lifetime of a complex and hence allowing more time for exchange. We recently reported a study of multiple exchange within a single complex. The study was in two parts. In the first we looked for double incorporation during a single reaction of allyl anions with D_2O (eq. 39). When the anion and D_2O are similar in basicity we

$$\overset{X}{\underset{|}{CH_2}}=\overset{}{C}-CH_2^- + D_2O \rightleftharpoons \overset{X}{\underset{|}{CH_2}}=\overset{}{C}-CHD^- + HOD \rightleftharpoons \overset{X}{\underset{|}{CH_2}}=\overset{}{C}-CD_2^- + H_2O \tag{39}$$

found that on the average over two exchanges per encounter occurred. More unexpectedly, we found that more than two exchanges occurred on the average even in a proton abstraction reaction which is 11 kcal/mole exothermic (eq. 40).

$$\overset{O}{\underset{||}{HC}}-\overset{CH_2}{\underset{||}{C}}-CH_3 + DO^- \longrightarrow \left[\overset{O}{\underset{||}{HC}}-\overset{CH_2}{\underset{||}{C}}-CH_2^- \cdot HOD \right] \longrightarrow \longrightarrow \overset{O}{\underset{||}{HC}}-\overset{CH_2}{\underset{||}{C}}-CHD^- + H_2O \tag{40}$$

These results suggest that the ion dipole complex formed after proton transfer has a relatively long lifetime, probably due to the many vibrational and rotational modes available in the carbanion for storing the excess energy. Thus even though proton transfers to and from carbon are relatively slow, the lifetime of the complex is long enough to permit multiple exchanges.

We have found support for this view in some recent studies of multiple exchange with the amide and hydroxide ion (16), as shown in eq. 41 and 42. The first of these is an almost

$$D_2N^- + NH_3 \quad \overset{73\%}{\longrightarrow} \quad H_2N^- + NHD_2 \tag{41}$$
$$\overset{27\%}{\longrightarrow} \quad HDN^- + NH_2D$$

$$H_2N^- + D_2O \quad \overset{83\%}{\longrightarrow} \quad DO^- + NH_2D \tag{42}$$
$$\overset{17\%}{\longrightarrow} \quad HO^- + NHD_2$$

thermoneutral proton transfer which should be relatively fast
since transfer is to and from nitrogen. Nevertheless only a
relatively small amount (27%) of multiple exchange is seen. Even
less (17%) multiple exchange is seen in the exothermic (13 kcal)
proton transfer from water to amide ion. It seems probable that
the small size of the molecules results in very short lifetimes
for the complexes, and hence of relatively little multiple
exchange.

Drift Studies

We have recently carried out a few preliminiary drift
studies to see how the chemistry of some anions will be modified
as their kinetic energy increases (17). So far our most useful
observation is that the effective basicity of HO^- and H_2N^- can be
increased in a drift field. For example as we pointed out
earlier, DO^- fails to exchange with ethylene under the usual
flowing afterglow conditions. However when a potential is
applied first exchange and then some proton abstraction is
observed.

$$DO^- + CH_2=CH_2 \quad \xrightarrow{\sim 0.5eV} \quad HO^- + CH_2=CHD \qquad (43)$$

$$\xrightarrow{\sim 1.5eV} \quad HOD + CH_2=CH^- \qquad (44)$$

Amide ion exchanges with ethylene at zero field, but it can
be converted completely to the vinyl anion when given kinetic
energy. This has given us the first opportunity to carry out

$$H_2N^- + CH_2=CH_2 \longrightarrow CH_2=CH^- + NH_3 \qquad (45)$$

studies of the chemistry of the vinyl anion. One interesting
example is given in eq. 47.

$$CH_2=CH^- + N_2O \longrightarrow HC\equiv C^- + N_2 + H_2O \qquad (46)$$

CONCLUSION

It is clear that recently developed experimental techniques
have greatly extended the range of ion-molecule experiments which
can be undertaken. We are now in a position to prepare anions
less acidic than ammonia, to synthesize many previously unknown
anions, particularly of second row elements, and to examine
ion-molecule reactions in a detail never before attainable. As a
result of these experiments we hope to be able to gain a deeper
understanding of how organic reactions occur in solution, and how
we might set out to design new synthetic methods which will be
useful to the organic chemist carrying out reactions in solution.

Acknowledgement. I would especially like to acknowledge Dr. Veronica Bierbaum who is a full partner in all of the work reported here. In addition I am grateful to my other collaborators whose names are given in the references. Financial assistance has been provided by the National Science Foundation, the U. S. Army Research Office and the Petroleum Research Fund administered by the American Chemical Society. We gratefully acknowledge the generous support of these agencies.

References

1. Adams, N. G., and Smith, D.: 1976, Int. J. Mass Spectrom. Ion Phys. 21, pp. 349-359.
2. DePuy, C. H., and Bierbaum, V. M.: 1981, Acc. Chem. Res. 14, 146-153.
3. McFarland, M., Albritton, D. L., Fehsenfeld, F. C., Ferguson, E. E., and Schmeltekopf, A. L.: 1973, J. Chem. Phys. 59, pp. 6610-6619.
4. DePuy, C. H., Bierbaum, V. M., Flippin, L. A., Grabowski, J. J., King, G. K., Schmitt, R. J., and Sullivan, S. A.: 1980, J. Am. Chem. Soc. 102, pp. 5012-5015.
5. Bierbaum, V. M., DePuy, C. H., and Shapiro, R. H.: 1977, J. Am. Chem. Soc. 99, pp. 5800-5802.
6. DePuy, C. H., Grabowski, J. J., and Bierbaum, V. M.: To be submitted to J. Phys. Chem.
7. Dillard, J. G., and Franklin, J. L.: 1968, J. Chem. Phys. 48, pp. 2353-2358. DePuy, C. H., and Bierbaum, V. M.: 1981, Tetrahedron Lett. 51, pp. 5129-5130.
8. Fehsenfeld, F. C., and Ferguson, E. E.: 1974, J. Chem. Phys. 61, pp. 3181-3193.
9. Bierbaum, V. M., Grabowski, J. J., and DePuy, C. H.: Unpublished results.
10. Anderson, D. R., Bierbaum, V. M., and DePuy, C. H.: Submitted to J. Am. Chem. Soc.
11. Damrauer, R., DePuy, C. H., and Bierbaum, V. M.: Organometallics, in press.
12. Stewart, J. H.; Shapiro, R. H., DePuy, C. H., and Bierbaum, V. M.: 1977, J. Am. Chem. Soc. 99, pp. 7650-7653.
13. Bartmess, J. E., and McIver, R. T., Jr.: 1979, in "Gas Phase Ion Chemistry" (Bowers, M. T., Ed.), Vol. 2, Academic Press, New York, Chapter 11, pp. 87-121.
14. Squires, R. R., DePuy, C. H., and Bierbaum, V. M.: 1981, J. Am. Chem. Soc. 103, pp. 4256-4258.
15. Grabowski, J. J., DePuy, C. H., and Bierbaum, V. M.: J. Am. Chem. Soc. in press.
16. Grabowski, J. J.: Unpublished results.
17. Grabowski, J. J., Bierbaum, V. M., and DePuy, C. H.: Unpublished results.

DETERMINATIONS OF GAS-PHASE ORGANIC ION STRUCTURES

G.S. Groenewold and M.L. Gross

Department of Chemistry, University of Nebraska,
Lincoln, Nebraska, U.S.A. 68588

Selected ion structure investigations are reviewed in this paper. The investigations illustrate the use of a variety of instrumental techniques, which include ICR mass spectrometry and several types of ion decomposition analyses, including unimolecular, collision induced and photo induced reactions. The isomeric ion systems described are $C_3H_6O^{+\cdot}$, $C_3H_6O_2^{+\cdot}$, $C_2H_5O^+$, $C_2H_2^{-\cdot}$, $C_7H_8^{+\cdot}$ and $C_6H_6O^{+\cdot}$. In addition, investigations of the $C_3H_6^{+\cdot}$, $C_3H_5^+$, $C_6H_6^{+\cdot}$, $C_8H_8^{+\cdot}$ and $C_4H_6^{+\cdot}$ structures have been carried out with a newer technique in which ions are derivatized in the gas phase and the ionic derivatives are analyzed by collision induced dissociation spectrometry.

INTRODUCTION

Gas-phase organic ion structure determinations have been actively pursued since the late 1960's. The subject has been a focus of research because gas-phase ion structure determines ion reactivity, and an understanding of reactivity is critical in both analytical and organic mass spectrometry. These studies are also of fundamental interest, because the structures of ions (and their reactions) are studied under conditions of very low pressure without being influenced by solvent. One consequence of this experimental condition is that ions which might be unstable in solution can be observed as stable structures in the mass spectrometer.

Although attempts have been made to do spectroscopy on gas-phase organic ions, the use of such techniques is not widespread. Alternatively, ion structures have been assigned

243

M. A. Almoster Ferreira (ed.), Ionic Processes in the Gas Phase, 243–265.
© 1984 by D. Reidel Publishing Company.

by comparing the reactivity of ions generated from isomeric precursors. This method depends on the assumption that ion reactivity is reflective of structure. As a result of the comparative nature of the experiment, structural analyses which rely on reactivity must be considered to be indirect.

There are four basic ways to compare ion structure and reactivity. The first three methods rely on monitoring ion decompositions which occur either unimolecularly (metastables), following collisional activation (CA) or photoactivation. The fourth method employs ion-molecule reactions for the determination of ion structure. Here the neutral molecule may be thought of as a derivatizing agent. This "derivatizing agent" serves to reveal the chemical properties of an unknown ion. These properties, which include both ion-molecule reaction pathways and reaction rates, are used to measure reactivity and to deduce structure.

Most magnetic sector mass spectrometers may be used for investigations of ion structure by monitoring unimolecular decompositions. A limitation of the unimolecular decomposition analysis is that it is influenced by variations in the amount of internal energy of the analyte ion. The structural and analytical sensitivity of the decomposition analysis may be enhanced by colliding the ion of interest with a neutral gas such as helium. The resulting collision induced dissociation (CID) spectra are more representative of the ion structure than are the unimolecular decompositions. In CID processes, effects of the internal energy of the ion are small, and the resulting spectra usually reflect the structure of the ion regardless of its internal energy.

Ion decompositions (CID and unimolecular) are most easily studied using the technique of mass spectrometry/mass spectrometery (MS/MS)(1). An instrument particularly well suited is a triple sector mass spectrometer which has the advantage of high mass resolution (> 1000) mass selection of the ion of interest prior to analysis of decomposition products(2). Recently, Frieser and Cody have also monitored collision induced decompositions using a Fourier transform mass spectrometer (FTMS), a technique which also may prove to be valuable in the determination of ion structure(3).

Ion structures, which have been determined by comparisons of their ion-molecule reactivity, have traditionally been studied using ion cyclotron resonance mass spectrometry. This type of spectrometers' unique double resonance feature allows structurally characteristic reaction pathways to be identified unequivocably. Furthermore, the spectrometers' capability to monitor ion intensity as a function of ion lifetime allows precise determination of ion-molecule reaction rate constants, which can be used as structural fingerprints.

Recently, ion structures have been studied using a high pressure chemical ionization (CI) source together with a tandem mass spectrometer(4). This technique combines features of ion structure determination by comparative ion-molecule reactions (now conducted in a CI source) and by ion decomposition (using the tandem mass spectrometer in an MS/MS mode). Carrying out the reaction at high pressure (1 torr) allows collisional stabilization of the ion-molecule adduct of interest, enabling structurally diagnostic ion-molecule derivatization reactions to be observed for distinguishing unique isomers which otherwise give similar decomposition spectra. The method may suffer as a result of competing side reactions, but the collisional stabilization feature makes it a valuable complement to ICR mass spectrometry for the study of ion-molecule reactions.

This report is a review of selected applications of the above mentioned techniques of ion structure determination. It is not a comprehensive review, and examples have been included because they are early and illustrative. Furthermore, many of the investigations described involve an interdisciplinary approach, which lends credibility to both the decomposition and the kinetic methods used. Finally, some applications are described of the new ion structure method which combines CID and ion-molecule reaction chemistry.

$C_3H_6O^{+\cdot}$

The question of the structures of the McLafferty rearrangement product ion and of the double McLafferty rearrangement product ion prompted Djerassi and coworkers to undertake an ion structure study of C_3H_6O radical cations using ICR mass spectrometry(5). They found that C_3H_6O cations originating from the ionization of acetone and the McLafferty rearrangement of methyl alkyl ketones underwent specific ion-molecule reactions which were consistent with proposed keto and enol structures. For example, the keto ion resulting from the ionization of acetone underwent a charge exchange with 2-hexanone (equation 1) and reacted with neutral acetone to form a condensation product which eliminated a methyl radical (equation 2). In contrast, the enol ion, resulting from the

$$\overset{O}{\underset{}{\Lambda}}{}^{+\cdot} \; + \; \overset{O}{\underset{}{\Lambda}}\!\!\diagdown\!\!\diagup \; \longrightarrow \; \overset{O}{\underset{}{\Lambda}}\!\!\diagdown\!\!\diagup{}^{+\cdot} \; + \; \overset{O}{\underset{}{\Lambda}} \qquad (1)$$

$$\text{(2)}$$

McLafferty elimination of propene from 2-hexanone, protonated neutral 2-hexanone (equation $\underline{3}$). It was found that the $C_3H_6O^{+\cdot}$ enol could also be formed from the elimination of C_2H_4 from 1-methyl-cyclobutan-1-ol. The enol ion then reacted with the neutral parent to form a cation condensation product which eliminated H_2O and C_2H_4 (equation $\underline{4}$) and H_2O and $\cdot CH_3$ (equation $\underline{5}$).

$$\text{(3)}$$

$$\text{(4)}$$

$$\text{(5)}$$

An important conclusion of this work was that only the hydrogen attched to the oxygen is capable of protonating a neutral ketone. This constitutes a powerful reaction for distinguishing between enol and keto structures. These techniques were also extended to the C_3H_6O radical cation resulting from the double McLafferty rearrangement of 5-nonanone. It was concluded that the ion had the enol structure. This work proved to be a benchmark example of the utility of diagnostic reactions studied by ICR for comparing ion structure.

$C_3H_6O_2^{+}$

Recently, Hemberger(6) (and what must be a record number of internationally prominent collaborators) used CID to establish the isomerization of methyl isobutyrate to an enol structure prior to methyl and ethylene eliminations (equation $\underline{6}$). The ion structure $\underline{a}$ was verified by comparison of its CID

$$\text{(6)}$$

spectrum with "authentic" keto and enol structures. The enol was the product of a McLafferty elimination of C_2H_4 from methyl butyrate and the keto was ionized methyl acetate.

The result was also verified by comparing the relative ion-molecule reaction rate constants for the reaction of the unknown and benchmark ions with a neutral ketone. Ion a and the "authentic" enol were observed to protonate acetone at the same dk/dE (as determined by double resonance, where k is the bimolecular rate constant and E is the translational energy of the ion). The protonation of a ketone by an ionic enol is consistent with Djerassi's earlier interpretations(5).

$C_2H_5O^+$

The investigations of the structures of isomeric C_2H_5O cations are illustrative of several structure elucidation techniques. Beauchamp and Dunbar employed ion-molecule reactions in an ICR mass spectrometer to identify reactions characteristic of the methoxymethyl (b) and protonated acetaldehyde (c) structures(7). They found that ion b (not c) underwent hydride and methyl cation transfer reactions with its neutral parent, methyl ethyl ether (equation 7). In addition,

$$CH_3-\overset{+}{O}=CH_2 \ + \ \text{\includegraphics{}} \ \longrightarrow \ \begin{cases} CH_3-O-CH_3 \ + \ CH_3-\overset{+}{O}=CH-CH_3 \\ \\ \text{\includegraphics{}} \ + \ H_2C=O \end{cases}$$

b

(7)

the protonated acetaldehyde ion c could be distinguished from b by its reactions with neutral 2-propanol, which yielded condensation products by eliminating H_2O and C_3H_6 (equation 8).

$$\overset{+OH}{\underset{H}{\text{\includegraphics{}}}} \ + \ \overset{OH}{\text{\includegraphics{}}} \ \longrightarrow \ \begin{cases} CH_3-CH=\overset{+}{O}-\text{\includegraphics{}} \ + \ H_2O \\ \\ CH_3-CH=O\cdots\overset{+}{H}\cdots OH_2 \ + \ C_3H_6 \end{cases}$$

c

(8)

The workers were not able to distinguish the protonated ethylene oxide structure d from the protonated acetaldehyde c using the above ion-molecule reactions. A later communication(8) showed that the two ions could be distinguished on the basis of their ion-molecule reactions with neutral phosphine (equation 9) and hydrogen sulfide. The ion d reacts readily with phosphine (or H_2S) to yield an adduct ion and neutral water. In contrast the protonated acetaldehyde ion structure is inert to both reagents.

$$\overset{+}{\underset{\underline{d}}{\triangle}}\text{OH} + \text{PH}_3 \longrightarrow \left[\text{HO} \overset{\cdot\cdot\cdot\text{H}\cdot\cdot}{\triangle} \text{PH}_2\right]^+ \longrightarrow \overset{+}{\triangle}\text{PH}_2 + \text{H}_2\text{O} \qquad (\underline{9})$$

A fourth $C_2H_5O^+$ isomer was generated by a "charge reversal" process and studied in a tandem mass spectrometer by CID spectroscopy(9). A $CH_3CH_2O^-$ ion was first formed in a high pressure source by H^+ abstraction from ethanol and was selected using MS-I of the tandem instrument. Collisional charge stripping of two electrons produced the C_2H_5O cation, which was then analyzed by CID. The CID spectrum generated by this experiment was significantly different from the spectra of conventionally produced methoxymethyl (b) and protonated acetaldehyde (c) $C_2H_5O^+$ isomers. The CID spectra of the new $C_2H_5O^+$ isomer contained important signals corresponding to the losses of O· and CH_2=O, which were consistent with the proposed $CH_3CH_2O^+$ ion structure e (equation 10). The charge reversal technique was the only way to generate and analyze this isomer, because the isomer rearranges too rapidly prior to analysis when formed by other means.

$$CH_3CH_2O^- \xrightarrow[\text{reversal}]{\text{charge}} \underset{e}{CH_3CH_2O^+} \begin{cases} \xrightarrow{\text{CID}} C_2H_5^+ + O\cdot \\ \\ \xrightarrow{\text{CID}} CH_3^+ + H_2C=O \end{cases} \qquad (\underline{10})$$

$C_2H_2^{\overline{\cdot}}$

The determination of the ion structure of the C_2H_2 radical anion was unique because the structure was inferred from evidence taken from the method of formation of the ion. The $O^{\overline{\cdot}}$ reagent ion was generated by EI and reacted with ethylene by abstracting two hydrogens to yield $C_2H_2^{\overline{\cdot}}$ and neutral $H_2O(10)$. Results of studies of deuterium and fluorine substituted neutral ethylene parents demonstrated that both of the leaving hydrogen atoms originate from the same carbon, establishing the structure of the product ion f (equation 11). This structural conclusion was supported by the later work of Nibbering and Dawson(11) who used MINDO/3 calculations to determine the electron affinities of $H_2C=C$: (EA = 0.0eV) and $HC\equiv CH$ (EA = -1.8eV). These indicate that acetylenic molecular anions should not be stable.

$$H_2C=CD_2 + O^{\overline{\cdot}} \longrightarrow \begin{cases} \longrightarrow H_2C=C^{\overline{\cdot}} + D_2O \\ \\ \longrightarrow {}^{\overline{\cdot}}C=CD_2 + H_2O \\ \\ \nrightarrow HC\equiv CD^{\overline{\cdot}} + DOH \end{cases} \qquad (\underline{11})$$

Additionally, the $H_2C=C^{\cdot-}$ undergoes further reaction with N_2O (equation 12) which the ion having the acetylenic structure would not be expected to do. This method has been used to form

$$H_2C=C^{\cdot-} + N\overset{+}{\equiv}N\overset{-}{-}O \longrightarrow H_2C=C=N^- + NO\cdot \qquad (12)$$

study examples of $R-CH=C^{\cdot-}$ ions(10), using olefins such as methoxyethylene and 1,1-difluoro ethylene (see equations 13 and 14). No $[M-H_2]^{\cdot-}$ was observed in the case of propene (equation 15).

$$D_3C-O-CH=CH_2 + O^{\cdot-} \longrightarrow D_3C-O-CH=C^{\cdot-} + H_2O \qquad (13)$$

$$F_2C=CH_2 + O^{\cdot-} \longrightarrow F_2C=C^{\cdot-} + H_2O \qquad (14)$$

$$H_3C-CH=CH_2 + O^{\cdot-} \overset{/}{\longrightarrow\!\!\!/} H_3C-CH=C^{\cdot-} + H_2O \qquad (15)$$

$C_7H_8^{+\cdot}$

The structures of various $C_7H_8^{+\cdot}$ ions constitute an intriguing problem whose investigation has required a variety of strategies and techniques. Based on a consideration of the possible mechanisms for the propene elimination from n-butylbenzene, two possible $C_7H_8^{+\cdot}$ isomers were suggested (equations 16 and 17). Bursey, Hoffman, and Benezra(12)

compared the $C_7H_8^{+\cdot}$ derived from n-butylbenzene to three isomeric radical cations from model compounds (toluene, cycloheptatriene, and norbornadiene) by reacting each of the cations with an alkyl nitrate in an ICR. Double resonance experiments were interpreted to indicate that this reaction was specific for C_7H_8 cations having the toluene structure (equations 18-20). The $C_7H_8^{+\cdot}$ from n-butylbenzene did not react with the alkyl nitrate (as did toluene), and the unknown ion was assigned the methylene cyclohexadiene structure g. Unfortunately, these double resonance experiments were misleading for reasons to be discussed below.

$$\text{(toluene radical cation, } CH_3) + CH_3ONO_2 \longrightarrow CH_3C_6H_5NO_2^+ + CH_3O\cdot \quad (\underline{18})$$

$$\text{(cycloheptatriene radical cation)} + CH_3ONO_2 \longrightarrow \text{No Reaction} \quad (\underline{19})$$

$$\text{(norbornadiene radical cation)} + CH_3ONO_2 \longrightarrow \text{No Reaction} \quad (\underline{20})$$

Because the ICR experiments only permitted analysis of low energy C_7H_8 cations and provided only negative information of the structure of the $C_7H_8^{\ddagger}$ derived from n-butylbenzene‡, various deuterated $C_7(H,D)_8^{\ddagger}$ isomers of higher energy were studied by examining unimolecular, collision induced and source-occurring decompositions(13). Analysis of partially deuterated $C_7(H,D)_8^{\ddagger}$ formed from 2-phenylethanol‡ showed the cation hydrogen atoms to be less scrambled at higher internal energies, indicating that methylene cyclohexadiene‡ $\underline{g}$ was initially formed. However, at longer (metastable) times, the deuterium and hydrogen atoms in the $C_7(H,D)_8^{\ddagger}$ product became completely scrambled. This was interpreted as methylene cyclohexadiene $\underline{g}$ isomerizing to cycloheptatriene and toluene, structures which also displayed complete H/D scrambling (equation $\underline{21}$).

$$\text{(methylenecyclohexadiene radical cation } \underline{g}) \longrightarrow \text{(toluene radical cation } CH_3) + \text{(cycloheptatriene radical cation)} \quad (\underline{21})$$

Photodissociation spectroscopy was used to further advance the understanding of the $C_7H_8^{\ddagger}$ system. Using this technique, the wavelength dependence of the disappearance (by fragmentation) of the non-decomposing C_7H_8 radical cations which are trapped in an ICR cell was determined(14). The cation spectra that were produced showed clearly that the $C_7H_8^{\ddagger}$ from n-butylbenzene and 2-phenylethanol were not the same as the toluene or cycloheptatriene radical cations. Although the experiment was not rigorously conclusive, the spectrum of the unknown $C_7H_8^{\ddagger}$ was consistent with the spectrum predicted for a triene such as methylenecyclohexadiene $\underline{g}$.

When Ausloos and Lias(15) reapplied the ICR double resonance technique to the alkyl nitrate reaction (equation $\underline{18}$), they found that the product $CD_3C_6D_5NO_2^+$ resulting when toluene-d_8 was used showed no dependence on the $C_7D_8^{\ddagger}$ (in contrast to the behavior of $CH_3C_6H_5^{\ddagger}$). The double resonance decrease in m/z 138 ($CH_3C_6H_5NO_2^+$) observed originally while irradiating m/z 92 ($C_7H_8^{+}$) was actually a harmonic effect (because 92 and 138 are multiples of 46) and would have been observed even in the absence of a m/z 92 ion. This work demonstrated that the NO_2 abstraction mechanism (equation 18) was incorrect. A more thorough investigation showed that the $C_7(H,D)_8NO_2^+$ products were actually formed by a solvent "switching" mechanism in which the NO_2^+ is considered to be solvated first by the CH_3ONO_2 and finally by the toluene (equations 22-24).

$$C_2H_5ONO_2^{+} \longrightarrow CH_2ONO_2^+ + CH_3 \cdot \qquad (22)$$

$$CH_2ONO_2^+ + C_2H_5ONO_2 \longrightarrow C_2H_5ONO_2 \cdot NO_2^+ + H_2CO \qquad (23)$$

$$C_2H_5ONO_2 \cdot NO_2^+ + CD_3C_6D_5 \longrightarrow CD_3C_6D_5NO_2^+ + C_2H_5ONO_2 \qquad (24)$$

The switching reaction was determined to occur only when a competing charge exchange reaction was not sufficiently exothermic. For example, toluene and benzene (which have ionization potentials (IP) values of 8.8 and 9.2 eV, respectively) underwent the switching reaction, but xylene (which has a substantially lower IP of 8.2 eV) participated only in charge exchange with the NO_2^+. Similarly, cycloheptatriene and norbornadiene underwent only charge exchange because they have lower IP values of 8.2 and 8.42 eV respectively.

Ausloos and Lias(15) also determined that the $C_7H_8^{+}$ structure could be investigated by looking at charge exchange reactions with the neutral molecule, n-butylbenzene. Since n-butylbenzene has an IP of 8.7 eV, any $C_7H_8^{+}$ with a higher IP will charge exchange with neutral n-butylbenzene. Using this technique, the toluene^{+} structure (IP = 8.8 eV) was easily distinguished from that of norbornadiene^{+} and cycloheptatriene^{+} (IP = 8.42 and 8.2 eV, respectively). The $C_7H_8^{+}$ formed in the rearrangement of n-butylbenzene^{+} charge exchanged much more slowly with neutral n-butylbenzene than did toluene^{+}. This result showed that: 1) >98% of the unknown $C_7H_8^{+}$ formed from the n-butylbenzene has the same structure; 2) this structure is not toluene^{+}; and 3) the structure has an IP which is similar to that of n-butylbenzene (8.7 eV). Furthermore, the methylene cyclohexadiene structure g was ruled out because it is expected to have a significantly lower IP based on a comparison with an analogous compound, methylene cyclopentadiene (IP = 8.36 eV). Alternatively, ring opened or protonated tolyl structures were suggested as possible structures for the $C_7H_8^{+}$. The precise structure of this radical cation remains an open question.

$C_6H_6O^{+\cdot}$

ICR studies, CID, and photodissociation have been used to gain insight into the structures of the $C_6H_6O^{+\cdot}$ isomers produced by ionic fragmentation. Tomer and Djerassi(16) used the neutral 1-methylcyclobutanol in the ICR as a structurally diagnostic reagent for the phenol$^{+\cdot}$ ("enol") structure (equation 25). The unknown $C_6H_6O^{+\cdot}$ structure resulting from the ketene elimination from bicyclo[2.2.2]oct-2-en-5,7-dione h underwent similar reactions, indicating that the ion was a mixture of the anticipated cyclohexadienone ("keto") structure i and phenol (equation 26). In addition, the 1-methylcyclobutanol reaction of deuterium labelled C_6H_6O cations indicated that the $C_6H_6O^{+\cdot}$ resulting from the ketene elimination from phenyl acetate$^{+\cdot}$ had the phenol$^{+\cdot}$ structure (equation 27).

$$(25)$$

$$(26)$$

$$(27)$$

ICR proton (deuteron) transfer reactions from $C_6H_5DO^{+\cdot}$ to the base 4-t-butylpyridine were used by Theissling and Nibbering(17) to show that the $C_6H_5DO^{+\cdot}$ produced from α-d_2-phenoxyethyl halides consists of two structures (equation 28). The phenol-OD$^{+\cdot}$ structure transfered only a deuteron to the

$$(28)$$

X = F,Cl,Br

base, in contrast to the cyclohexadienone which transfered similar amounts of D and H. The phenoxyethyl flouride proved to be the most revealing of the halides since it was the only compound of the series not to undergo a phenoxy-halogen positional interchange.

The structures of the $C_6(H,D)_6O^{+}$ resulting from the same compounds were also analyzed by CID spectroscopy(18). The spectra of the $C_6H_6O^{+}$ resulting from the ionization of phenol and the decomposition of bicyclo[2.2.2]oct-2-en-5,7-dione h showed clearly that both the phenol^{+} and the cyclohexadienone^{+} i exist as stable species. The study also showed that in the case of phenoxyethyl flouride, the $C_6H_6O^{+}$ derived from the transfer of a hydrogen atom originating from the beta position (that is, adjacent to the oxygen atom, see j) was identical to the authentic phenol radical cation based on a comparison of their CID spectra. In the case of the $C_6H_6O^{+}$ derived from the transfer of a hydrogen atom originating from the alpha position, the CID spectrum appeared to be a mixture of a trace of cyclohexadienone^{+} i and predominantly phenol^{+}. The chloride and bromide differed from the fluoride in that the CID spectra of the $C_6H_6O^{+}$ from these compounds included a significantly enhanced proportion of the cyclohexadienone^{+} admixed with the phenol structure. Although there were quantitative differences between the interpretation of the ICR and the CID data, the CID experiments offered good qualitative support for the conclusions drawn from proton transfer ion-molecule reactions.

More recently, $C_6H_6O^{+}$ was investigated using the photodissociation ICR technique(19). Photodissociation spectra were generated from determinations of the rate constant for equation 29 versus the irradiation wavelength. The spectra

$$C_6H_6O^{+} + h\nu \longrightarrow C_5H_6^{+} + CO \qquad (29)$$

of $C_6H_6O^{+}$ from four different sources showed clearly the existence of two unique isomers. Additional information was obtained by plotting the percentage of photodissociating ions versus the irradiation time. The C_6H_6O ions from phenetole and phenol were homogeneous populations of the phenol^{+} structure. In contrast, $C_6H_6O^{+}$ resulting from bicyclo[2.2.2]oct-2-en-5,7-dione and 2-phenoxyethyl chloride were demonstrated to be mixtures of the phenol and cyclohexadienone structures in accord with the interpretations of results from ion-molecule reactivity comparisons and from CID. Although there are minor differences in the relative proportions of the two structures, interpretations based on all the techniques converge on the conclusion that a mixture of structures are formed in the dissociation of phenoxy ethyl halides.

$C_3H_6^{\ddagger}$

A large research effort has been focussed of the question of whether cyclopropane‡ ring opens or remains ring intact. The research has employed specific ion-molecule reactions, collisional charge stripping and molecular orbital calculations. Ion-molecule reactions studied by ICR were used to show that the cyclopropane‡ reacted with neutral ammonia to form condensation products by the elimination of ethylene and ethyl· from the activated complex (equation $\underline{30}$)(20). The cyclopropane‡ structure was differentiated from the propene‡, which did not react with ammonia (equation $\underline{31}$). C_3H_6 radical

$$\triangleright^{\ddagger} + NH_3 \rightleftharpoons \left[\triangleright^{+NH_3}\right] \xrightarrow[\quad]{\quad} \begin{array}{l} \xrightarrow{-C_2H_4} \; {}^{+}H_3N{-}CH_2\cdot \\[2mm] \qquad\qquad \downarrow -H \\[2mm] \xrightarrow{-C_2H_5\cdot} \; {}^{+}H_2N{=}CH_2 \end{array} \qquad (\underline{30})$$

$$\diagup\!\!\diagdown^{\ddagger} + NH_3 \rightleftharpoons \left[\triangle^{+NH_3}\right] \longrightarrow \text{No Reaction} \qquad (\underline{31})$$

cations which were generated from ionized tetrahydrofuran, oxepane and cyclohexanone also reacted as ionized cyclopropane. Deuterium labelling of the tetrahydrofuran precursor was used together with the ammonia reaction to prove that the reacting $C_3H_6^{\ddagger}$ had retained the cyclopropane structure (equation $\underline{32}$). The $C_3H_4D_2^{\ddagger}$ transferred both CH_2 and CD_2 to ammonia in a ratio of 2:1, which was predicted for the cyclopropane‡ structure. The alternative trimethylene radical cation produced directly (i.e., without equilibration with the cyclic structure) would be expected to show a corresponding ratio of 1:1.

$$\text{(diagram)} \qquad (\underline{32})$$

The isomeric $C_3H_6^{\ddagger}$ ions were also studied in terms of their reactivity with NH_3 using photoionization to form the ions(21). This work showed that a fraction to the of the ionized cyclopropane radical cations protonated ammonia in addition to undergoing a condensation reaction (equation $\underline{30}$). Since the protonation of ammonia was shown to be a characteristic reaction of propene‡, these results indicated that some of the $C_3H_6^{\ddagger}$ from cyclopropane existed in a ring-opened structure, even at the lowest ionizing energies studied. Further, the fraction of ring-opened structures increased with increasing ionizing energies.

Photodissociation ICR spectrometry was used by van Velzen and van der Hart to investigate the structures resulting from cyclopropane radical cation(22). Most of the radical cations were found to photodissociate by loss of ·H. The results again indicated that cyclopropane‡ exists as a mixture of ring-intact and ring-opened (propene) structures, and that at higher internal energies propene becomes the major component of the mixture.

The structures of the $C_3H_6^{\ddagger}$ isomers were further probed by collisional charge stripping using tandem mass spectrometers. The results again were in accord with the interpretation that the cyclopropane‡ and the propene‡ were unique structures(23), but the phenomenon of structure populations varying with internal energy was not addressed. This aspect of the problem has recently been attacked by studying the doubly charged spectra at low ionizing energy and at high MS-I resolution(24). The high resolution was used to resolve the ^{13}C isotope of the $C_3H_5^{+}$ (i.e., $^{13}CC_2H_5^{+}$) from the $C_3H_6^{\ddagger}$ mass. To achieve a reasonable ion production at low ionizing voltage, the source of the mass spectrometer was modified by placing grids between the filament and the cage and between the trap and the cage. The results of these experiments showed clearly that: 1) $C_3H_6^{\ddagger}$ from cyclopropane and propene are unique at all ionizing energies; 2) the structure of $C_3H_6^{\ddagger}$ from propene is independent of ionizing energy; and 3) the spectra of $C_3H_6^{\ddagger}$ ions from cyclopropane vary consistently with ionizing energy, suggesting two structures, neither of which is the propene‡. The observation of at least two structures whose proportions vary with ionizing energy is in accord with the photoionization(21) and photodissociation(22) work cited previously. The ab initio calculations of Collins and Gallup(25) support an interpretation that two of the observed structures resulting from ionized cyclopropane are cyclopropane‡ and trimethylene‡, ·CH$_2$CH$_2$CH$_2$+ .

Similar structures have been observed in the investigations of substituted cyclopropanes and heterocyclic compounds. The heterocyclic cyclopropane radical cation investigations constitute another example of the use of ion-molecule reactions studied by ICR spectrometry for the characterization of ion structures(26). Phosphirane and thiirane k radical cation transfer heteroatoms and methylene groups to ammonia (equations 33 and 34). Aziridine, on the other hand, undergoes neither of these reactions (equation 35). The results of these reactions were interpreted in terms of the thiirane and phosphirane retaining their cyclic structures whereas the aziridine radical cation undergoes ring opening.

$$\text{\raisebox{0.5ex}{$\triangleright$}}S_k^{\dot{+}} + NH_3 \longrightarrow SNH_3^{\dot{+}} + C_2H_4 \tag{33}$$

$$\text{\raisebox{0.5ex}{$\triangleright$}}S_k^{\dot{+}} + NH_3 \longrightarrow H_2C=NH_2^+ + CH_3S\cdot \tag{34}$$

$$\text{\raisebox{0.5ex}{$\triangleright$}}NH \xrightarrow{e^-} C_2H_5N^{\dot{+}} + NH_3 \longrightarrow \text{No } CH_2 \text{ Transfer} \tag{35}$$

$C_3H_5^+$

C$_3$H$_5$ cations from a variety of parent compounds have been studied to determine if various $C_3H_5^+$ isomers existed and if they can be distinguished. CID analyses of C$_3$H$_5$ cations from ionized C$_3$H$_5$Br parents demonstrated that two stable C$_3$H$_5$ isomers exist, which were assumed to be the allyl ($\underline{1}$) and 2-propenyl ($\underline{m}$) structures(27). CID spectra determined by other workers were interpreted to show that the 1-propenyl cation ($\underline{n}$) was unstable and existed as a mixture of the first two structures $\underline{1}$ and $\underline{m}$(28). Radiolysis and proton transfer reactions were interpreted in terms of a third, stable cyclic $C_3H_5^+$ structure $\underline{o}$(29).

$$\underline{1} \qquad\qquad \underline{m} \qquad\qquad \underline{n} \qquad\qquad \underline{o}$$

The differences in the CID spectra of the $C_3H_5^+$ isomers are small; hence the analysis is not as convincing as one would hope. In addition, interpretations of CID and theoretical studies(30) of the cyclic $C_3H_5^+$ do not agree with those drawn from the proton transfer and radiolysis experiments mentioned above. For these reasons, a more unequivocal analysis technique was sought. Initially, possible structure specific ion-molecule reactions were investigated in an ICR spectrometer(31). Double resonance showed that $C_3H_5^+$ reacted with neutral aromatics to transfer a CH group. The reactions, however, were not diagnostic for distinguishing the $C_3H_5^+$ isomers because the $C_3H_5^+$ ions generated from allyl halides, cyclopropyl halides, and 1- and 2-propenyl halides all reacted with benzene to give $C_7H_7^+$.

A potentially more structurally characteristic alternative experiment involves a "derivatization" of the gas-phase C$_3$H$_5$ cations with neutral reagent gases, followed by collisional

stabilization of the derivative in a high pressure source, and
finally CID analysis by tandem mass spectrometry (MS/MS)(32).
The experimental strategy was first suggested by Dymerski and
McLafferty(33). The expectation is the technique will enhance
CID differences of isomeric ions which, if underivatized, result
in nearly identical CID spectra. The technique may be compared
to a laboratory synthesis, in which the high pressure chemical
ionization source functions as the reaction vessel, MS-I serves
to isolate the product ion, and MS-II produces a structurally
characteristic mass spectrum of the product ion.

The advantage of running the reactions in a high pressure
chemical ionization source rather than a low pressure ICR is
that the ion-molecule adducts formed in exothermic reactions may
be collisionally stabilized to the extent that they can be
observed as stable ions in the mass spectrometer. This feature
allows ion structure assignments to be made based on
comparisons with synthetically generated model compounds.
Thus, the approach should yield more direct information on ion
structure than simple comparisons of ion reactivity.   The
technique . is limited by the lack of the double resonance
experiment which is unique to ICR spectrometry.

The ion derivatization chosen for the C_3H_5 cations
involved reacting them with neutral benzene (equation $\underline{36}$) to
give a $C_9H_{11}^+$ adduct. This adduct was not observed under the
lower pressure conditions of the ICR; instead only $C_7H_7^+$ was
detected.   The CID spectra of the stabilized $C_9H_{11}^+$ adducts

$$C_3H_5^+ + C_6H_6 \longrightarrow C_9H_{11}^+ \qquad\qquad (\underline{36})$$

originating in reactions of $C_3H_5^+$ from the four isomeric C_3H_5X
precursors showed enhanced spectral differences relative to the
spectra of the underivatized cations. Based on the CID spectra
of the derivatives, it was possible to demonstrate more
convincingly that the allyl and 2-propenyl cations could be
distinguished, and the 1-propenyl cation appears to be a mixture
of the allyl$^+$ and 2-propenyl$^+$ isomers, although the existence of
a third, stable $C_3H_5^+$ isomer cannot be rigorously ruled out.
In addition, the CID spectrum of the derivatized cyclopropyl
cation shows that it is very similar to both of the derivatized
allyl cations, indicating that the cyclopropyl cation has ring
opened prior to reaction with neutral benzene.

Although the above comparisons are useful for
distinguishing the $C_3H_5^+$ isomers, they do little to reveal the
structures of the isomers. A more direct structure assignment
was made by comparing the CID spectra of the $C_9H_{11}^+$ adducts to
those of protonated C_9H_{10} model compounds. For example, the CID
spectrum of the [2-propenyl cation + C_6H_6] adduct is very

similar to that of protonated 2-propenyl benzene. The lack of an intense C_3H_4 loss to give m/z 79 in the CID spectra of other protonated model isomers eliminates them as possible structures. The presence of the C_3H_4 loss identifies the protonated α-methylstyrene structure p, and is strong evidence for the 2-propenyl$^+$ structure (equation 37). The [allyl cation + C_6H_6] adduct is similarly identified as protonated allyl benzene q (equation 38). In all cases, the cyclopropyl cation behaved identically to the allyl cation. The CID spectra of the ion-molecule reaction products of the $C_3H_5^+$ from the 1-propenyl halides reacting with C_6H_6, C_6D_6 and phenol all indicate that this isomer exists as approximately a 2:1 mixture of the allyl and 2-propenyl cation structures.

$$\text{(37)}$$

$$\text{(38)}$$

$C_6H_6^{+}$

ICR spectrometry showed that the reaction of $C_6H_6^{+}$ with 2-propyl iodide was characteristic of ionized benzene. Ionized acyclic C_6H_6 isomers, dimethylenecyclobutene, fulvene and benzvalene all do not react with neutral 2-propyl iodide. The technique of obtaining CID spectra of ion derivatives was also applied to the investigation of the ion-molecule reaction adducts formed from ionized benzene and 2-propyl iodide(34). The work demonstrated that 2-propyl iodide formed a $C_9H_{13}^+$ adduct with ionized benzene by eliminating an iodine radical. Comparisons of the CID spectrum of the $C_9H_{13}^+$ adduct with the protonated model compounds, 1- and 2-propylbenzene, methyl ethyl benzenes and trimethyl benzenes, showed unequivocally that the structure of the adduct was protonated 2-propylbenzene (equation 39). This result identifies the reacting $C_6H_6^{+}$ as a benzene radical cation.

$$\text{(39)}$$

A small amount of the initially formed $C_9H_{13}I^{\ddot{+}}$ was also collisionally stabilized in the high pressure source. Consecutive reaction monitoring(35), a feature of the triple analyzer mass spectrometer, was used to obtain a CID spectrum of the $C_9H_{13}^+$ produced by metastable iodine radical elimination from the $C_9H_{13}I^{\ddot{+}}$ in the first field free region of the instrument. The experiment showed unambiguously that the protonated 2-propylbenzene structure was the product of the decomposing $C_9H_{13}I^{\ddot{+}}$ and removed any question about the origin of the $C_9H_{13}^+$ produced in the source.

Other products of the decomposing $C_9H_{13}I^{\ddot{+}}$ adducts were shown by deuterium labelling and metastable ion analysis to correspond to $C_6H_5I^{\ddot{+}}$, $C_3H_7I^{\ddot{+}}$, $C_9H_{13}^{\ddot{+}}$ and $C_6H_5^+$ (equation 40). This decomposition pattern was analogous to that of a biphenyl iodonium cation (r) sputtered into the gas phase by fast atom bombardment (equation 41). A mechanism which involves the

$$
C_9H_{13}I^{\ddot{+}} \longrightarrow
\begin{cases}
C_6H_5I^{\ddot{+}} + C_3H_8 \\
C_3H_7I^{\ddot{+}} + C_6H_6 \\
C_9H_{13}^+ + I\cdot \\
C_6H_5^+ + C_3H_8I\cdot
\end{cases}
\tag{40}
$$

$$
\underset{\underline{r}}{C_6H_5\overset{+}{-}I-C_6H_5} \longrightarrow
\begin{cases}
C_6H_5I^{\ddot{+}} + C_6H_5\cdot \\
C_{12}H_{10}^{\ddot{+}} + I\cdot \\
C_6H_5^+ + C_6H_5I
\end{cases}
\tag{41}
$$

initial formation of a "Wheland type" intermediate (structures r and s) followed by a reductive elimination of an iodine radical derives from this comparison (equation 42).

$$
\tag{42}
$$

$C_8H_8^{\ddot{+}}$

The technique of obtaining CID spectra of derivatized ions has also been a valuable tool for the investigation of ion-molecule cycloadditions. Certain cations possessing a 1,3-diene moiety might be expected to undergo cycloadditions with olefinic neutrals by analogy with neutral chemistry. Such cycloadditions would be extremely diagnostic for the characterization of those cations.

Three $C_8H_8^{\ddagger}$ isomers, styrene, cyclooctatetraene and o-quinodimethane (alternatively o-xylylene, structure t) can be easily distinguished by their respective reactions with neutral deuterated styrene(36). The deuterated styrene is necessary in the present study to separate the desired cross chemistry from any internal ion-molecule reactions. Examination of the CID spectra of the collisionally stabilized $C_{16}(H,D)_{16}^{\ddagger}$ adducts and appropriately labelled model compounds shows that the [o-quinodimethane + styrene]‡ adduct is identical to a [4+2] model cycloadduct, 2-phenyltetralin structure u (equation 43).

$$(43)$$

This cycloaddition reaction yielding ion u is consistent with behavior anticipated from a cis- diene t. In contrast to this behavior, the [styrene + styrene]‡ adduct is at present unidentified, although recent investigations suggest that its structure is a 1,4 ion dipole structure v (equation 44). The cyclooctatetraene‡ is unreactive with neutral styrene. Therefore, it is easily distinguished from the other isomeric C_8H_8 radical cations.

$$(44)$$

These ion-molecule reactions were used to probe the structure of a fourth $C_8H_8^{\ddagger}$ isomer, the radical cation resulting from the ionization of benzocyclobutene w. A ring intact and a ring-opened structure are logical possibilities for this molecular ion. The CID experiments showed that ionized benzocyclobutene, w, reacted with styrene to give an intermediate which is identical to that formed from o-quinodimethane‡ (t) and neutral styrene. This behavior is consistent with at least partial ring opening prior to condensation with styrene (equation 45). While a ring closing of the o-quinodimethane‡ structure cannot be rigorously ruled out, that possibility seems unlikely from a consideration of the probable structure of the ion-molecule reaction product (ion u).

$$(45)$$

$C_4H_6^+$

Another ion structure which may be characterized by ion-molecule cycloadditions is 1,3-butadiene$^+$. ICR spectrometry of deuterium labelled reagents was used to demonstrate that 1,3-butadiene$^+$ and methyl vinyl ether underwent a cycloaddition at pressures of approximately 10^{-5} torr(37). This conclusion was based on the observation that both the [1,1,4,4-d$_4$-1,3-butadiene + methyl vinyl ether]$^+$ adduct and 2,6,6-d$_3$-$\overline{4}$-methoxycyclohexene eliminated CH$_3$OD (equations $\underline{46}$ and $\underline{47}$). Furthermore, CID experiments indicated that the $\overline{C_6H_8}^+$ product of the methanol elimination from the ion-molecule adduct is cyclic.

$$\text{(structure)} \longrightarrow C_6H_5D_3^+ + CH_3OD \qquad (\underline{46})$$

$$\text{(structure)} \longrightarrow C_6H_6D_2^+ + CH_3OD \qquad (\underline{47})$$

This chemistry was subsequently studied using CID to investigate the [1,3-butadiene + methyl vinyl ether] adduct formed and stabilized at a pressure of approximately 1 torr(38). The CID spectrum of the adduct most closely resembled that of 1-methoxy-2,4-hexadiene, leading to the conclusion that the adduct must be acyclic (equation $\underline{48}$).

$$\text{(structure)} \xrightarrow{1\ Torr} \text{(structure)} \qquad (\underline{48})$$

The technique of obtaining CID spectra of ion-molecule adducts was also used to investigate the reaction of ionized 1,3-butadiene with neutral 1,3-butadiene(38). This work showed that the CID spectrum of the [butadiene + butadiene]$^+$ adduct formed and stabilized at 0.1 torr matches that of the [4+2] cycloadduct, 4-vinylcyclohexene (equation $\underline{49}$). However, the CID spectrum of the [butadiene + butadiene]$^+$ intermediate formed and stabilized at higher pressure (1 torr) matched that of an acyclic compound, 5-methyl-1,3,6-heptatriene. These results are entirely analogous to those observed for the reaction of ionized 1,3-butadiene and methyl vinyl ether. That is, a cycloaddition is observed at lower pressures and an acyclic adduct is observed at higher pressures.

(49)

At present these results are best interpreted in terms of a thermodynamically more stable cycloaddition product which dominates the low-pressure ion-molecule chemistry. At high pressures, where extensive collisional stabilization is available, only the more kinetically favored (less thermodynamically stable) acyclic products are observed.

One extension of this interpretation is that only the 1,3-butadiene ions in the cis- conformation are involved in cycloaddition. At higher pressures, where more collisional stabilization is available, the acyclic products of the reactions of the more abundant trans- ion dominate the observed chemistry. If this view proves to be accurate, then the ion-molecule reactions allow us not only to characterize the C_4H_6 cations but also to investigate their stereochemistry.

CONCLUSIONS

The investigations discussed above highlight the powerful mass spectrometric ion structure determination techniques currently used and bring into focus several efforts in the field. While many of the experimental techniques have been developed to a high state-of-the-art, no single technique is capable of solving all ion structure problems. Because no instrumental panacea exists, the solution of difficult problems by making use of two techniques which are complementary, such as acquiring the CID of ions derivatized in a high pressure source using reactions characterized by ICR double resonance, should be a fruitful approach.

REFERENCES

1. Cooks, R.G. and Glish, G.L., Chem. Eng. News, November 30, 1981, p. 40.

2. a) Maquestiau, A., Van Haverbeke, Y., Flammang, R., Abrassart, M., and Finet, D., Bull. Soc. Chim. Belg., 1978, 87, 765; b) Maquestiau, A., Van Haverbeke, Y., de Meyes, C., Duthoit, C., Meyerant, P. and Flammang, R., Nouv. J. Chim., 1978, 3, 517; c) McLafferty, F.W., Todd, P.J., McGilvery, D.C., and Baldwin, M.A., J. Amer. Chem. Soc., 1980, 102 3360; d) McLafferty, F.W., Acc. Chem. Res., 1980, 2 33; e) Gross, M.L., Chess, E.K., Lyon, P.A., Crow, F.W., Evans, S. and Tudge, H., Int. J. Mass Spectrom. Ion Phys., 1982, 42, 243.

3. Cody, R.B., and Freiser, B.S., Anal. Chem., 1982, 54 1431.

4. a) Wolfshutz, R. and Schwarz, J., Int. J. Mass Spectrom. Ion Phys., 1980, 33, 285; b) Chess, E.K., and Gross, M.L., J. Org. Chem., in press.

5. a) Diekman, J., MacLeod, J.K., Djerassi, C., and Baldeschwieler, J.D., J. Amer. Chem. Soc., 1969, 91 2069; b) Eadon, G., Diekman, J., and Djerassi, C., J. Amer. Chem. Soc., 1969, 91 3986; c) Eadon, G., Diekman, J., Djerassi, C., J. Amer. Chem. Soc., 1970, 92 6205.

6. Hemberger, P.H., Kleingeld, J.C., Levsen, K., Mainzer, N., Mandelbaum, A., Nibbering, N.M.M., Schwarz, H., Weber, R., Wiesz, A., and Wesdemiotis, C., J. Amer. Chem. Soc., 1980, 102 3736.

7. Beauchamp, J.L., and Dunbar, R.C., J. Amer. Chem. Soc., 1970, 92 1477.

8. Staley, R.H., Corderman, R.R., Foster, M.S., and Beauchamp, J.L., J. Amer. Chem. Soc., 1974, 96 1260.

9. Bursey, M.M., Hass J.R., Harvan, D.J., and Parker, C.E., J. Amer. Chem. Soc., 1979, 101 5485.

10. Goode, G.C., and Jennings, K.R., Adv. Mass Spectrom., 1974, 6 797.

11. Dawson, J.H.J., and Nibbering, N.M.M., J. Amer. Chem. Soc., 1978, 100 1928.

12. a) Hoffman, M.K., and Bursey, M.M., Tet. Letters, 1971, 27 2539; b) Bursey, M.M., Hoffman, M.K., and Benezra, S.A., Chem. Commun., 1971, 1417.

13. Levsen, K., McLafferty, F.W., and Jerina, D.M., J. Amer. Chem. Soc., 1973, 95 6332.

14. Dunbar, R.C., and Klein, R., J. Amer. Chem. Soc., 1977, 99 3744.

15. Ausloos, P., and Lias, S.G., Chem. Phys. Lett., 1977, 47 495.

16. Tomer, K.B. and Djerassi, C. Tetrahedron, 1973, 29, 3491.

17. Theissling, C.B., and Nibbering, N.M.M., Adv. Mass Spectrom., 1977, 1287.

18. Borchers, F., Levsen, K., Theissling, C.B., and Nibbering, N.M.M., Org. Mass Spectrom., 1977, 12 746.

19. van Velzen, P.N.T., van der Hart, W.J., van der Greef, J., Nibbering, N.M.M., and Gross, M.L., J. Amer. Chem. Soc., 1982, 104 1208.

20. Gross, M.L., and McLafferty, F.W., J. Amer. Chem. Soc., 1971, 93 1267.

21. Sieck, L.W., Gordon, Jr., R., and Ausloos, P., J. Amer. Chem. Soc., 1972, 94 7157.

22. Van Velzen, P.N.T. and Van der Hart, W.J., Chem. Phys., 1981, 61, 335.

23. a) Bowen, R.D., Barbalas, M.P., Pagano, F.P., Todd, P.J., and McLafferty, F.W., Org. Mass Spectrom., 1980, 15 51; b) Holmes, J.L., Terlouw, J.K., Burgers, P.C., and Rye, R.T.B., Org. Mass Spectrom., 1980, 15 149.

24. Miller, D.L. and Gross, M.L., J. Amer. Chem. Soc., in press.

25. Collins, J.R., and Gallup, G.A., J. Amer. Chem. Soc., 1982, 104 1530.

26. a)Profous, Z.C., Wanczek, K.-P., and Hartman, H., Z. Naturforsch., 1975, 30A 1470; b) Baykut, G., Wanczek, K.-P., and Hartman, H., Adv. Mass Spectrom., 1980, 8A 186.

27. Bowen, R.D., Williams, D.H., Schwarz, H., and Wesdemiotis, C., J. Amer. Chem. Soc., 1979, 101 4681.

28. Bowers, M.T., Shuying, L., Kemper, P., Stradling, R., Webb, H., Aue, D.H., Gilbert, J.R., and Jennings, K.R., J. Amer. Chem. Soc., 1980, 102 4830.

29. a) Aue, D.H., Davidson, W.R., and Bowers, M.T., J. Amer. Chem. Soc., 1976, 98 6700; b) Colosimo, M. and Bucci, R., J. C. S. Chem. Commun., 1981, 659.

30. a) Radom, L., Hariharan, P.C., Pople, J.A., and Schleyer, P.v.R., J. Amer. Chem. Soc., 1973, 95 6531; b) Raghavachari, K., Whiteside, R.A., Pople, J.A., and Schleyer, P.v.R., J. Amer. Chem. Soc., 1981, 103 5649; c) Merlet, P., Peyerimhoff, S.D., Buenker, R.J., and Shih, S., J. Amer. Chem. Soc., 1974, 96 959; d) Halgren, T.A., and Lipscomb, W.N., Chem. Phys. Letters, 1977, 49 225.

31. Lay, J.O., Doctoral Dissertation, University of Nebraska, 1982.

32. Lay, J.O., and Gross, M.L., J. Amer Chem. Soc., in press.

33. Dymerski, P.P., and McLafferty, F.W., J. Amer. Chem. Soc., 1976, 98 6070.

34. Miller, D.L., and Gross, M.L., J. Amer. Chem. Soc., in press.

35. Burinsky, D.J., Cooks, R.G., Chess, E.K. and Gross, M.L., Anal. Chem., 1982, 54, 295.

36. Chess, E.K., and Gross, M.L., J. Org. Chem., in press.

37. van Doorn, R., Nibbering, N.M.M., Ferrer-Correia, A.J.V., and Jennings, K.R., Org. Mass Spectrom., 1978, 13 729.

38. Groenewold, G.S., Lay, J.O., and Gross, M.L., American Society for Mass Spectrometry, "30th Annual Conference on Mass Spectrometry and Allied Topics, Honolulu, Hawaii".

ACKNOWLEDGEMENT

Preparation of this review was made possible by support of the United States National Science Foundation (Grant CHE80-08008) and by the Midwest Center for Mass Spectrometry, an NSF regional instrumentation facility (Grant CHE78-18581).

CARBON SKELETAL REARRANGEMENTS VIA PYRAMIDAL CARBOCATIONS

Helmut Schwarz, Helga Thies and Wilfried Franke

Institut für Organische Chemie der Technischen Universität, D-1000 Berlin 12, Straße des 17. Juni 135, W. Germany

Abstract — The scrambling of carbon atoms in gaseous carbocations is readily explicable by invoking transition states or intermediates of pyramidal structure, which are now be seen as the logical bridge between organic and organometallic chemistry. The electronic structure of this species is treated with respect to some of their physical and chemical properties (energy, charge distribution, geometry). Specific examples discussed include the following hydrocarbon ions, the gas phase chemistry of which proceeds via pyramidal intermediates: $C_5H_5^+$, $C_5H_9^+$, $C_6H_{11}^+$, $C_6H_5^+$. Reasons are provided why, in contrast, the carbon scrambling in some other systems, as for example tropylium ion $\rightleftharpoons$ benzyl cation, does not involve pyramidal cations but proceeds via a sequence of orbital symmetry allowed isomerization, or why in the case of simple saturated hydrocarbon ions, as for example $C_4H_9^+$, the C-skeleton reorganization is achieved by the well-known Wagner/Meerwein type rearrangement. The carbon scrambling in ionised cyclopentadiene, which precedes the formation of allene cation radical and acetylene, is due to the intermediacy of ionised bicyclo[2.1.0]pent-2-ene, a species which is also involved in the phototransposition of carbon atoms in neutral cyclopentadiene. Pyramidal-like structures are much too high in energy to play a role in the degenerate isomerization of $C_5H_6^{+\cdot}$.

Carbocations and in particular those with a pyramidal

M. A. Almoster Ferreira (ed.), Ionic Processes in the Gas Phase, 267–286.
© *1984 by D. Reidel Publishing Company.*

structure are now seen as the logical bridge between organic
and organometallic chemistry (1 - 4). Although as hydrocarb-
bons, carbocations are unquestionably organic, nonetheless
their valence properties and structural features, not least of
which are the coordination number of carbon and the electron
deficient 3 centre/2 electron bonding properties, owe much to
organometallic compounds. The structural similarity between
carbocations and organometallic compounds is even more apparent in the pyramidal carbocations, e. g. $(CH)_5^+$ (1) and $(CH)_6^{2+}$
(2). In these species the C-atoms of interest are not only five-
and six-coordinated, respectively, but the structural topology
of the $(CH)_n^+$ cations finds it exact counterpart in organometallic chemistry. Consequently, 1 and 2 are analogous to ferro-
cene (3), since they both have semi-sandwich structures. Following the more general approach to cluster theory formulated by
Wade (5), Williams (6) and Rudolph (7) or a more recent graph-
theoretical approach (8) the cations 1 and 2 could be described
as nido-clusters, since they have the same topology and are
isoelectronic with the carbaboranes (5 - 10). For example, the
carbaborane 4 can be generated from the dication 2 by replacement of the apical and one basal C^+ atom by two isoelectronic
B atoms.

Pyramidal cations such as 5, can formally be considered as being the result of the interaction of the CH^+ molecular fragment (protonated carbon) with an appropriate 4 electron
system. In the case of 1 the 4e system corresponds to cyclobutadiene, whereas 2 may be described as a nido-cluster resulting
from combination of CH^+ and the antiaromatic cyclopentadienyl
cation. The only requirements are that the interacting orbitals

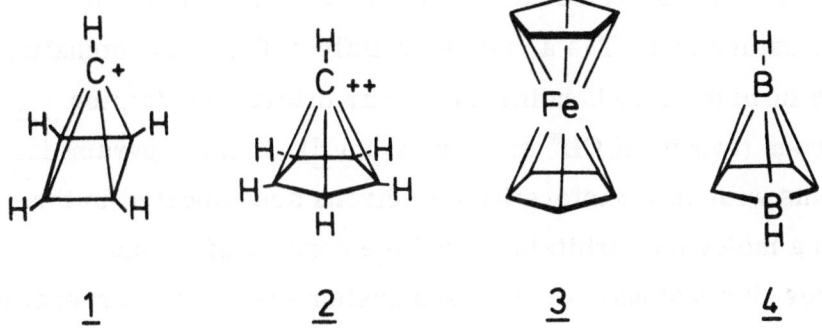

<u>1</u>   <u>2</u>   <u>3</u>   <u>4</u>

of the molecular fragments be of similar energy and of appropriate symmetry. Furthermore, the segment forming the base
must supply <u>four</u> electrons while the CH^+ fragment provides
<u>two</u> electrons and two empty orbitals (2, 3).

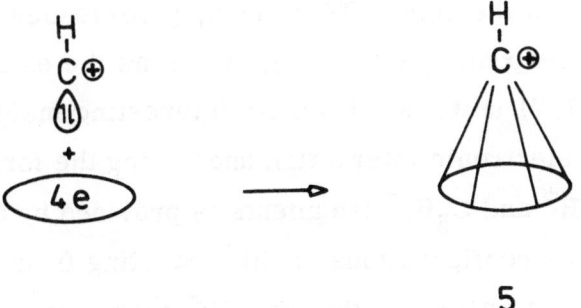

<u>5</u>

The interaction diagram (Fig. 1) which should not be mistaken for a correlation diagram for the central linear approach of CH^+ to cyclobutadiene (6) clearly describes the electronic nature of 1. The a_1 and e orbitals of CH^+ on conbination with 6 furnish three bonding molecular orbitals containing six electrons (two from CH^+ and four from 6). In fact, pyramidal systems with 4n + 2 interstitial electrons accomodated in three bonding molecular orbitals extend the concept of aromaticity to three-dimensional, delocalised systems which are generally expected to be stable (11).

A brief comment on the charge distribution in pyramidal cations is, perhaps, appropriate. The charge is not located at the apical carbon as may be concluded from the structural notation used in the formulae. On the very contrary, for pyramidal carbocations it has been found computationally (4) that the apical carbon atoms bear only a small positive charge or, in some cases, even a slight negative charge. For example, the apical C atom in 1 carries only 3.8 % of the total charge, whereas the basal C atoms carry 35 %. For the di-cation 2 the charge distribution was computed by ab initio methods (12) and it was found that the apical CH group only possesses 0.29 charge units (corresponding to 14.9 %), whereas the entire basis possesses 1.71 units (85.5 %). An interesting insight into the nature of the electronic interaction underlying the formal combination of CH^+ and $C_5H_5^+$ fragments is provided by analysis of the valence configurations of CH^+. Starting from a $\sigma_{CH}^2 \pi^2 \sigma^0$ configuration for the free CH^+ fragment, a $\sigma_{CH}^2 \pi^{1.76} \sigma^{0.96}$ configuration is obtained for the CH-partial structure in 2. This can be interpreted as follows. The five-membered ring donates 0.96 electrons to the empty σ-orbital of CH^+, whereas 0.24 electrons from the π-orbital of CH^+ are donated back to the five-membered ring. This type of bonding is commonplace for organometallic compounds (13). For many other pyramidal carbocations charge distributions have been calculated and seems to be a quite general feature that most of the charge is residing in the basal part of the species, whereas the carbyne carbon atom is practically neutral, with the charge being strongly deflected to the periphery. In fact, it is quite often found that it is the "hydrogen periphery" which carries a substantial part of the positive charge. This is an electrostatically favourable state of affairs which is a characteristic of the polyhedral structure. The analogy with the distribution of charge on a spherical surface is obvious.

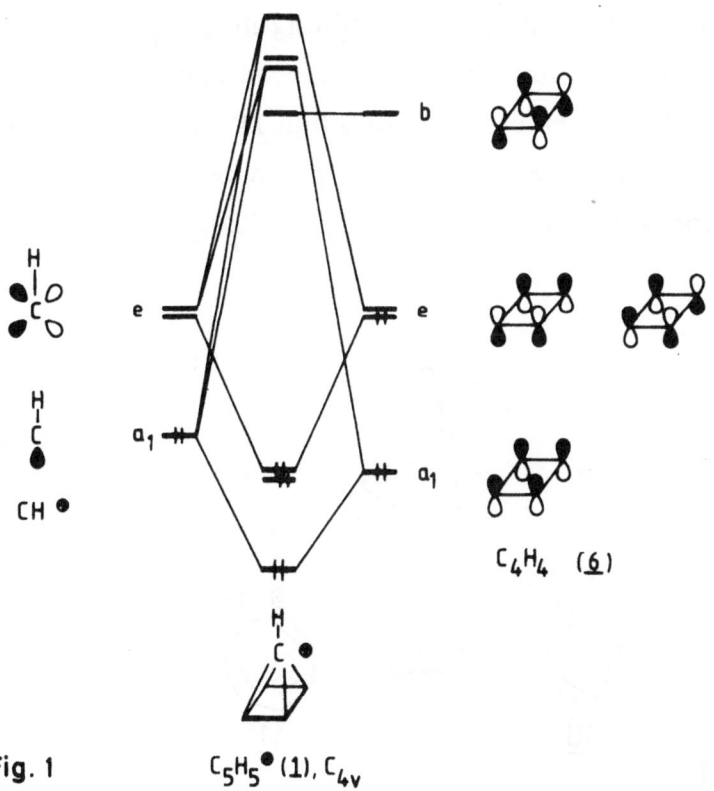

Fig. 1 $C_5H_5^{\bullet}$ (1), C_{4v}

The foregoing treatment also accounts for the fact that certain carbocations and carbanions cannot be stable if pyramidal geometry is adopted. Thus, the pyramidal $C_7H_7^+$ cation (7) is unstable (14), although it has been postulated (15) in the carbon scrambling of benzyl cation (8) and tropylium ion (9), which precedes acetylene loss from $C_7H_7^+$ precursors (15,16). 7 contains eight electrons (six from benzene and two from the CH^+ fragment) two of which have to be placed in orbitals of high energy. In fact, 7 may be regarded as a prototype of a three-dimensional antiaromatic (4n electron) system. Nature arranges things differently. Carbon atom scrambling between 8 and 9 does not involve 7. According to extensive MINDO/3 calculations (14) the minimal energy requirement path (MERP) for carbon atom scrambling in $C_7H_7^+$ proceeds via a sequence of orbital symmetry allowed isomerizations (Scheme 1). The barrier for 9 $\longrightarrow$ 8 (via 11 and 10) has been calculated to 57 kcal/mol (14) which is substantially lower than the activation energy for C_2H_2 loss from $C_7H_7^+$ (95 kcal/mol; 17).

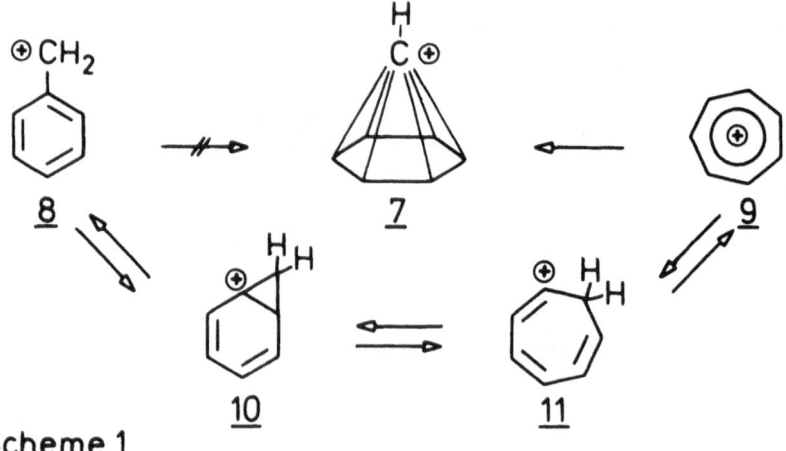

Scheme 1

In the gas phase chemistry of $C_nH_{2n+1}^+$ cation, as for example $C_4H_9^+$, carbon scrambling is often found to proceed the unimolecular dissociation of the species, and the question is whether this is due to the intermediacy of pyramidal ions. The answer is no, because it is impossible to derive from $C_nH_{2n+1}^+$ any energetically feasible 4 electron fragment $C_{n-1}H_{2n}$ of correct symmetry, which is mandatory for the combination with CH^+ to form pyramidal cations. In fact, the energetically most favoured pathway for carbon skeleton reorganization in $C_nH_{2n+1}^+$ ions can be explained in terms of the well-known Wagner/Meerwein type rearrangements, i.e. sequences of [1.2]-hydride and alkyl shifts.

On the other hand, carbon atom scrambling in $C_5H_5^+$ prior to acetylene loss (16,18), is easily explicable in terms of an intermediate pyramidal cation 1. This follows directly from the fact that the symmetry-allowed isomerisation of the bent cyclopentadienyl cation 12 to the thermochemically less stable (19) pyramidal ion 1 ($\Delta\Delta H_f^O = 14.4$ kcal/mol) affords less energy than the elimination of C_2H_2 from $C_5H_5^+$ (57 versus 80 kcal/mol). Thus, prior to fragmentation a complete reorganization of the carbon skeleton is likely to occur via the equilibrium 12 $\rightleftharpoons$ 1.

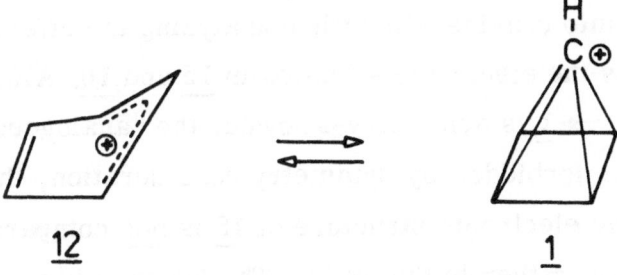

12   1

Derivatives of 1 has been studied both experimentally and computationally (4). The influence of benzannelation on the relative stability of classical versus pyramidal structures has been discussed in detail by Olah and Schleyer et al. (20). Numerous MINDO/3 calculations clearly show that the pyramidal structures 1, 14, and 16 are destabilized upon increasing annelation compared to the classical structures 12, 13, and 15 ($\Delta\Delta$ H_f^o = 14.4, 49.1 and 73.0 kcal/mol, respectively). There are two reasons for this substantial increase in energy. Firstly the antiaromatic character of the classical structures is diminished upon increasing benzannelation, i.e. 15 is more stable than 13, which is in turn more stable than 12. Secondly, the stability of the pyramidal structures decreases in the order 1 > 14 > 16. The differences are readily apparent from the "complexation energies" arising from the hypothetical reaction olefin + $CH^+ \longrightarrow$ pyramidal cation. The relevant values for the formation of 1, 14 and 16 are: -176.2, -153, and -139.9 kcal/ mol, respectively. As the complexation energy arises from the interaction of the doubly degenerate p-LUMO of the CH^+ fragment with the π-orbital of the diene, the observed trend becomes understandable increasing annelation diminishes the π energy and at the same time decreases the coefficients of the carbon centres involved in coordination. Another factor one has to take into consideration when analyzing the effect of benz-annelation is the electronic situation in 15 and 16. Although the process 12 $\longrightarrow$ 1 is symmetry-allowed, the "analogous" reaction 15 $\longrightarrow$ 16 is forbidden by symmetry consideration, the reason being that the electronic structure of 15 is not comparable to that of 12, but rather to that of 17. The latter, which is a lu-momer of 12, cannot isomerize to 1 for symmetry reasons (2).

With other words, carbon scrambling in benzannelated derivatives of 1 via pyramidal cations is only expected to take place
in those cases, in which the decomposition of the respective ions
is extremely high in energy. So far, no system is known which
meets this requirement.

Similar effects are predicted by theory (21) for phenyl
substituents, which stabilize the cyclopentadienyl structure
over the pyramidal form. Consequently, isomerization is likely

to take place only partially, if at all. Indeed, analyses of the unimolecular losses of disubstituted acetylenes (X-C≡C-X, X-C≡C-Y, and Y-C≡C-Y, respectively) from cations $\underline{18}$ provide experimental support for the computational prediction in that the results are not compatible with the assumption of a complete carbon skeleton rearrangement of the 5-membered ring (22). Moreover, depending on the substitutents X, Y (X, Y = C_6H_5, p-$CH_3C_6H_4$, p-FC_6H_4) and the actual substituent pattern the distributions for the losses of disubstituted acetylenes varies slightly. However, it is important to note, that all systems $\underline{18}$ studied so far eliminate X-C≡C-X. This requires a substantial reorganization of the five-membered ring systems. Both the pyramidal cation $\underline{19}$ and the edge CH^+-complexed tetrahedrane derivative $\underline{20}$ (or isomers thereof) can principally serve as intermediates in these rearrangements whereas the polycyclic cation $\underline{21}$ cannot.

$\underline{18}$ $\underline{19}$ $\underline{20}$ $\underline{21}$

Pyramidal cations do, however, play a decisive role in the gas phase chemistry of $C_5H_9^+$ and $C_6H_{11}^+$ ions (23, 24). For example, metastable $C_5H_9^+$ ions generated from several precursors by dissociative ionization, lose ethylene in a uni-molecular fashion to give an allyl cation. This elimination is preceded by a complete scrambling of all five carbon atoms as shown by the investigation of $^{13}C_2$-labelled substrates (i.e. iso-topomeric bromocyclopentanes, 5-bromo-1-pentenes, and cyclo-butylmethyl bromide, respectively). Model studies and quantum mechanical calculations show that C atom scrambling in the cyclopentyl ion (22) can be accounted for in two ways: either by a bisected cyclobutylcarbinyl cation (23) or by the pyramidal cation (24). 24 can be viewed as a complex between CH^+ and two ethylene molecules, the latter provide the 4 electron. The alternative, that the four basal atoms in 24 interact with each other in such a way to form a cyclobutane ring has not been con-firmed by molecular orbital calculations. In fact, the distance between the opposing C_2 units is 2.63 Å (STO-3G basis set) and 1.44 Å within a C_2 unit. The latter value corresponds to a lenghtened CC-double bond of a diene moiety and the former excludes a cyclobutane ring. Ethylene elimination occurs only with the pyramidal cation 24. Electronic reorganization in 24 leads to a transition state TS_3 (Fig. 2), whose structure corre-sponds to a partially opened cyclopropyl cation "solvated" by an ethylene molecule. Alternative mechanisms for the process $C_5H_9^+ \longrightarrow C_3H_5^+ + C_2H_4$, as for example the direct cyclore-version 23 $\longrightarrow$ 26 or various two-step reactions involving acyclic $C_5H_9^+$ ions, have been found by calculation to be ener-getically unfavourable (23).

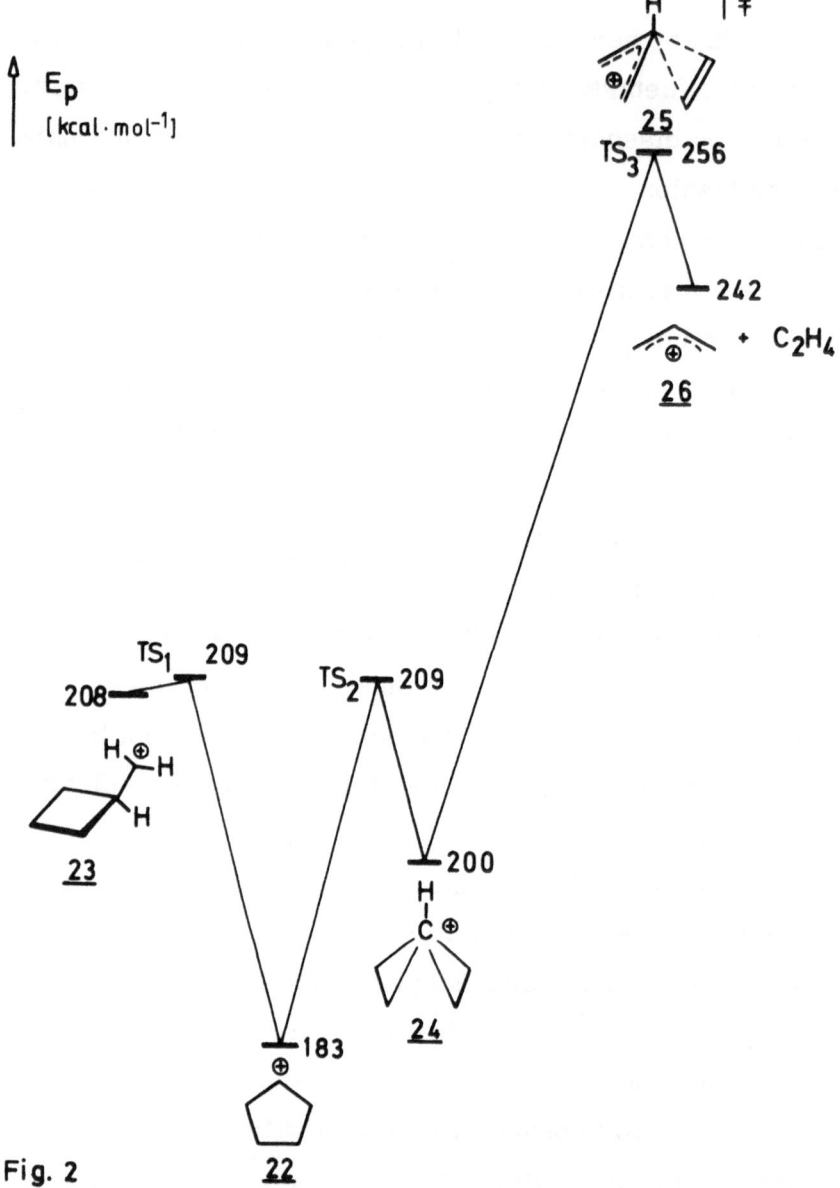

Fig. 2

Similar results were obtained for the homologous
$C_6H_{11}^+$ ion which decomposes unimolecularly to the 1-methyl-
allyl cation and ethylene (24). The investigation of several
$^{13}C_2$-labelled cyclohexane and methylcyclopentane derivatives

proves that the dissociation of $C_6H_{11}{}^+$ is preceded by complete carbon atom scrambling irrespective of the constitution of the molecular ions (five- or six-membered ring systems) and the nature of the neutral X to be eliminated from $C_6H_{11}X$ (X = COOH, Br). The scrambling is best described in terms of an equilibrium between the cyclohexyl (27) and the 1-methylcyclopentyl cation (28), involving both protonated cyclopropane derivatives (29) (25) and the methyl-substituted pyramidal cations 30 and 31 (Scheme 2).

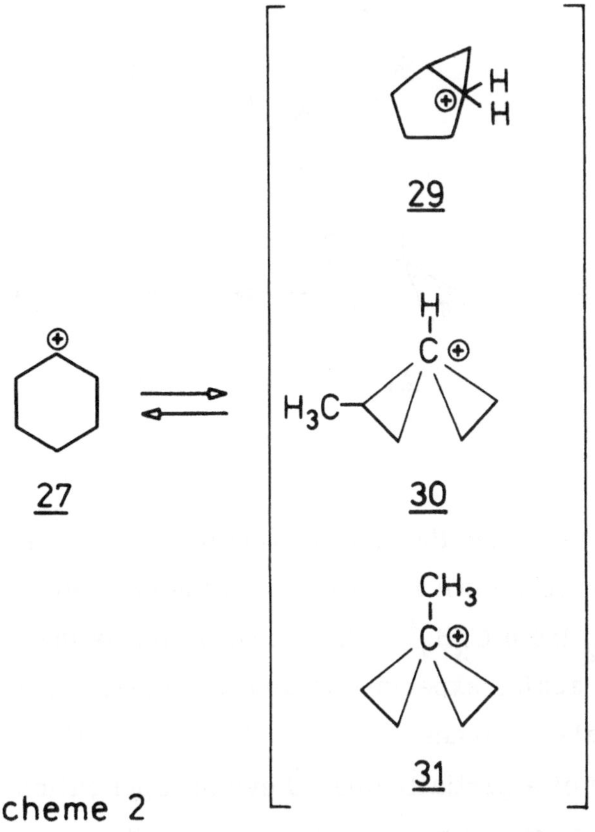

Scheme 2

Ethylene elimination, however, proceeds most favourably from the transition state 32. According to MNDO calculations, 32 is at least 11 kcal/mol more stable than its isomeric

form 33. 32, which can be viewed as a partially open 2-methyl-
cyclopropyl cation solvated by interaction with C_2H_4, decomposes to C_2H_4 and $C_4H_7{}^+$. The latter has in agreement with
experimental results the structure of 1-methylallyl cation (34)
and not that of the isomeric 2-methylallyl cation (35), which
could be formed only from 33 (Scheme 3).

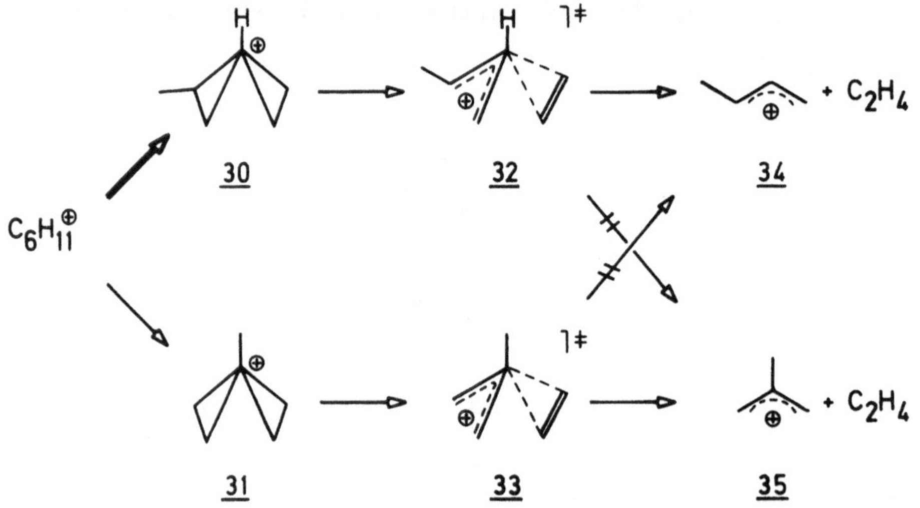

Scheme 3

It should be mentioned that elimination of C_2H_4 from
$C_5H_9{}^+$ and $C_6H_{11}{}^+$ is mechanistically similar to the unimolecular elimination of H_2 from $C_3H_7{}^+$, which also produces the
allyl cation (26). Once again, experiments and calculations both
show that the pyramidal-like transition state (36) (X = H, R = H)
exhibits the properties of a partially opened cyclopropyl cation
which is "solvated" by a σ-ligand, i.e. H_2. Indeed, the reaction $C_3H_7{}^+ \longrightarrow C_3H_5{}^+ + H_2$ is the first example of two coupled
symmetry-allowed processes, namely the opening of a corner-
protonated cyclopropyl cation to the allyl ion and the simulta-

neous cheletropic elimination of hydrogen (26).

	R	X	Reaction	Ref.
	H	H	$C_3H_7^{\oplus} \longrightarrow C_3H_5^{\oplus} + H_2$	(26)
	H	CH_2	$C_5H_9^{\oplus} \longrightarrow C_3H_5^{\oplus} + C_2H_4$	(23)
36	CH_3	CH_2	$C_6H_{11}^{\oplus} \longrightarrow C_4H_7^{\oplus} + C_2H_4$	(24)

The gas phase chemistry of $C_6H_5^+$ ion has been studied both experimentally and computationally. There exist two pathways for the degenerate isomerization of phenyl cations (37). One involves a set of [1.2]-hydrogen migration (37 $\longrightarrow$ 38 $\longrightarrow$ 37; 27), the other proceeds via the pyramidal cation (39); 39 can be described as a complex between a carbon atom and the 4 electron donating cyclopentadienyl cation. For the pathway involving 38 an activation energy of 44 kcal/mol has been calculated (28), whereas the reaction via 39 affords at least 69 kcal/mol (29). The lowest activation energy for the loss of C_2H_2 from $C_6H_5^+$ has been calculated to be 108 kcal/mol (29); thus, it is obvious that $C_6H_5^+$ ions are expected to undergo both carbon and hydrogen scrambling (independent of each other) prior to dissociation, which is, indeed, observed experimentally (30). For phenyl ions (37) having internal energy sufficient for automerization

but not high enough for dissociation it has been shown experimentally (27) that the energetically preferred pathway for the
degenerate isomerization proceeds via $\underline{38}$ and not $\underline{39}$ in complete
agreement with the computational results.

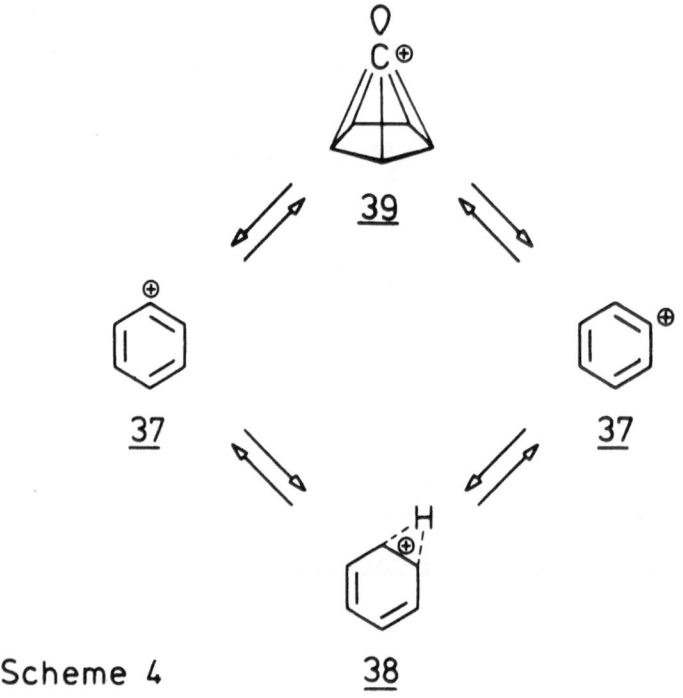

Scheme 4 $\underline{38}$

A complete scrambling of carbon atoms has been re-
ported (16, 31) for the molecular ion of cyclopentadiene ($\underline{40}$),
which decomposes to C_2H_2 and $C_3H_4^{+\cdot}$. The latter has been
shown by collisional activation mass spectrometry to have the
structure of ionised allene and not that of propyne (31). MINDO/3
molecular orbital calculations (31) indicate that the minimal
energy requirement path for carbon atom scrambling in ionised
$\underline{40}$ proceeds via the bicyclic cation radical $\underline{41}$ (Scheme 5), in
very close similarity to the well-known photochemical trans-

position of C atoms in neutral 40 (32). According to the calculations the pyramidal cation 42 (a complex between ionised
$CH_2^{+\cdot}$ and cyclobutadiene) does not exist at all on the potential
energy surface. It undergoes spontaneous rearrangement to the
polycyclic cation 43, the heat of formation of which, however,
is substantially higher than that of 41 (ΔH_f^o = 44.9 kcal/mol).
As 41 is only 20.7 kcal/mol less stable than 40, it is likely that
prior to dissociation carbon atom scrambling will take place as
sketched in Scheme 5.

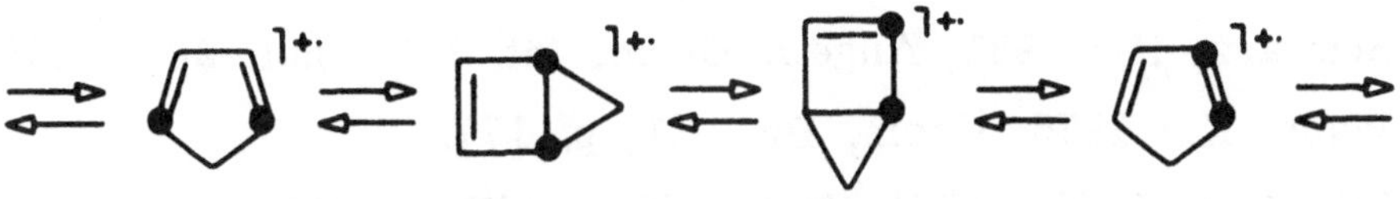

40a   41a   41b   40b

42   43

Scheme 5

ACKNOWLEDGEMENTS

The generous financial support by the Fonds der Chemischen Industrie, the Deutsche Forschungsgemeinschaft (projects Schw 221/6-1, 6-2) and the Gesellschaft von Freunden der Technischen Universität Berlin is gratefully acknowledged.

REFERENCES

(1) Hogeveen, H., and Kwant, P. W.: 1975, Acc. Chem. Res. 8, p. 413.

(2) Stohrer, W. D., and Hoffmann, R.: 1972, J. Am. Chem. Soc. 94, p. 1661.

(3) Masamune, S.: 1975, Pure Appl. Chem. 44, p. 861.

(4) Schwarz, H.: 1981, Angew. Chem. Int. Ed. Engl. 20, p. 991.

(5) Wade, K.: 1975, Chem. Br. 11, p. 177.

(6) Williams, R. E.: 1971, Inorg. Chem. 10, p. 210.

(7) Rudolph, R. W.: 1976, Acc. Chem. Res. 9, p. 446.

(8) Balaban, A. T., and Rouvray, D. H.: 1980, Tetrahedron 36, p. 1851.

(9) Grimes, R. N.: 1970, Carbaboranes, Academic Press, New York.

(10) Lipscomb, W. N.: 1973, Acc. Chem. Res. 7, p. 257.

(11) Jemmis, E. D., and Schleyer, P. v. R.: 1982, J. Am. Chem. Soc., in the press.

(12) Jonkman, H. T., and Nieuwport, W. C.: 1973, Tetrahedron Lett., p. 1671.

(13) Elian, M., and Hoffmann, R.: 1975, Inorg. Chem. 14, p. 1058.

(14) Cone, C., Dewar, M. J. S., and Landman, D.: 1977, J. Am. Chem. Soc. 99, p. 372.

(15) Rinchardt, K. L. Jr. , Buchholz, A. C. , van Lear, G. E. , and Cantrill, H. L. : 1968, J. Am. Chem. Soc. 90, p. 2983.

(16) Davidson, R. A. , and Skell, P. S. : 1973, J. Am. Chem. Soc. 95, p. 6843.

(17) McLafferty, F. W. , and Winkler, J. : 1974, J. Am. Chem. Soc. 96, p. 5182.

(18) Thies, H. , and Schwarz, H. : unpublished results.

(19) Dewar, M. J. S. , and Haddon, R. C. : 1973, J. Am. Chem. Soc. 95, p. 5836.

(20) Olah, G. A. , Prakash, G. K. S. , Liang, G. , Westerman, P. W. , Kunde, K. , Chandrasekhar, J. , and Schleyer, P. v. R. : 1980, J. Am. Chem. Soc. 102, p. 4485.

(21) Hehre, W. J. , Schleyer, P. v. R. : 1973, J. Am. Chem. Soc. 95, p. 5837.

(22) Thies, H. , and Schwarz, H. : manuscript in preparation.

(23) Franke, W. , Schwarz, H. , Thies, H. , Chandrasekhar, J. , Schleyer, P. v. R. , Hehre, W. J. , Saunders, M. , and Walker, G. : 1981, Chem. Ber. 114, p. 2808.

(24) Franke, W. , Frenking, G. , Schwarz, H. , and Wolfschütz, R. : 1981, Chem. Ber. 114, p. 3878.

(25) Saunders, M. , Vogel, P. , Hagen, E. L. , and Rosenfeld, J. : 1973, Acc. Chem. Res. 6, p. 53.

(26) Schwarz, H. , Franke, W. , Chandrasekhar, J. , and Schleyer, P. v. R. : 1979, Tetrahedron 35, p. 1969.

(27) Speranza, M. : 1980, Tetrahedron Lett. , p. 1983.

(28) Dill, J. D. , Schleyer, P. v. R. , Binkley, J. S. , Seeger, R. , Pople, J. A. , and Haselbach, E. : 1976, J. Am. Chem. Soc. 98, p. 5428.

(29) Tasaka, M. , Ogata, M. , and Ichikawa, H. : 1981, J. Am.

Chem. Soc. 103, p. 1885.

(30) Dicinson, R. , and Williams, D. H. : 1971, J. Chem. Soc.
     B, p. 249.

(31) Thies, H. , Halim, H. , and Schwarz, H. : manuscript in
     preparation.

(32) Andrews, G. D. , and Baldwin, J. E. : 1977, J. Am. Chem.
     Soc. 99, p. 4851.

LOW RATE FRAGMENTATION REACTIONS OF GASEOUS IONS

M.A. Almoster Ferreira

Chemistry Department, Faculty of Sciences
University of Lisbon, Portugal.

Abstract: The importance of the relatively slow decomposition of gaseous metastable ions is reviewed. Experimental results are presented illustrating slow decaying ionic processes and, when possible, comparisons are made about the use of different techniques. Various approaches to the interpretation of these processes as well as several applications are discussed with the purpose of showing how relevant these studies became for the understanding of the structure of small molecules in the ionised state as well as of the kinetics and energetics involved in these phenomena.

INTRODUCTION

Some of the ions formed in the ionisation chamber of a conventional mass spectrometer have lifetimes of the order of 10^{-6} - 10^{-5}s. These ions with observable lifetimes are called metastable ions. They live long enough to cross the ion accelerating region and decompose outside the ion source, very often in the first field free region, sometimes further down on their flight towards the collector.

They usually are a small fraction of the ions with the same mass and charge that decompose inside the ionisation chamber. However, the fragmentation of normal ions occurs so quickly after ionisation that the dissociation process can not be observed directly. One important feature of the metastable ions is precisely the fact that their dissociation is delayed by several microseconds as compared with the decomposition of normal ions.

These slow fragmentation reactions, usually referred to as

M. A. Almoster Ferreira (ed.), Ionic Processes in the Gas Phase, 287–302.
© *1984 by D. Reidel Publishing Company.*

metastable transitions, are more often thought of in connection
with large organic molecules since they are very useful to
determine reaction paths or to elucidate the mechanism of their
fragmentation under electron impact or other ionising processes.
However, the study of the delayed spontaneous dissociation of
ions is equally important in the study of small molecular or
fragment ions arising from diatomic, triatomic or tetratomic
gaseous molecules.

Experimental results on unimolecular and collision induced
decomposition of (metastable) N_2^+, CO^+ and CO_2^+ ions have been
reported (1) (2) long before Hipple, Fox and Condon (3) have
pointed out the existence of metastable peaks in the mass spectra
of some organic compounds. In fact, they had been observed earlier
still by Aston (4) who distinguished between primary peaks re-
presenting ions formed inside the ion source and diffuse peaks
(or bands) due to collisional processes with the residual gas in
the instrument and known as Aston bands (5), or to spontaneous
fragmentation of metastable ions. Friedlander (2) also reported
unimolecular decay of metastable CO_2^{2+} and other doubly charged
metastable ions.

Metastable ions have been extensively applied in fundamental
studies in mass spectrometry. Electron impact techniques have
been more widely used but field ionisation (6) and chemical
ionisation techniques have also become important in such studies,
especially the latter which, when comparing similar processes
taking place under electron impact and chemical ionisation, may
provide useful information on the energy transfer during
collisional processes occurring inside the ion source (7).

The discovery of the unique usefulness of metastable ions on
mass analysis has led to the development of methodologies and
instrumental improvements and modifications that make it possible
to detect them under various conditions (8) whatever the ionising
technique used.

Metastable studies have recently known an unusual application
in addition to an already wide range of chemical applications:
the quantitative determination of deuterium in water (9) which is
of vital importance for the understanding of many biochemical
processes. The method has the advantage of not requiring any
particular sample purification, provided that adequate calibration
curves can be obtained in the same instrumental conditions as the
unknown sample.

ENERGETICS AND KINETICS OF FRAGMENTATION OF METASTABLE IONS

A metastable ion is always formed in an excited state. It

is important to know how long these ions can live, how they decompose and how their excess energy dissipates. The answer to these questions involves the knowledge of the kinetics of ionic decomposition and its energetics.

Electron impact techniques have been considered to have advantages in the study of these subjects, especially when compared with photoionisation, not only because it is an easier technique, but also because the corresponding selection rules are less restrictive. Nevertheless the difficulty in obtaining a monoenergetic electron beam to ionise gaseous molecules and all inherent consequences should not be minimized.

The internal energy of a metastable ion is converted, upon decomposition, into different forms of energy of the dissociation products. For the comprehension of the decomposition of metastable ions, it is particularly important to consider the part that is released as kinetic energy which is easily detectable due to the broadening of the corresponding metastable peak. Beynon (10) has shown that the average value of the kinetic energy released, T, is related to the width of the peak by the expression $d = (4m_2^2/m_1)(\mu T/eV)^{1/2}$ where d is expressed in m.u., m_1 and m_2 are the precursor and daughter ion masses, V the full accelerating voltage the metastable ion crosses before decomposing, and $\mu = (m_1 - m_2)/m_2$.

The kinetic energy released is also important to determine the exact position on the spectrum of the peak associated with the dissociation of the metastable ion. In fact the ionic fragment resulting from the decomposition of a single charged metastable ion has a smaller momentum than the ion with the same mass and charge formed inside the ion source, since its formation occurs after full acceleration of the parent ion. As a consequence it will not be registered at the corresponding mass to charge ratio, but the so called metastable peal will appear on the spectrum with apparent mass (11) given by

$$m^* = \frac{m_2^2}{m_1}\left[1 + \frac{\mu T}{eV} + 2\left(\frac{\mu T}{eV}\right)^{1/2}\right]$$

when decomposition occurs after full acceleration. T-values have usually been calculated at half-height ($T_{0.5}$) for spectra run in conventional mass spectrometers.

The necessity for a better knowledge about the released kinetic energy was soon understood. For instance, it is not easy to correlate it with thermochemical or kinetic data, mostly because correct values for threshold heats of formation of a large number of ions are not available. Another difficulty met with is the lack of a theory relating the shape of a metastable peak to the rate processes involved in its formation.

An important contribution towards the comprehension of these problems has been given by Holmes and al. and Terwilliger and al. who developed computational methods to evaluate the profiles of metastable peaks taking into account the effects of instrumental parameters (12)(13). Using also a mathematical model, a method has been proposed for obtaining the distribution of the kinetic energy released in a metastable decomposition (14). Values for T_{av} and $T_{0.5}$ are calculated and it was possible to establish that, since $T_{0.5}$ is not characteristic of a given process, the relation $T_{0.5}:T_{av} = 1:2.16$ is valid only for the rare truly Gaussian peaks.

The photoion-photoelectron coincidence technique (PIPECO) has proved that the average kinetic energy released in any fragmentation reaction depends on the internal energy of the parent ion. Stockbauer (15) has constructed an instrument that can select parent ions with known internal energy which possesses the necessary conditions to measure the kinetic energy released on fragmentation as a function of the internal energy of the parent ion. So far it has been mostly used to study processes which do not yield metastable ions, for which electron bombardment methods continue to be applied (16).

Experimental results can be checked by other theoretical methods such as the Monte-Carlo method that reproduces a metastable peak from a kinetic energy distribution (17), and the Monte-Carlo calculation that simulates the shape of a metastable peak in the more complex situation of consecutive metastable reactions (18).

The first measurement of the kinetic energy release on the fragmentation of an ion formed as a consequence of an ion--molecule reaction in the ion source of a chemical ionisation mass spectrometer has been reported by Huntress (19). As regards the kinetic aspects of the fragmentation of metastable ions, they are important because they give an idea of how a molecule can store energy above its dissociation limit and what determines its breakdown.

The unimolecular decomposition of excited ions under electron impact and how they depend on time is a fundamental problem which must be considered in different perspectives, whether we are dealing with a large molecule or a small one, i.e., a di- or tri--atomic molecule. As to the latter, the conventional quasi--equilibrium theory (QET) is not likely to apply, since small ions have such simple structures that application of statistical methods does not seem to be appropriate. Instead the process can be well described in terms of potential energy surfaces.

As it has already been mentioned, ions that decompose outside

the ion source are long-lived excited species. An excited ion
can spontaneously loose part or all of its excitation energy
through either a radiative transition to a lower state or through
a process of internal conversion, for instance, a predissociation
process. The latter has been studied, especially through its
consequences on the width and intensity of emission and absorption
bands. However, since each predissociated level originates
fragments that separate with discrete amounts of energy, the
measurement of such small quantities of energy results in a
precise knowledge of predissociated states. This is one reason why
mass spectrometric measurements can approach such a problem and
use the results to interpret the meaning of the occurrence of
metastable peaks in mass spectra. Several sophisticated techniques
have been developed that, through the mass analysis of ions obtained
by electron impact, make it possible to measure energies with an
accuracy of 10^{-4}eV or better (20).

The occurrence of a mechanism of predissociation can be
confirmed establishing the dissociation products, determining
their appearance energies, measuring the kinetic energy released
and the lifetimes of the predissociated levels.

It is worth mentioning that since Hipple's work on metastables,
attempts have been made to measure metastable lifetimes, very
often without great success. Recently Klapstein (21) has accurately
measured lifetimes of N_2O^+ and COS^+ in selected vibronic levels,
following some previous work that used an electron impact method
(22)(23) and also photo-electron-photon coincident measurements
(24-26). Cooks and al. (27) measured the kinetic energy release
for the unimolecular fragmentation of metastable ions as a function
of ion lifetime, varying the ion accelerating voltage, assuming
statistical distribution of the energy of the activated complex,
with the aim of estimating the energy dependence of the rate
constant. They concluded that the energy dependence of the unimolecular rate constant was considerably larger than when
established through ion abundance measurements, at least for
reactions involving the loss of HCN and H from the molecular ions
of respectively benzonitrile and benzene.

Usually some difficulties arise to estimate the correct
values of appearance energies of the fragment ions formed in the
decomposition of a metastable ion. Several methods have been tried
and different standards have been used to calibrate the energy
scale but the accuracy of the results is seldom satisfactory.
Burger and Holmes (28) have recently proposed an accurate and
reproducible method for measuring the appearance energy of the
daughter ion in a metastable decomposition using as calibrant a
metastable peak of a carefully chosen reference reaction, the
loss of CH_3^+ by $(CH_3CH_2)_2O$. Normalization of the efficiency curves
for both reactions was performed before calculating the difference

of appearance energies for the pair of selected peaks. Almoster
Ferreira and Santos (29) have used a similar process (without
normalization due to parallelism of both efficiency curves used)
to determine the appearance energy of the metastable reaction of
HCl loss by 1,2-dichloroethane using as calibrant the metastable
peak at apparent mass 20.51 in the mass spectrum of 2,3-dimethyl-
butane ($86^+ \rightarrow 42^+$), for which the appearance energy has been
accurately measured. The same calibrant has also been used to
determine appearance energies of daughter ions resulting from
the decomposition of small molecular metastable ions such as
CS_2^+, COS^+, CO_2^+ and CO^+.

Burger's method is supposed to apply even when the metastable
peak is broad and diffuse, in which case the kinetic energy
released is a function of the excited lifetime and the width of
the metastable peak is considered to be due to the occurrence of
a kinetic shift.

EXPERIMENTAL RESULTS: DISCUSSION AND INTERPRETATION

Our attention will be mainly focused on experimental results
obtained with small gaseous ions. They are excellent species to
give information about the details of fragmentation mechanisms,
because their electronic excited states are well separated and,
very often, thermochemical data for some of these states are
available.

Through the study of the unimolecular and collision induced
dissociation of ions formed in an excited or metastable state it
is possible not only to follow the mechanisms of their fragmentation but also to determine the thermochemistry involved in the
process. Collision induced dissociation of small molecular metastable ions can also be used to identify the electronic states of
the products that are formed following high energy collisional
excitation. Studies of this kind have yielded a large number of
publications due to their increasing interest (30-43).

As already mentioned, the internal energy of a metastable
ion is converted, upon dissociation, into different forms of
energy of the dissociation products. A method to identify energy
states of the products of ion dissociations using mass-analysed
ion kinetic energy spectrometry (MIKES) has been described by
Kim (42). It makes it possible to measure not only the excitation
energy of the reactant ion but also the kinetic energy released
on fragmentation and, subtracting the latter from the former, the
energy of the products can be calculated. Collision induced dissociations being a particular type of ion-molecule reactions can
be studied in a double focusing mass spectrometer if the translational energy is not too high. According to Kim (42) collision

induced dissociation (CID) involves a collisional excitation followed by unimolecular fragmentation. It is a process that resembles the unimolecular dissociation of a metastable ion decomposing in a field free region of a mass spectrometer so that it is possible to study it using the same methods. This shows that metastable ion decompositions play an important role on the identification of electronic states of ionic and neutral species.

Frequently each ion formed in the decomposition of small polyatomic molecular ions can be associated with a metastable transition. This is the case, for instance, of CS_2^+ in which evidence of four metastable ions was found through the corresponding decomposition reaction (39). An interesting situation occurs with S^+ ions which were found to be formed from three different metastable ions as shown in fig. 1.

The behaviour of low rate decaying metastable ions, such as CS^+, CO^+ and H_2S^+, decomposing according to the following reactions: $CS_2^+ \rightarrow CS^+ + S$ (41), $CO^+ \rightarrow O^+ + C$ (40) and $H_2S^+ \rightarrow HS^+ + S$ (44) under

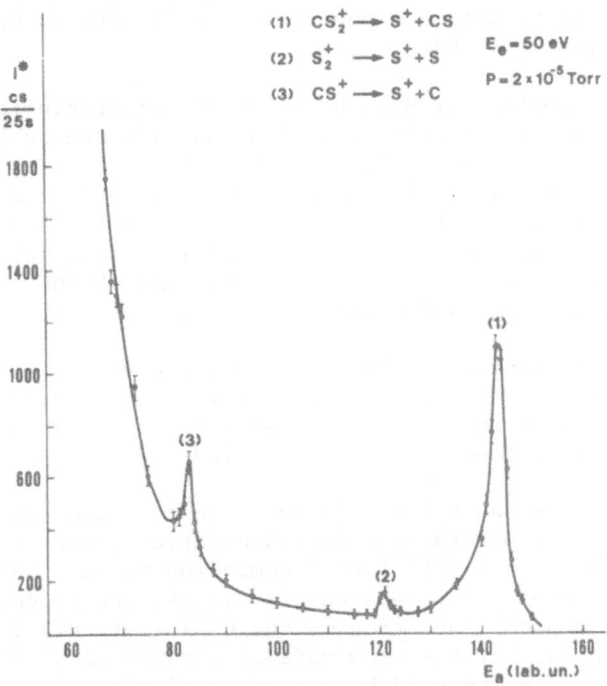

Fig. 1: S^+ daughter ions due to the decomposition of different metastable ions formed in the electron impact fragmentation of CS_2.

different conditions of pressure and ionising energies, has been
studied through measurements of the respective metastable peak
intensity as a function of the ion source pressure and of the
energy of ionising electrons. In each case the product ion results
from two different processes, a spontaneous unimolecular reaction
and a reaction induced by collision with the residual gas. A common
feature is that higher electron energies always enhance the unimolecular process, and a typical situation is shown in fig. 2.
Higher electron energies favour the process of generation of long-
-lived ions that are subsequently excited and decompose outside
the ion source, a situation analogous to that of a collision
induced dissociation (42). Since fragmentation following electron
impact is very similar to fragmentation due to collision with a
neutral species, the energies involved in both processes, even if
not known, must distribute among the same degrees of freedom of
the molecule. This may account for the enhancement of unimolecular
metastable processes by high electron energies.

It is noteworth that Cody and Freiser (45) have recently
developed a technique in which electron impact excitation
constitutes an alternative to collision induced dissociation
giving to the ICR technique capabilities equivalent to collision
induced dissociation (CID).

The slow unimolecular decomposition of metastable ions in
diatomic and other small polyatomic ions usually occurs by predissociation (46)(29-34) as suggested by Momigny in 1961 (47)(48).
To determine a predissociation mechanism involved in the dissociation of a particular excited ion it is necessary to know the correct
potential energy surfaces of the precursor ion. These are very
often calculated by *ab initio* calculations through the configuration interaction method which can provide the necessary data (49-51).

It should be mentioned that the kinetic energy released to
the fragments in the dissociation of small ions whenever repulsive
states are involved can also provide some information concerning
potential energy surfaces of the parent ion (42).

A few examples can illustrate the interpretation of slow
fragmentation reactions through predissociation processes. For
instance, for metastable CS_2^+ ions decomposing to give $CS^+ + S$ (39)
the maximum kinetic energy released is 0.4 eV, the process being
interpreted by a predissociation of the $C^2\Sigma_g^+$ state via the $^2\Sigma^-$
state. In a similar way the slow rate appearance of S^+ from metastable CS_2^+ can be interpreted by a predissociation of the B $^2\Sigma_u^+$
state by the $^4\Sigma^-$ state. A strong spin orbit coupling due to the
presence of sulphur atoms in the molecule favours such processes,
thus justifying the violation of the selection rule that forbids
$\Sigma^+ \leftrightarrow \Sigma^-$ correlations (52).

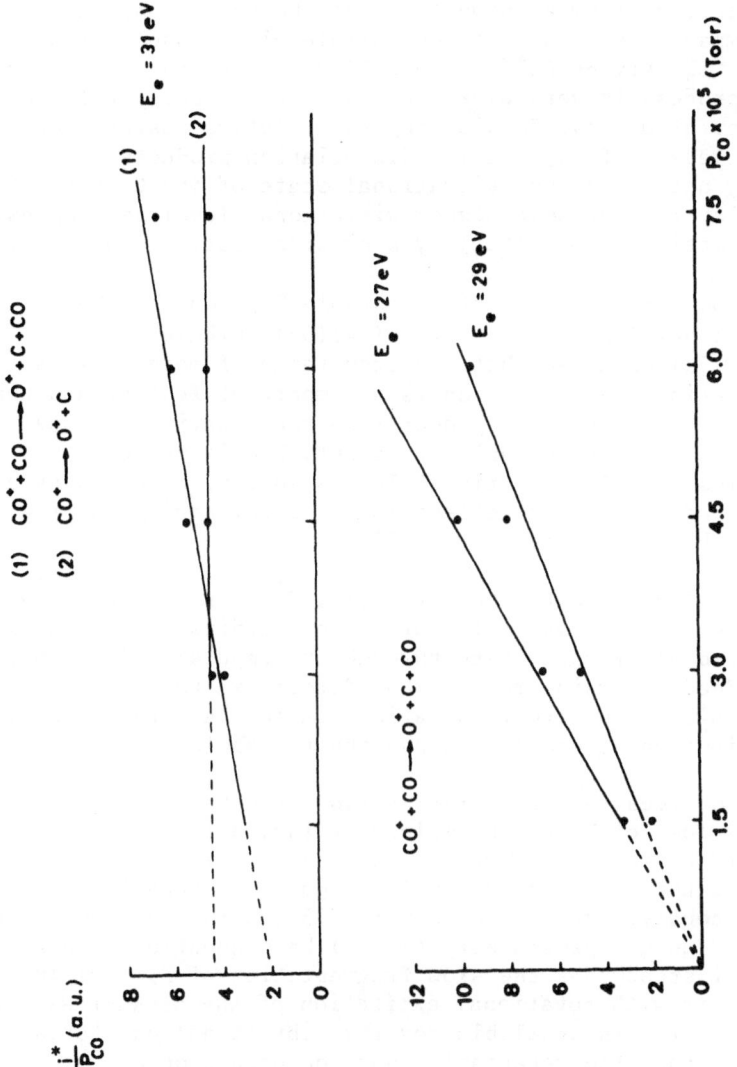

Fig. 2: Decomposition of CO^+ metastable ions: dependence of O^+ ions abundance both on the ion source CO pressure and on the energy of ionising electrons. Up to 29 eV only a collision induced process occurs, whilst at 31 eV collision induced and unimolecular decompositions occur simultaneously.

It has been shown by photoion-photoelectron coincidence spectroscopy (35) and photoion kinetic energy analysis that the CO_2^+ (C $^2\Sigma_g^+$) predissociates to $O^+(^4S_u) + CO$ ($^4\Sigma_g^+$). Again a pre-dissociation mechanism is involved and the first of these two methods makes it possible to determine the states of the products starting from a known state of the parent species. An electron impact study (40) of the unimolecular dissociation of the CO_2^+ (C $^2\Sigma_g^+$) giving $CO^+(^2\Sigma_g^+)$ and $O(^3P)$ led to the conclusion that this process is very weak and the kinetic energy release is of the order of 0.4 eV. Considering the relative positioning of the energy levels of CO_2^+ and its dissociation products, it is clear that, not the ground vibrational state of the C state of the molecular ion but some higher vibrational level is responsible for the formation of CO^+ ($^2\Sigma_g^+$) by a predissociation mechanism.

The slow decomposition of metastable SH_2^+ ions has been studied by several authors (44)(53)(54)(55). Dibeler and Rosenstock (53) observed that the occurrence of metastable de-compositions in a triatomic ion is an important feature and that when hydrogen was replaced by deuterium the behaviour of the molecular ion was considerably different. Predissociation probabilities for all vibrational levels must be affected as a consequence of the isotope effect in the predissociation of a bound state by a repulsive state (56).

At energies close to the threshold, S^+ ions are due to a slow unimolecular decomposition of HS^+ by predissociation of its first excited electronic state through the repulsive $^4A_2$ state. This is indeed the main process, but direct excitation to the $^2B_2$ state followed by predissociation via the $^4A_2$ state can give some contribution at energies above the threshold (54).

It is obvious that the dissociation of H_2S^+ to give S^+ also implies the loss of H_2 by the molecular ion, a fact which, according to Herzberg (52) may be the reason for the lack of sharpness of the predissociation limit due to a tunneling process. Recently, Jarrold and Bowers (55), using mass analysed ion kinetic energy spectrometry (MIKES) have questioned this, suggesting instead that the slow fragmentation of H_2S^+ yielding S^+ ions occurs with rotational excitation of the product H_2 since not enough energy is available for its vibrational excitation. This implies that the metastable reaction occurs on a $^2A_1 - {}^2B_1 - {}^4A_2$ surface and not on a $^2A_1 - {}^2B_2 - {}^4A_2$ surface according to calculations by Hirsch and Bruna (57).

From H_2S^+, metastable HS^+ ions are also formed which, in turn, decompose into S^+ + H. Three ionisation thresholds were found (44) for this reaction, at energies of 14.0 ± 0.3 eV, 15.7 ± 0.7 and 21.0 ± 1.0 eV. The shape of the metastable peak indicates that it consists of the overlapping of two peaks:

a nearly Gaussian peak and a flat top one. If a diagram showing
the energy levels for the neutral H_2S and for the radical $HS^{\cdot}$ is
drawn, it is easily understood that the processes occurring at
14.0 eV and at 15.7 eV are not related to metastable HS^+ ions
arising from the dissociative ionisation of H_2S, but to $HS^{\cdot}$
radicals, probably formed by pyrolysis of hydrogen sulphide in
the filament. Pyrolysis of molecules in the filament of an
electron impact instrument has also been considered appropriate
to describe some experimental results obtained in the study of
the dissociative ionisation of other small molecules (58).

The occurrence of the usually called anomalous metastable
peaks, that is, peaks due to the decomposition of a metastable
ion which appear in the spectrum localized at an apparent mass
to charge ratio higher than the calculated value, is an established
fact observed in several metastable decomposition reactions. A
typical example is the peak associated with the loss of a hydrogen
atom by methanol ions, due to a reaction induced by the collision
of the parent ion with the walls of the flight tube inside the

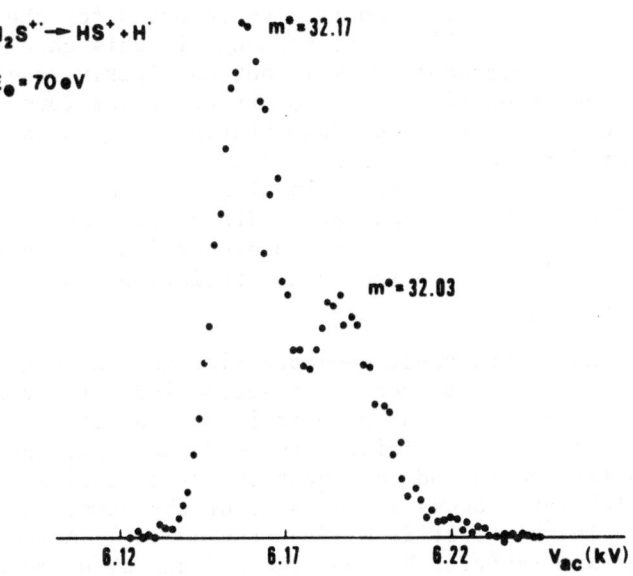

Fig. 3: The two components of the metastable peak associated
with the reaction $H_2S^+ \rightarrow HS^+ + H$ (Ref. 44).

magnet (59)(60).

For the slow decomposition reaction $H_2S^+ \rightarrow HS^+ + H$, an apparently anomalous experimental observation has also been reported but not explained (44). The shape of its metastable peak recorded in a double focusing electron impact mass spectrometer operated on the Barber and Elliot mode (61) shows an approximately Gaussian form. But if the resolution of the instrument and the ionising electron energy are both increased, a second component appears shifted to a lower scanning ion accelerating voltage, which is more important than the component corresponding to the calculated apparent mass, as shown in fig. 3.

The characteristics of the second peak made visible as a consequence of the modification of the operating conditions of the instrument, resemble those reported by Cooks (59) for SID peaks: it is a peak due to a loss of H reaction, showing linear dependence on the source pressure and an apparent gain in momentum. However, one important difference is that the decomposition of the metastable H_2S^+ occurs in the first field free region of the instrument, which seems to be ruled out for the mentioned SID peaks. A similar situation is also observed for the metastable decomposition $CS_2^+ \rightarrow CS^+ + S$ (58) although in this case the structure of the metastable peak is not as clearly resolved as for the H_2S^+ decomposition. Yet, no structure was found in the metastable peaks for the slow fragmentation of metastable D_2S^+ giving DS^+ or for the reaction $HDS^+ \rightarrow DS^+ + H$. It is possible that, as a consequence of intermolecular isotope effects (56)(62)(63), both dissociations occur after direct excitation to a repulsive state of the respective molecular ion, and not by a predissociation mechanism after a non radiative transition at a curve crossing.

For the H_2S^+ metastable decomposition the influence of increasing pressure of H_2S upon the reaction indicates that the peak at the calculated apparent mass is the result of a unimolecular process and a collision induced one occurring simultaneously, whilst the second component observed results from a truly unimolecular process independent of the sample pressure taking place in the first field free region (44). Although there are striking similarities between the SID anomalous metastable peak reported by Cooks and the results observed on the H_2S^+ decomposition leading to HS^+, the latter must correspond to a different process.

On the above grounds one would expect the existence of a long-lived highly excited H_2S^+ ion which is able to cross the acceleration region of the ionisation chamber and which decomposes in the first field free region, that is, long before it could reach the flight tube inside the magnet.

The analysis of the photoelectrons spectrum and of the emission spectrum (64) of H_2S^+ clearly shows that the dissociation of H_2S^+ into $HS^+ + H$ should preferentially involve the predissociation of the $^2B_2$ state by the $^4A_2''$ state, this explaining the normal component of the composite metastable peak. However, the predissociation of the $^2A_1$ state by the same repulsive quartet state is not impossible, but it should occur by a third order interaction via the $^2B_1$ state. Considering that the latter situation would only involve high vibrational levels of the $^2A_1$ state, considering also that high vibrational levels play a smaller role in collision processes, and that for the $^2A_1$ state predissociation can favourably compete with radiative deexcitation, the appearance of the unimolecular process could be explained. This is consistent with the fact that a smaller amount of kinetic energy is released in the process involving the truly unimolecular decomposition and which gives rise to the second component observed in the metastable peak, in comparison with the kinetic energy release calculated for the peak at the expected mass to charge ratio. The apparent gain in momentum is most probably due to the fact that part of the excitation energy of the parent ion is converted into translational energy of the daughter ion, which is statistically possible.

A matter of considerable interest is the influence of isotopic substitution on the fragmentation of isotopically labelled metastable ions. It was pointed out that, when the loss of H by metastable H_2S^+ and the loss of D by metastable D_2S^+ were compared, a different behaviour of the two molecular ions was observed (53). Since the mass of deuterium is approximately twice the mass of hydrogen, hydrogen/deuterium substitution provides a good insight into the isotope effects in general. Moreover, it will be particularly suitable to compare intramolecular and intermolecular isotope effects (65).

When small metastable ions decomposition is considered, the effect of isotopic substitution on predissociation must simultaneously be taken into account. It is usually admitted that potential energy surfaces of a molecular or molecule ion system are not altered as a consequence of isotopic substitution, which, nevertheless, modifies vibrational frequencies and moments of inertia. The zero-point energy is lower for the ion which has the heavier isotope than for the corresponding unlabelled ion. Therefore, for the two parent ions, which are isotopically distinct but chemically identical, the critical energies for dissociation are different, which is a relevant factor in the occurrence of intermolecular kinetic isotope effects. If a predissociation mechanism is called for to explain the decomposition of an isotopically homogeneous metastable precursor ion, since isotope effects in predissociation are strongly dependent on the shape of both surfaces involved in the predissociation, it is

possible that the equivalent reaction in the isotopically
labelled reactant ion does not occur by the same mechanism.

As regards intramolecular effects depending only on the
properties of the common transition state for the two decomposition reactions under comparison, it is usually very difficult to
decide what will be the influence of the isotopic substitution,
unless kinetic studies to determine rates of decomposition are
undertaken.

On studying the relatively low rate ionic processes occurring in gases, namely, the slow decomposition of metastable ions,
one readily arrives to the conclusion that sophisticated improvements in instrumentation are enabling particularly good conditions
of research, and also that good theoretical support is being
offered by theoreticians. However, even if an understanding of
these processes is steadily improving, an enormous amount of
work is still required in order to provide the necessary knowledge
of intramolecular dynamics leading to the fragmentation of ions
in the gas phase.

ACKNOWLEDGEMENTS:

Financial support from the Instituto Nacional de Investigação Científica, Portugal, is gratefully acknowledged.

REFERENCES

1 - Smyth, H.D.: 1931, Rev. Mod. Phys. 3, p.347.
2 - Friedlander, E., Kallmann, H., Lasareff, W., and Rosen, B.:
     1932, Z. Physik. 76, p.70.
3 - Hipple, J.A., Fox, R.E., and Condon. E.U.: 1946, Phys. Rev.
     69, p. 347.
4 - Aston, F.W.: 1920, Proc. Cambridge Phil.Soc. 19, p. 317.
5 - Mc Gowan, W., and Kerwin, L.: 1963, Can.J.Phys. 41, p. 316.
6 - Derrick, P.J., and Burlingame, A.L.: 1974, Acc. Chem. Res.
     7, p. 328.
7 - Cameron, D., Clark, J.E., Kruger T.L., and Cooks, R.G.:1977,
     Org.Mass Spectrom.12(2), p. 112.
8 - Jennings, K.: 1984, this book, and references therein.
9 - Schmit, J.P., and Baulay, G.: 1981, Anal.Chemistry, 53,
     p. 2359.
10 - Cooks, R.G., Beynon, J.H., Caprioli, R.M., and Lester, G.R.:
     1973, "Metastable Ions", Elsevier Scient. Pub. Co., Amsterdam.
11 - Lester, G.R.: 1971, "Recent Topics in Mass Spectrometry",
     Ed. R.I.Reed, Gordon and Breach Sci. Pub., London.

12 - Holmes, J.L., Osborne, A.D., and Weese, G.M.: 1975, Int. J. Mass Spectrom. Ion Phys., 19, p. 207.

13 - Terwilliger, D.T., Elder, J.F., Beynon, J.H., and Cooks, R.G.: 1975, Int. J. Mass Spectrom. Ion Phys. 16, p. 225.

14 - Holmes, J.L., and Osborne, A.D.: 1977, Int. J. Mass Spectrom. Ion Phys. 23, p. 200.

15 - Stockbauer, R.: 1977, Int. J. Mass Spectrom.Ion Phys. 25, p. 89.

16 - Rosenstock, H.M., Buff, R., Almoster Ferreira, M.A., Lias, S.G., Parr, A.C.,Stockbauer, R.L., and Holmes, J.L.: 1982, J. Am. Chem. Soc. 104, p. 2337.

17 - Holmes, J.L., and Osborne, A.D.: 1978, Int. J. Mass Spectrom. Ion Phys. 27, p. 271.

18 - Holmes, J.L., Cartledge, K., and Osborne, A.D.: 1979, Int. J. Mass Spectrom. Ion Phys. 29, p. 171.

19 - Huntress Jr., W.T., Sharma, D.K.S., Jennings, K.R. and Bowers, M.T.:1977, Int.J.Mass Spectrom.Ion Phys. 24, p. 25.

20 - Schopman, J., Fournier, P.G., and Los, J.: 1973, Physica 73, p. 518.

21 - Klapstein, D., and Maier, J.P.: 1981, Chem. Phys. Letters 83, p. 590.

22 - Dayton, I., Dalby, F., and Bennet, R.G.: 1960, J. Chem. Phys. 33, p. 179.

23 - Fink, E.H., and Welge, K.H.: 1968, Z. Naturforsch 23a, p. 358.

24 - Bloch, M., and Turner, D.W.: 1975, Chem. Phys. Letters 30, p. 344.

25 - Maier, J.P., and Thommen, F.: 1980, Chem. Phys. 51, p. 319.

26 - Frey, R., and al.: 1978, Chem. Phys. Letters 54, p. 411.

27 - Cooks, R.G., Kim, K.C., Keough, T., and Beynon, J.H.: 1974, Int. J. Mass Spectrom. Ion Phys. 15, p. 271.

28 - Burger, P.C., and Holmes, J.L.: 1982, Org. Mass Spectrom. 14, p. 123.

29 - Almoster Ferreira, M.A., and Santos, J.A.: 1972, 6th Mass Spectroscopy Group Meeting, Swansee.

30 - Fiquet-Fayard, F., and Guyon, P.M.: 1966, Mol. Phys. 11, p. 17.

31 - Haugh, M.J., and Bayers, K.D.: 1971, J. Chem. Phys. 75, p. 1472.

32 - Lorquet, A.J., Lorquet, J.C., Wankenne, H., and Momigny, J.: 1971, J. Chem. Phys. 55(8), p. 1971.

33 - Wells, W.C., Borst, W.L., and Zipf, E.C.: 1972, J. Geophys. Res. 77, p. 69.

34 - Mathieu, G., Wankenne, H., and Momigny, J.: 1972, Chem. Phys. Let. 17(2), p. 260.

35 - Eland, J.H.D.: 1972, Int. J. Mass Spectrom. Ion Phys. 9, p. 397.

36 - Brehm, B., Eland, J.H.D., Frey, R., and Küstler, A.: 1973, Int. J. Mass Spectrom. Ion Phys. 12, p. 213.

37 - Eland, J.H.D.: 1973, Int. J. Mass Spectrom. Ion Phys. 12, p. 389.

38 - Brehm, B., Eland J.H.D., Frey R., and Küstler, A.: 1973, Int. J. Mass Spectrom. Ion Phys. 12, p. 197.

39 - Momigny, J., Mathieu, G., Wankenne, H., and Almoster Ferreira,
     M.A.: 1973, Chem. Phys. Letters, 21(3), p. 606.
40 - Almoster Ferreira, M.A., and Vaz Pires, M.M.: 1974, "Advances
     in Mass Spectrometry" vol. 6, ed. A.R. West, App. Science
     Pub., p. 939.
41 - Momigny, J., Mathieu, G., Wankenne, H., Flame, J.P., and
     Almoster Ferreira, M.A.: 1974, "Advances in Mass Spectrometry,
     vol. 6, ed. A.R. West, App. Science Pub., p. 923.
42 - Kim, K.C., Uckotter, M., Beynon, J.H., and Cooks, R.G.: 1974,
     Int. J. Mass Spectrom. Ion Phys. 15, p. 23.
43 - Olivier, J.L., Locht, R., and Momigny, J.: 1982, Chem. Phys.
     68, p. 201.
44 - Vaz Pires, M.M.: 1979, Ph.D. Thesis, University of Lisbon.
45 - Cody, R.B., and Freiser, B.S.: 1979, Analytical Chem. 51(4),
     p. 547.
46 - Wankenne, H., and Momigny, J.: 1971, Int. J. Mass Spectrom.
     ion Phys. 7, p. 227.
47 - Momigny, J.: 1961, Bull. Soc. Chim. Belg. 70, p. 291.
48 - Momigny, J.: 1966, Mem. Soc. Roy. Sci. Liège, 13, p. 138.
49 - Lorquet, A.J., Lorquet, J.C., Wankenne, H., and Lefebre-
     -Brion, H.: 1971, J. Chem. Phys. 55, p. 4053.
50 - Gallway, C., and Lorquet, J.C.: 1978, Chem. Phys. 30, p. 169.
51 - Lorquet, J.C.: 1982, this book and references therein.
52 - Herzberg, G.: 1967, "Molecular Spectra and Molecular
     Structure", III Electronic Spectra and Electronic Structure
     of Polyatomic Molecules, Van Nostrand, N.Y.
53 - Dibeler, V.H., and Rosenstock, H.M.: 1963, J. Chem. Phys.
     39(11), p. 3106.
54 - Jones, E.G., Beynon, J.H., and Cooks, K.G.: 1972, J. Chem.
     Phys. 57(8), p. 3207.
55 - Jarrold, M.F., Illies, A.J., and Bowers, M.T.: 1982, Chem.
     Phys. 65, p. 19.
56 - Fiquet-Fayard, F.: 1970, J. Chim. Phys. Special Issue 57-63.
57 - Hirsch, G. and Bruna, P.J.: 1980, Int. J. Mass Spectrom. Ion
     Phys. 36, p. 37.
58 - Almoster Ferreira, M.A.: unpublished results.
59 - Cooks, R.G., Ast, T., and Beynon, J.H.: 1975, Int. J. Mass
     Spectrom. Ion Phys. 16, p. 348.
60 - Cooks, R.G., and Terwilliger, D.T., Ast, T., and Beynon, J.H.:
     1975, J. Am. Chem. Soc. 97, p. 1583.
61 - Barber, M., and Elliot, R.M.: 1969, ASTM Committee E-14, 12th
     Annual Conference on Mass Spectrometry, Montreal.
62 - Trone, M., Goursaud, S., Azria, R., and Fiquet-Fayard, F.:
     1973, J. Phys.(Paris) 34, p. 381.
63 - Fiquet-Fayard, F., Sizun, M., and Abgrall, H.: 1976, Chem.
     Phys. Letters 37(1), p. 72.
64 - Dixon, R.N., Duxbury, G., Horani, M., and Rostas, J.: 1971,
     Mol. Phys. 22, p. 977.
65 - Derrick, P.: 1982, to be published.

APPLICATIONS OF THE FRANCK-CONDON FACTORS TO POLYATOMIC
MOLECULES

R. BOTTER, J. CARLIER

Département de Physico-Chimie, CEN/Saclay
91191 Gif-sur-Yvette Cedex (France)

ABSTRACT
Franck Condon factors (FCF) for polyatomic molecules will be
discussed. Elaborate potential functions can only be used in a
quasi diatomic approximation. Most of the polyatomic calculations
of FCF are made with a harmonic potential. Various methods of
calculations using this approximation will be briefly presented.
The problem of transformation of normal coordinates from one
state to the other will be discussed. Approximations in this
transformation may lead to large errors. Three types of applications will be developed for the ionization of polyatimoc molecules.

1) INTRODUCTION

Calculations of Franck Condon factors (FCF) for diatomic molecules
have been largely developed using elaborated potential functions:
Morse $\lceil 1 \rfloor$ or Rydberg, Klein, Reese $\lceil 2 \rfloor$. Very accurate results
are obtained for the overlap integral. The differences observed
between experimental and calculated vibronic transition probabilities may be ascribed to the variation of the electronic
transition moment with internuclear distance. This variation is
represented by the r centroid $\lceil 1 \rfloor$.

For polyatomic molecules the problem is more complicated not
only because of the increase of the dimension of the representation (3 N - 6 vibriational modes) but because the exact description of the potential hypersurface is no more possible. Morse
potential functions or RKR numerical potentials for polyatomic
systems may only be used in a quasi diatomic approximation. This

303

M. A. Almoster Ferreira (ed.), Ionic Processes in the Gas Phase, 303–325.

is valid only if there is one totally symmetric vibrational mode
in the molecule like in linear symmetric triatomic molecules:
CO_2 $[3]$. In all other cases, reducing the polyatomic vibrational
problem to a quasi diatomic one may lead to larger errors than by
using the exact vibrational treatment in the harmonic approximation $[4, 5]$.

The normal vibrations are expressed in terms of normal coordinates
and the frequencies are calculated with the well known GF matrix
method developed by WILSON $[4]$. The normal coordinates are
related to the internal symmetry coordinates by the relation
S = L Q (1) where S and Q are the column matrices of the internal
symmetry and the normal coordinates and L a square matrix which
defines the transformation between the two sets of coordinates.
The symmetry coordinates are combinations of internal displacement
coordinates (variation of distances and angles). Therefore a
normal vibration can generally not be related to a simple variation of distances and (or) angles as we will see below. This makes
the quasi diatomic approximation non valid. However in some cases
like Van der Waals molecules the mixing of the low energy coordinates between the molecules and the internal coordinates of the
initial molecules is so low that the new (Van der Waals) coordinate
may be considered as a quasi diatomic bond.

After a brief survey of the F.C.F. calculation in the harmonic
approximation, the Duchinsky effect (mixing of normal modes) will
be discussed. Three examples of application of the F.C.F. will
then be given.

II) FRANCK CONDON FACTORS CALCULATIONS FOR POLYATOMIC MOLECULES

Various methods have been developed for the calculation of the
F.C.F. in the harmonic approximation in polyatomic molecules.
D.C. MOULE $[6]$ has given recently a general review on some of
these methods. We will briefly recall the general development
of SHARP and ROSENSTOCK $[7]$ and only mention the other methods.

These authors were the first to take into account the exact
Duchinsky transformation of the normal coordinates $[8]$ between
the initial and final states. They used the generating function
method for the integration of the product of the Hermite poly-
nomials.

The vibronic transition probabilities in the Born Oppenheimer
approximation are given by:

$$P_{e', n', e'', n''} = |<\Psi'_{n'}(Q') |R_{e'e''}(Q)| \Psi''_{n''}(Q'')>|^2 \qquad (2)$$

where $\Psi'_{n'}(Q')$ and $\Psi''_{n''}(Q'')$ are the vibrational wavefunctions for the final and initial states respectively with n' and n" vibrational quanta, Q' and Q" being the normal coordinates (harmonic approximation).

$$R_{e'e''}(Q) = \langle\Psi'_{e'}(r',Q') \,|M_e|\,\Psi''_{e''}(r'', Q'') \,\lambda\rangle \tag{3}$$

$\Psi'_{e'}(r', Q')$ and $\Psi''_{e''}(r'', Q'')$ are the electronic wave functions of the final and initial states, r' and r" the electronic coordinates, M_e the electronic dipole moment.

If we assume that $R_{e'e''}(Q)$ is a slowly varying function of Q we may expand the electronic transition moment about the equilibrium position of one of the states:

$$R_{e'e''}(Q') = R_{e'e''}(0) + \sum_{i=I}^{3N-6} \left(\frac{\delta R_{e'e''}}{\delta Q_i}\right)_0 Q_i +$$

$$\frac{1}{2} \sum_i \sum_j \left(\frac{\delta^2 R_{e'e''}}{\delta Q_i \, Q_j}\right)_0 Q_i \, Q_j + \ldots \tag{4}$$

If only the first term is important (small Q dependence of $R_{e'e''}(Q)$) then the transition probability will be given by the classical formula

$$P_{e'n'e''n''} = |R_{e'e''}|^2 |\langle\Psi'_{n'}(Q') |\Psi''_{n''} (Q'')\rangle|^2 \tag{5}$$

$$I_{n'n''} = \langle\Psi'_{n'}(Q') \,|\Psi''_{n''}(Q'') \rangle \tag{6}$$

is the overlap integral. The F.C.F. are given by

$$P_{n'n''} = |I_{n'n''}|^2 = |\langle\Psi'_{n'}(Q') \,|\Psi''_{n''}(Q'') \rangle|^2 \tag{7}$$

The selection rules for such transitions have been given [9]. If the initial state is totally symmetric (ground vibrational state) only totally symmetric vibrations of the final state may be excited.

In order to take into account the forbidden transitions, the higher terms of equation (4) have to be considered. The vibrational wave function in the harmonic approximation is

$$\Psi_n(Q) = \prod_{i=1}^{3N-6} \Psi_{n_i} (Q_i) \tag{8}$$

$$\Psi_{n_i}(Q_i) = N \exp\left(-\frac{1}{2} \gamma_i Q_i^2\right) H_{n_i} (\gamma_i^{1/2} Q_i) \tag{9}$$

N = normalisation factor

$$\gamma_i = \frac{4\Pi^2 \nu_i}{h} \quad \nu_i = \text{vibrational frequency.}$$

H_{ni} = the n_i^{th} Hermite polynomial.

The generating function method has been used by SHARP and ROSENSTOCK to perform the integration of equation (7). A general relation has been given for the transition probabilities from any vibrational level of the initial state to all the vibrational levels of the final state. A simplified expression has been given (9) for the simultaneous excitation of two normal modes in the final state for a transition from the ground vibrational state.

Without going into more details this method has been used and further developed by many authors.

COON et al. [10] developed the first the calculation of F.C.F. in polyatomic molecules in the harmonic approximation. They used the simplified transformation of the normal coordinates (see equation (15)) and F.C.F. are just given by square one dimensional overlap integrals. As will be shown this may not always lead to correct results.

WARSHEL and KARPLUS [11] and LUCAS [12] have worked on the normal coordinate transformations. LUCAS has discussed the effect of the transformation matrix between internal coordinates and of anharmonicity.

As pointed out above, MOULE [6] has given a general review on the polydimensional Franck Condon method.

Many other calculations using SCHRÖDINGER wave mechanics have been given. BARANOV et al. [13] have developed a general algorithm for the numerical calculation of the F.C.F. for all initial and final vibrational states for allowed as well as for forbidden transitions. They also calculate the vibrational overlap integral using Gaussian wavefunctions.

FAULKNER and RICHARDSON [14] have also developed a method of direct transformation of the normal coordinates.

In all these calculations the equilibrium geometries of the two states have to be known in order to calculate the F.C.F. However, the geometry of the final state in photoabsorption or photo-ionization is generally not known. It can be determined by fit-ting the calculated and experimental F.C.F. This method of

introducing the coordinates of the final state as parameters
in the calculation may be lengthy.

DOCTOROV et al. $\lceil$ 15 $\rfloor$ have developed a method for the direct
evaluation of the equilibrium geometry of the final state from
the vibrational transition probabilities. These calculations
have been made in the harmonic approximation using SCHRÖDINGER
wave mechanics and the second quantization method. The forbidden
transitions have also been evaluated by these authors.

CEDERBAUM and DOMKE $\lceil$ 16 $\rfloor$ developed a very elegant and powerful
method of calculation of the vibronic spectra. These authors use
the second quantization approach and the Green's function method
to calculate the potential energy of the final state in normal
coordinate space at the equilibrium geometry of the initial state.
The transition probability is expressed in terms of the first and
second derivative of the potential energy along the normal modes
of the initial state

$$\kappa_i = \frac{1}{\sqrt{2}} \left(\frac{\delta E(Q)}{\delta Q_i}\right) \quad (10) \qquad \gamma_{ij} = \frac{1}{4} \left(\frac{\delta^2 E(Q)}{\delta Q_i \delta Q_j}\right)_0 \qquad (11)$$

This method does not necessitate the knowledge of the final
state equilibrium geometry and it is less dependent on the
anharmonic corrections in the case of large geometry changes
between the two states. The authors also state that their method
is much simpler. It must however be remembered that it needs a
fairly accurate ab initio molecular orbital calculation in order
to obtain a good precision on the derivative of the potential
surfaces. Vibronic interactions may explicitly be taken into
account in this development.

This last method avoids also the problem of transformation of
normal coordinates from one state to the other given by
DUCHINSKY. Let us briefly discuss this problem.

III) THE DUCHINSKY EFFECT

The overlap integral is calculated in normal coordinate space
of one of the two states. Therefore one of the vibrational
wavefunctions has to be expressed in terms of the normal coordinate of the other state.

The general transformation of the normal coordinate has been
given by DUCHINSKY $\lceil$ 8 $\rfloor$

$$\underset{\sim}{Q'} = \underset{\sim}{J} \, \underset{\sim}{Q''} + \underset{\sim}{K} \qquad (12)$$

$\underset{\sim}{Q'}$ and $\underset{\sim}{Q''}$ normal coordinate vector matrices of the final and
initial state.

J is a square matrix corresponding to the rotation of the coordinates between the two states,

K a column matrix corresponding to the translation of the coordinates.

S = LQ (1) gives the transformation of internal to normal coordinates.

J and K may be calculated from the relation $\begin{bmatrix} 9 \end{bmatrix}$

$$S' = AS'' + R \qquad (13)$$

S' and S'' the internal symmetry coordinates are directly related to the cartesian displacement coordinate $\begin{bmatrix} 4 \end{bmatrix}$ and A and R can be deduced from the equilibrium geometry of the two states.

J and K are given by

$$J = (L')^{-1} AL'' \quad (14) \qquad K = (L')^{-1}R$$

The simplified transformation

$$Q' = Q'' + K \qquad (15)$$

has been used by COON et al. $\begin{bmatrix} 10 \end{bmatrix}$, assuming that there is a one to one correspondence between the normal coordinates. It is possible in this case to use a quasi diatomic approximation, the F.C.F. will be given by one dimensional square overlap integrals. However such an approximation is only valid if the equilibrium geometry of the two states is nearly identical. Once the rotation of the coordinate becomes important, the A matrix is far from being diagonal. If we take the $NH_3^+(X)$ $\tilde{N}H_3(X)$ transition the A matrix is given for the totally symmetric coordinates in the C_{3v} pyramidal representation.

$$\begin{vmatrix} S'_1 \\ \\ S'_2 \end{vmatrix} = \begin{vmatrix} 0.895 & -0.445 \\ \\ 0.335 & 0.793 \end{vmatrix} \begin{vmatrix} S''_1 \\ \\ S''_2 \end{vmatrix} + \begin{vmatrix} R_1 \\ \\ R_2 \end{vmatrix}$$

The non diagonal terms represent about 50%. The internal and normal coordinates are therefore mixed through this transformation.

In the case of C_2H_4, the L^{-1} matrix elements for the transformation $Q = L^{-1}$ are given in table 1 for Q_2 and Q_3. It can be seen that the normal coordinates are a mixture of stretching and bending coordinates. Therefore during the transition $C_2H_4^+(X) \leftarrow$ $C_2H_4(X)$ which changes mostly the C-C bond length, both the stretching and bending modes can be excited. Only the harmonic approximation introduces presently such a mixture of coordinates.

Three applications of the F.C.F. calculation will be given below and illustrated.

IV) VIBRATIONAL STRUCTURE OF PHOTOELECTRON SPECTRA

Although the vibronic structure of excited electronic states of polyatomic molecules has been the subject of many comparisons between experimental and calculated spectra [13, 14, 15, for example] we will restrict the present paper to ionized states.

High energy photoelectron spectra (HeI, HeII) give generally the vibrational transition probabilities for the various ionic states of a molecule (below the photon energy [17]). At this high energy, autoionization from a superexcited neutral state may not occur, because no neutral state, in coincidence with the discrete photon line may exist. Therefore, the intensity of the various peaks in an electronic band, after correction for the apparatus function, represents the vibrational transition probabilities. This however is not the case for the threshold photoelectron spectra where autoionization generally dominates over direct ionization [18, 19]. Another important problem may arise from resonances between neutral excited and ionic states. These resonances [20] not only enhance the ionization cross sections, but the transition probabilities to the various vibrational levels in the perturbed state may depart significantly from that for the direct ionization. Moreover they are very broad (about 1 eV) and may appear a few eV above the ionization potential of the state: these resonances have only been studied in a few molecules (N_2, O_2, C_2H_2) and we will assume that they don't appear in the examples given below.

TABLE I – L^{-1} matrix elements for C_2H_4 and C_2D_4 and for the corresponding ions.

C_2H_4	S''_1	S''_2	S''_3	C_2D_4	S''_1	S''_2	S''_3
Q''_2	-0.15	-0.96	0.25	Q''_2	-0.30	-0.95	0.08
Q''_3	0.01	0.89	0.46	Q''_3	-0.09	0.69	0.78
$C_2H_4^+$	S'_1	S'_2	S'_3	$C_2D_4^+$	S'_1	S'_2	S'_3
Q'_2	-0.14	-0.94	0.31	Q'_2	-0.23	-0.97	0.03
Q'_3	0.02	0.91	0.42	Q'_3	-0.10	0.63	0.77

$$S_1 = \frac{1}{2}\left(\Delta r_3 + \Delta r_4 + \Delta r_5 + \Delta r_6\right)$$

$$S_2 = \Delta R$$

$$S_3 = \frac{1}{2}\left(\Delta T_3 + \Delta T_4 + \Delta T_5 + \Delta T_6\right)$$

$C_2H_4 \; - \; C_2D_4$

$C_2H_4' \; - \; C_2D_4'$

$r \;\rightarrow\; C - H$

$R \;\rightarrow\; C - C$

$T \;\rightarrow\; H - C - H$

$\varepsilon = 0°$

$\varepsilon = 10°$

$R = 1.339 \; \overset{o}{A}$

$R = 1.41 \; \overset{o}{A}$

ε = half angle between the H–CH planes

$\alpha = \alpha'$

The calculations of the F.C.F. in photoelectron spectra of polyatomic molecules have been made by the various methods described above. The precision of the results depends on the methods

a) $H_2O^+ (X^2B_1)$ $H_2O (X^1A_1)$ transition.

The F.C.F. for the three isotopic species H_2O, HDO and D_2O have been calculated $/ 21 /$ with the method of SHARP and ROSENSTOCK and compared with the experimental results. The variation of the geometry and the vibrational frequencies between the two states are not important. Therefore, the F.C.F. may be calculated with a good accuracy in the harmonic approximation. The results are given in table II. The agreement between the calculated and the experimental values is good. Only the probability for the (2, 0, 0) vibrational overtone is outside the estimated error limits.

b) $NH_3^+ (X^2A_1")$ $NH_3 (X^1A_1)$ transition.

This transition has been widely studied experimentally and theoretically. The interest is that the ground state is pyramidal (symmetry C_{3v}) whereas the ionized as well as the Rydberg states are planar (symmetry D_{3h}). The large change in angle induces a long vibrational progression corresponding mostly to the ν_2 symmetric bending mode. It must be remembered at this point that a variation of the geometry between the two states induces large excitation of the vibrational modes in contrast to a change of the vibrational frequencies which has little effect on the vibrational excitations $/ 9 /$.

DOMCKE et al. $/ 16 b /$ have made a detailed comparison of the various methods of calculation. Some of their conclusiions will be briefly mentioned. The use of more elaborate potential functions for the calculation of the F.C.F. $/ 22, 23 /$ with separation of the two normal coordinates Q_1 and Q_2 and vibrations ν_1 and ν_2 does not lead to better results as the harmonic approximation $/ 24, 25 /$. All these calculations fail to represent correctly the vibrational sequence of the ionic band. DOMKE et al. $/ 16 b /$ using their method were able to reproduce with a remarkable precision the vibrational structure of the experimental spectrum of RABALAIS $/ 26 /$.

Extended calculations have also been performed by M. BEFFERT-FORGES $/ 27 /$ using the method of SHARP and ROSENSTOCK. Anharmonicities have been introduced using first order perturbation theory :

$$\Psi_n = \Psi_n^o \sum_{m \neq n} \frac{W_{m\,n}}{E_m^o - E_n^o} \qquad (16)$$

TABLE II - Comparison between calculated and measured transition probabilities for the ionization of H_2O, D_2O and HDO.

H_2O

Peak number	1	2	3	4	5
Assignment	000	001	100 (+ 002)	101 (+ 001)	200 (+020)(+102)
Experimental F.C.F.	0.736±0.03	0.081±005	0.146±0.004	0.018±0.005	0.021±0.005
FCF Brundle (Ref. 42)	0.757±005	0.069±0.005	0.143±0.005	0.013±0.002	0.018±0.002
Calculated FCF r=1.001±0.01 Å Φ=110.2±0.2°	0.737±0.012	0.080±0.005	0.152±0.007	0.022±0.002	0.006±0.001

D_2O

Assignment	000	001	100 (+002)	101 (+003)	200 (+0203)(+102)
Experimental FCF	0.676+0.003	0.101+0.005	0.157+0.04	0.037+0.005	0.026+0.005
F.C.F. Brundle (Ref. 42)	0.202+0.005	0.987+0.005	0.148+0.005	0.034+0.002	0.025+0.002
Calculated FCF r=0.9995+0.001 Å φ=109.8+0.2°	0.680+0.020	0.101+0.006	0.172+0.012	0.031+0.004	0.013+0.002

where Ψ_n and Ψ_n^0 represent respectively the perturbed and unperturbed vibrational wave functions with n vibrational quanta;
E_m and E_n the energy of the unperturbed states and

$$W_{m\ n} = \{\Psi_m^0\ W\ \Psi^0\ dq \qquad\qquad (17)$$

are the elements of the perturbation function W.

A perturbation function of the form

$$W = - fx^3 + gx^4$$

has been chosen.

f and g are determined from $w_e x_e$ and $w_e y_e$. The x^3 term has been
used for the stretching mode $\bar{v}_1$. In order to take into account
the positive anharmonicity for the bending mode (i.e. the energy
difference increases with vibrational quantum number) only the
term in x^4 has been used for v_2. f and g have been determined
and a fairly good agreement has been obtained with the experimental results of MENES [28] (Fig. 1). This figure shows that
the best results are obtained with a slightly reduced N-H bond
length in the planar ion, taking into account the combination
bands $(v_1 v_2)$ and with $v_1 = 3v_2$.

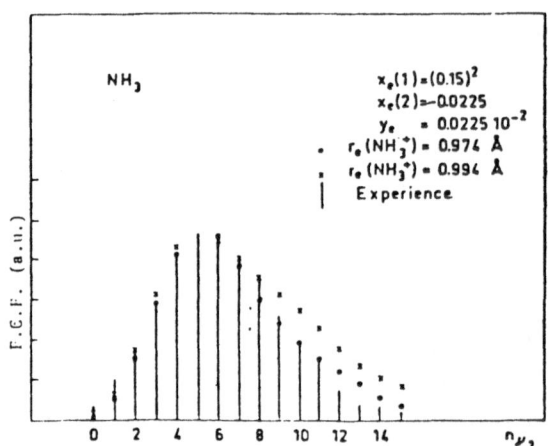

Figure 1 : Calculated and experimental F.C.F.
for two interatomic distances.

N - H = 1.014 $\overset{o}{A}$ NH$_3$ N - H = 0.974 $\overset{o}{A}$ NH$_3^+$

AVOURIS et al. [40] using an analogous treatment and a quad-
ratic force field arrived to the opposite conclusion. The N-H
bond length should be longer in NH$_3^+$ by about 0.06 $\overset{o}{A}$. This
discrepancy has not been explained.

The anharmonic approximation seems to give good results in the case of large variations of the equilibrium coordinates between initial and final states - and extended vibrational excitation.

c) $H_2O^+(A^2A_1) \leftarrow H_2O(X^1A_1)$ transition.

This is a very interesting case of a linear to bent transition.

$$H_2O^+(A^2A_1) \rightarrow D_{\infty h} \qquad\qquad H_2O(X^1A_1) \rightarrow C_{2v}$$

There are three vibrational modes in H_2O and four in H_2O^+ (A^2A_1). Taking into account the rotational coordinates and the Eckart conditions it is possible to find a correct transformation matrix A for the internal symmetry coordinates. This problem has been solved exactly by J. CARLIER $\underline{/\ 29\ /}$. A simplified transformation has been used by M. BAFFERT-FORGES $\underline{/\ 27\ /}$ in which one of the degenerate bending mode of H_2O^+ (A) corresponds to a rotation (zero frequency vibration $\underline{[4\ ]}$) in $H_2O(X)$. The results are given in figure 2. The combination bands $\nu_1\nu_2$ have to be included in order to obtain a reasonable fit. The assumption $\nu_1 = 4\nu_2$ has been made in these calculations (ν_1 = streching mode, ν_2 = bending mode). These calculations should be done with an exact transformation.

Although the calculations in the harmonic approximation with a correct transformation matrix of the coordinates and taking into accoung anharmonicities give a reasonable agreement with experiment, the method developed by CEDERBAUM and DOHMKE give better results for the F.C.F. calculation.

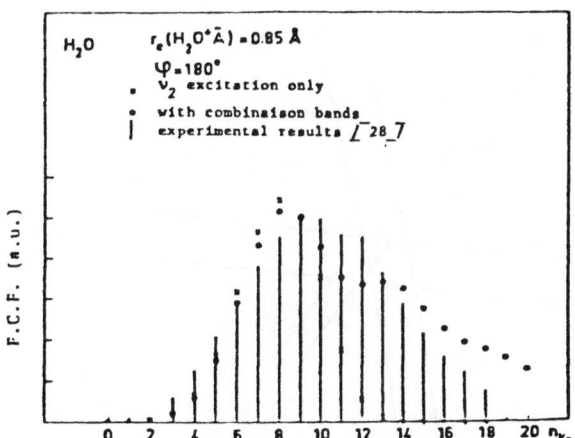

Figure 2 : Comparison of the photoelectron spectrum with the calculated F.C.F. for the transition $H_2O^+(A) \leftarrow H_2O(X)$

V) EQUILIBRIUM GEOMETRY OF THE FINAL STATE.

As we have seen previously most calculations of the F.C.F. necessitate the knowledge of the equilibrium geometry of the two

states. Conversely, the equilibrium geometry of the final state
may be deduced from the F.C.F. which depend essentially on the
coordinates.

In the classical harmonic calculation of SHARP and ROSENSTOCK
$\underline{/\,7\,\underline{/}}$ the coordinates of the final state are introduced as adjustable parameters and the best fit with experimental results
determines the values. DOCTOROV et al. $\underline{/\,15\,\underline{/}}$ have developed a
method, in the harmonic approximation, for the direct calcula-
tion of the equilibrium geometry of the final state from a few
experimental transition probabilities. These values are then
used by the authors to calculate the whole vibrational distribution in the corresponding electronic band. The method has been
applied to various examples : SO_2, BS_2, $ZnTl_2$, C_6H_6 which will
not be discussed here.

a) Equilibrium geometry of H_2O^+, $D_2O^+(X^2B_1)$ and $HDO^+(X^2B')$

The distances r (O-H), (O-D) and the angle Ψ(H-O-H) have been
deduced from the F.C.F. for the three species $\underline{/\,21\,\underline{/}}$. In figure
3, the calculated transition probabilities are given as a func-
tion of r and Ψ for three vibrational levels of the H_2O^+ ion.
The overlap of the three curves is well defined and the geometry

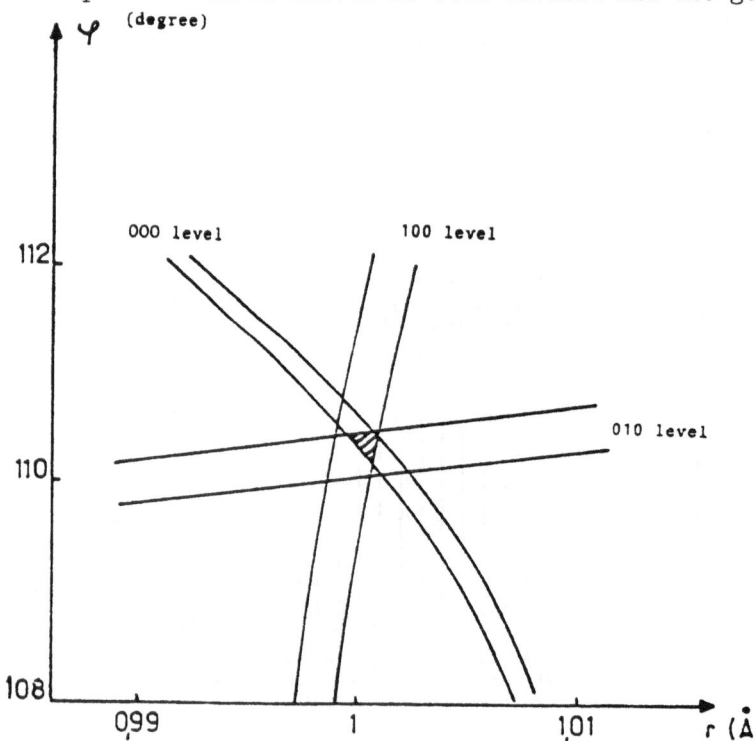

Fig. 3 - Equilibrium geometry of the H_2O^+(X) ion.
Surfaces of coincidence between calculated and
experimental F.C.F. for the given vibrational
levels (000, $1v_1 00$, $01v_2 0$).

of the final state can be determined with a high accuracy. The
values of the coordinates for H_2O^+, D_2O^+ and HDO^+ are given
in table III. The results are in very good agreement with those
of a rotational analysis by LEW and HEIBER $/\!\!\_30\_/$. The precision
is of the order of a few times 10^{-3} on both the O-H distance
and the HOH angle.

b) Equilibrium geometry of the $C_2H_{4-n}D_n^+$ ground state ions
   $(0 \leqslant n \leqslant 4)$.
The ejected $1b_{3u}$ electron is located on the Π double bond, therefore one expects a variation of the C = C bond length. Moreover
an intense peak corresponding to the out of plane ν_4 vibration
with 2 quanta is observed in the experimental spectrum as can be
seen in figure 4. This indicates that the ionic state must no
more be planar. The results of J. CARLIER $/\!\!\_31\_/$ and of KOPPEL
et al. $/\!\!\_37\_/$ are given in table IV together with the value for
the ground state. The two series of results for $C_2H_4^+$ are in good
agreement and have been obtained respectively with the methods
of SHARP and ROSENSTOCK $/\!\!\_7\_/$ and of CEDERBAUM and DOMCKE $\lceil16\_\rfloor$.
J. CARLIER has adjusted the parameters on all the isotopic spe-
cies, KOPPEL et al. on C_2H_4 and C_2D_4 taking into account the
vibronic coupling.

As will be shown, although the angle α does not vary significantly,
the influence of the F.C.F. is dramatic.

TABLE III - Equilibrium geometry of $H_2O(X)$ and $H_2O^+(X)$

	Molecule (a)	Ion FCF calculation $/\!\!\_31\_/$	Ion Rotational analysis $/\!\!\_30\_/$
$\underline{H_2O}$			
r (Å)	0.9572	1.001 ± 0,001	0,999
ϕ	104.52°	110.2 ± 0.2	110.5°
$\underline{D_2O}$			
r (Å)	0.9575	0.9995 ± 0.0015	
ϕ	104.47°	109.8 ± 0.2	
$\underline{HDO}$			
r (Å)	0.9571	1.000 ± 0.001	
ϕ	104.53°	110.1 ± 0.2	

(a) W.S. BENEDICT, N. GAILAR and E.K. PLYLER,
    J. Chem. Phys. 24, 1139 (1956).

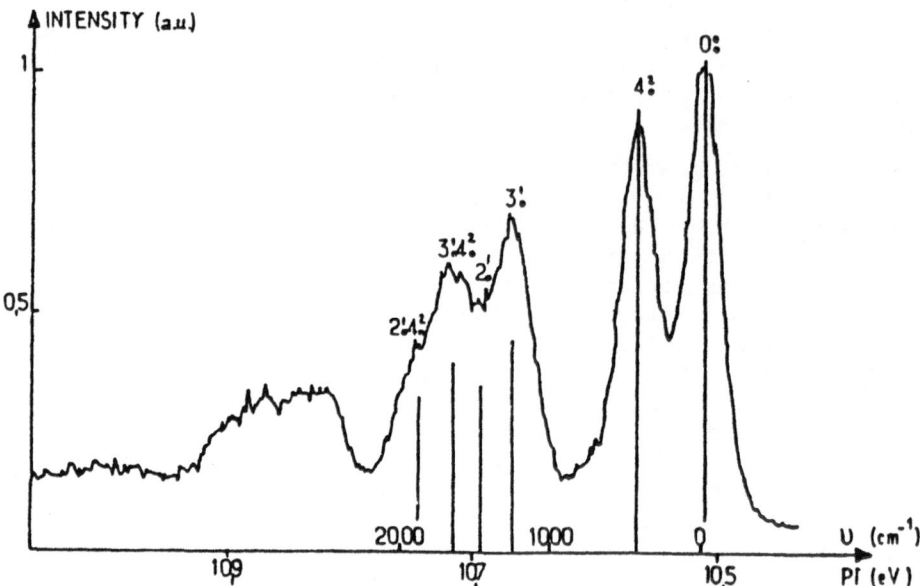

Figure 4 : First band of the HeI photoelectron spectrum of C_2H_4 and calculated transition
probabilities for : R = 1405 Å ε= 8.5° α =α' - 0.5° (vertical bars)

TABLE IV : Equilibrium geometry of C_2H_4 and $C_2H_4^+$.

	$r_{(C-H)}$ (Å)	$R_{(C=C)}$ (Å)	$2\alpha(H-\hat{C}-H)$	ε
C_2H_4	1.085	1.339	117.80°	0°
$C_2H_4^+$ $\underline{/317}$	1.085	1.40	116.30°	16°
$\underline{/377}$	1.091	1.405	117.80°	25°

VI) ASSIGNMENT OF THE VIBRATIONAL MODES IN THE PHOTOELECTRON SPECTRA.

In photoelectron spectroscopy the assignment of the vibronic peaks is made generally in an empirical way $/$ 17, 32$_/$. The ejected electron corresponds to a non bonding, a bonding or an antibonding orbital. In the first case, the bond length should not vary and no vibrational structure should be observed. In the two last cases the bond corresponding to the ejected electron will vary (lengthen or shorten) and a vibrational structure of the normal mode corresponding to this bond should be observed. This simple picture should however not be very representative in a polyatomic molecule and may lead to errors. Two examples will be given below.

a) $C_2H_4^+$ (X^2B_{3u}).

As can be seen in table IV the largest variation of the coordinates between C_2H_4 and $C_2H_4^+$ is the $C = C$ bond length. Therefore one expects that the $C = C$ streching mode ν_2 should be excited. This led TURNER et al. $/$ 17$_/$, STOCKBAUER et al. $/$ 33$_/$ and CVITAS et al. $/$ 34$_/$ to assign the most intense peak around 1500 cm^{-1} in the first band (see fig. 4) of $C_2H_4^+$ to the ν_2 vibration and the next peak to ν_3 (symmetric bending mode). In $C_2D_4^+$ the intensity of the two peaks is reversed and so is the assignment. A tentative explanation of this effect has been given by CVITAS et al. $/$ 34$_/$. The isotopic shift for the two frequencies is different in the molecule and ion if this attribution is correct (table V).

TABLE V : Isotopic shift of the ν_2 and ν_3 frequencies in C_2H_4 and $C_2H_4^+$.

	$C_2H_4(X)$	$C_2H_4^+(X)$	
		Stockbauer $/$ 33$_/$	Carlier $/$ 31$_/$
$\dfrac{\nu_2\,(C_2H_4)}{\nu_2\,(C_2D_4)}$	1.074	0.953	1.070
$\dfrac{\nu_3\,(C_2H_4)}{\nu_3\,(C_2D_4)}$	1.362	1.52	1.350

Let us now consider the F.C.F. calculations. The $\underset{\sim}{L}^{-1}$ matrix corresponding to the transformation $\underset{\sim}{Q} = \underset{\sim}{L}^{-1}\underset{\sim}{S}$ given in table I

shows that there is a strong mixing of the stretching and bending
internal coordinates in the two normal modes ν_2 and ν_3. This
mixing is quite different for C_2H_4 and C_2D_4. So one should expect
that a variation of $\Delta R(C = C)$ can produce an excitation of both
ν_2 and ν_3 but different for C_2H_4 and C_2D_4. The calculations show
that for $\Delta\alpha = 0$, ν_2 and ν_3 have almost equal intensities in
C_2H_4 but are quite different ($I_{(\nu_2)} > I_{(\nu_3)}$ in C_2D_4). If we now
consider the variation of $I_{(\nu_2)}$ and $I_{(\nu_3)}$ as a function of α ,
we see (figure 5) that the excitation probabilities vary drasti-
cally with $\Delta\alpha$. They are just reversed for a variation $\Delta\alpha$ of 6°
in $C_2H_4^+$. The equal probabilities for the two modes do not correspond to the same value of α in $C_2H_4^+$ ($\alpha' = \alpha''$) and C_2D_4
($\alpha' = \alpha'' - 3°$) The best agreement between experimental and
calculated values is obtained for $\alpha' = \alpha'' - 1°$ for all the 7
substituted ethylene molecules. This leads to an inverse assignment of ν_2 and ν_3 in $C_2H_4^+$ from the one given before (Table VI).
This assignment leads to an identical isotopic shift in the ion
and in the molecule for the two vibrational modes (Table V).

It seems that the new ordering $\lfloor 31 \rfloor$ of ν_2 and ν_3 in $C_2H_4^+$
explains better the experimental results than the one given by
previous authors $\lfloor 17, 33, 23 \rfloor$.

b) CH_3OH^+.

This example has been calculated by FLAMENT et al. $\lfloor 35 \rfloor$ using
the method of CEDERBAUM and DOMCKE $\lfloor 16 \rfloor$.

The geometries and the total energies of the neutral and ionized
states have been calculated with an ab initio method using a
STO-6G basis. The first and second derivatives of the energy
(vertical I.P.) with respect to the normal coordinate of the
neutral molecule have been determined. The results for CH_3OH,
CD_3OD and CH_3OD have been compared with the photoelectron
spectra. First, the authors notice that the isotopic substitution modifies considerably the composition of the normal modes
(L matrix). Second that the previous assignment of the peaks in
the first band of the photoelectron spectra, are not correct
$\lfloor 36 \rfloor$. The second peak has been assigned to ν_{C-O} using the
empirical rules given by TURNER $\lfloor 17 \rfloor$. In fact the ab initio
calculation of the geometries shows that in the ground ionic
state, corresponding to the removal of a n electron of the oxygen
atom, not only r (C - O) changes but also the C - O - H angle.
The F.C.F. calculations show that the second peak should be
assigned to the $\nu(\delta\ COH)$ in CH_3OH and to a mode corresponding to
a mixture of δ (COH) and r (C - H_2, H_3). The other peaks in the
band have also been assigned.

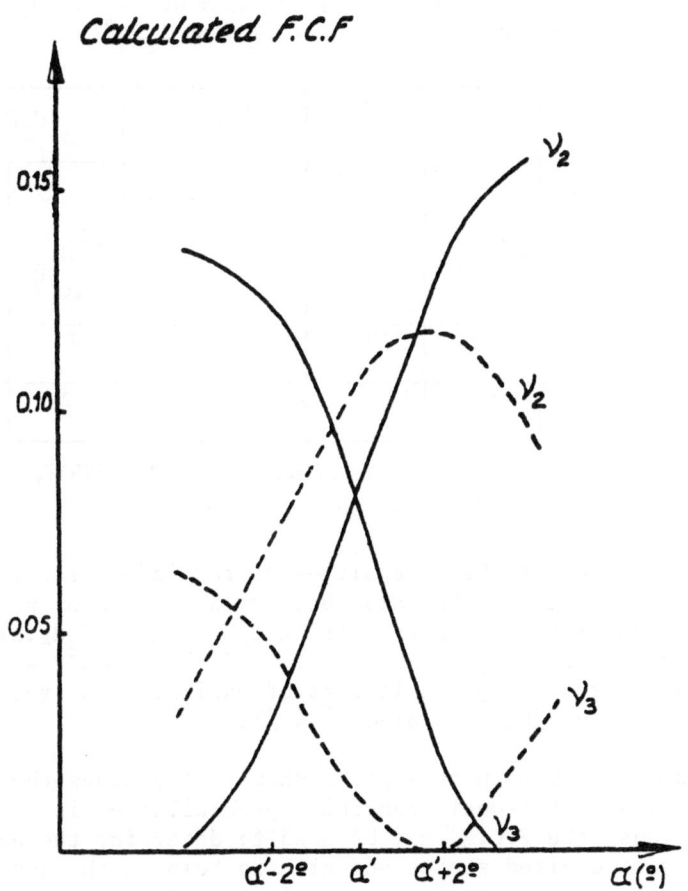

Figure 5 : Excitation probabilities of the ν_2 and ν_3 normal
modes as a function of a in $C_2H_4^{+2}$ (———)
and C_2D_4 (---)
R = 1.41 Å ε = 8°.

These examples show that the F.C.F. may be very useful for
interpretation of the photoelectron spectra.

VII) CONCLUSION.

Franck Condon factors are of invaluable help for many problems
in which an electronic transition is involved. They are used in
the calculations of the probability of predissociation $/\overline{\ 41\ }/$
as well as in reactive collision experiments in order to define
the type of interaction. Bound-free transitions have been the
subject of numerous works $/\overline{\ 38\ }/$. These problems are generally
developed in the diatomic approximation.

TABLE VI : Vibrational frequency assignment of the ν_2 and ν_3
modes in C_2H_4, $C_2H_4{}^+$, C_2D_4 and $C_2D_4{}^+$.

C_2H_4	$C_2H_4{}^+$		C_2D_4	$C_2D_4{}^+$	
Ground state (a)	X^2B_{3u} [33]	X^2B_{3u} [31]	Ground state (a)	X^2B_{3u} [33]	X^2B_{3u} [31]
ν_2 1630	1210	1375	1518	1340	1285
ν_3 1342	1450	1225	985	1050	905

a) J.L. DUNCAN, D.C. Mc KEAN and P.D. MALLINSON,
   J. Mol. Spectros. 45, 221 (1973).

The calculation of the intensities of forbidden transitions in
polyatomic molecules, has also been treated in many publications.
Vibronic coupling in the transition $C_2H_4{}^+(X) \leftarrow C_2H_4(X)$ has been
taken into account by KOPPEL et al.[37]. DOCTOROV et al. [15]
have calculated the probabilities of excitation of forbidden
transitions in C_6H_6 (see also [14]).

The examples given in this paper show that besides the calculation of the vibrational transition probabilities in electronic
transitions, the F.C.F. may be used to determine the equilibrium
geometry of excited states and also to help in the assignment of
the vibrational structure in the electronic spectra.

The F.C.F. calculations may also become important for the
interpretation of high resolution spectra obtained with lasers in
supersonic beam cooled molecules where the vibrational spectra
for compounds as large as tetracene are well resolved [39].

REFERENCES

[1] Nicholls, R.W.,:1961, J. Res. Nat. Bur. Stand.,65 A, p. 451.

[2] Benesch, W., Vanderslice, J.T., Tilford, S.G., and
       Wilkinson, P.G.,: 1966, Astrophys. J., 143, p. 236.

[3] Petropoulos, B., and Botter, R.: 1968, C.R. Acad. Sc.
       Paris, 266, p.1.

[4] Wilson, E.B., Decius, J.C., and Cross, P.C.,: 1955,
       Molecular Vibrations, Mc Graw Hill, N.Y.

[5] Woodward, L.A.: 1972, "Introduction to the theory of molecular vibrations and vibrational spectroscopy", Oxford University Press.

[6] Moule, D.C.: 1977, in "Vibrational state and structure", Ed. J.R. During, Elsevier, Amsterdam, Netherlands, 6, p. 227.

[7] Sharp, T.E., and Rosenstock, H.M.: 1964, J. Chem. Phys., 41, p. 3453.

[8] Duchinsky, F.: 1930, Acta Physicochem., USSR, 36, p. 410.

[9] Botter, R.: 1968, Thèse d'Etat, Fac. des Sciences d'Orsay.

[10] Coon, J.B., de Wames, R.E., and Loyd, C.M.: 1962, J. Mol. Spect., 8, p. 285.

[11] Warshel, A., and Karplus, M.:1972, Chemical Physics Let. 17, p.7.

[12] Lucas, N.J.D.: 1973, J.Phys. B. 6, p. 155.

[13] Baranov, V.I., Gribov, L.A., and Novosadov, B.K.:1981, J. Mol. Struct., 70, p.1; 1981, J. Mol. Struct., 70, p. 31.

[14] Faulkner, T.R., and Richardson, F.S.: 1979, J. Chem. Phys., 70, p. 1201.

[15] Doktorov, E.V., Malkin, I.A., and Man'ko, V.I.: 1975, J. Mol. Spectros., 56, p. 1; 1977, J. Mol. Spectros, 64, p. 302; 1977, Chem. Phys. Letters, 46, p. 183.

[16] a) Cederbaum, L.S., and Domcke, W.: 1977, Adv. Chem. Phys., 36, p. 205.

b) Domcke, W., Cederbaum, L.S., Koppel, H., and Von Nissen, W.: 1977, Mol. Phys. 34, p. 1759.

[17] Turner, D.W., Baker, C., Baker, A.D., and Brundle, C.R.: 1970, Molecular Photoelectron Spectroscopy, Wiley, London.

[18] Baer, T.: 1979, in "Gas Ion Chemistry", ed. M.T. Bowers, Academic Press, N.Y.

[19] Delwiche, J., Hubin-Franskin, M.J., Guyon, P.M., and Nenner, I.: 1981, J. Chem. Phys., 74, p. 4219.

[20] Parr, A.C., Ederer, D.L., West, J.B., Holland, D.M.P.,
     and Dehmer, J.L.: 1982, J. Chem. Phys., 76, p. 4349.

[21] Botter, R., and Carlier, J.: 1977, J. Elect. Spect. Rel.
     Phen. 12, p. 55.

[22] Harschbarger, W.R.: 1972, J. Chem. Phys. 56, p. 177;
     1970, J. Chem. Phys., 53, p. 903.

[23] Durmaz, S., Murrell, J.N., Taylor, J.M., and Suffold, R.:
     1970, Mol. Phys., 19, p. 533.

[24] Smith, W.L., and Warsop, P.A.: 1968, Trans. Faraday Soc.
     64, p. 1165.

[25] Botter, R., and Rosenstock, H.M.: 1968, Adv. Mass
     Spectrometry, 4, p. 579.

[26] Rabalais, J.W., Karlsson, L., Werme, L.O., Bergmark, T.,
     and Siegbahn, K.: 1973, J. Chem. Phys., 58, p. 3370.

[27] Baffert-Forges, M.: 1973, Thèse 3ème Cycle, Orsay.

[28] Menes, F., unpublished results.

[29] Carlier, J., to be published.

[30] Lew, L., and Heiber, I.: 1973, J. Chem. Phys., 58, p. 1246.

[31] Carlier, J.: 1979, These d'Etat, Fac. des Sciences PARIS VI.

[32] Eland, J.H.D.: 1974, Photoelectron Spectroscopy, Butter-
     works, (London).

[33] Stockbauer, R., and Inghram, M.G.: 1975, G. Elect. Spect.
     Rel. Phen., 7, p. 492.

[34] Cvitas, T., Gusten, H., and Klasinc, L.: 1979, J. Chem.
     Phys., 70, p. 57.

[35] Flament, J.P., Hopilliard, Y., Jaudon, P., and
     Youkharibache, P.: 1981, Nouveau J. Chimie, 5, p. 61.

[36] Mc Neil, K.A.G., and Dixon, R.N.: 1977, J. Elect. Spect.
     Rel. Phenom., 11, p. 315.

[37] Koppel, H., Domcke, W., Cederbaum, L.S., and Von Niessen,
     W.: 1978, J. Chem. Phys., 69, p. 4252.

[38] Gislason, E.A., Klein, A.W., and Los, J.: 1981, Chem.
     Physics, 59, p. 91.

[39] Amirav, A., Even, U., and Jortner, J.: 1981, J. Chem. Phys., 75, p. 3770.

[40] Avouris, P., Rossi, A.R., and Albrecht, A.C.: 1981, J. Chem. Phys., 74, p. 5516.

[41] Atabeck, O., Beswick, J.A., Lefevre, R., Mukamel, S., and Jortner, J.: 1976, J. Chem. Phys., 65, p. 4035.

[42] Brundle, C.R., and Turner D.W.: 1972, Proc. Roy. Soc. A. 326, p. 181.

ION CLUSTERS:

SUMMARY OF THE PANEL DISCUSSION

A. W. Castleman, Jr.

Department of Chemistry
The Pennsylvania State University
University Park, PA 16802

INTRODUCTION

Investigations of the dynamics of formation, structure,
and properties of small cluster ions formed by the attachment of
molecules to anions and cations comprises a very active area of
research. Growing interest in cluster ions is evident in both
the basic, as well as the applied literature. In terms of basic
chemical physics, investigations of the bonding of ion clusters
provides information on the well depth and forces between ions
and neutral molecules. Studies of the dynamics of cluster ion
formation and dissociation, as well as investigations of their
photodissociation, provide information on energy transfer, the
kinetics of association reactions, and on unimolecular dissociation processes. Although there is a paucity of detailed information, problems involved in ion-ion recombination can be further
elucidated through investigations of ion cluster-ion cluster
interactions (Smith and Adams, 1980)

Recognition that clusters in general represent an aggregated
state of matter, having properties midway between the gaseous
and the condensed phase, provides further impetus for research
in this field. Investigations into the formation and properties
of increasingly larger clusters offer a way of studying the molecular details of the course of change between the gaseous and the
condensed phase, commonly termed nucleation phenomenon. In this
regard it is known that such studies can further elucidate the
molecular details of the collective effects responsible for phase
transitions, the development of surfaces, and ultimately solvation phenomenon. A natural extension of work on large ion
clusters is provided by attempts of various investigators to

M. A. Almoster Ferreira (ed.), Ionic Processes in the Gas Phase, 327–354.
© *1984 by D. Reidel Publishing Company.*

connect their properties to ions in liquid or solids. This
research enables a further understanding of bonding and reactivity
of complexes known to exist in the solution phase that have gas
phase analogies. Studies of the attachment of electrons to neutral
clusters provide a technique for investigating polarons (the
solvated electron). Both ion and fast atom bombardment techniques
are utilized in surface chemistry and solid state physics to
investigate surfaces and reactivity at surface sites. Data on
gas phase cluster ions is useful in interpreting the results.
Furthermore, small clusters are currently being used as proto-
types of surface sites in theoretical efforts to model electronic
states and properties of the surface state (Surface Science,
1981). Exciting prospects are offered in gaining insight into
the molecular details of catalysis (Beauchamp, this conference).

Exciting new results and experiments involving both ions and
neutral clusters suggest that relatively few molecules must be co-
clustered in order for either the ion or the neutral system to
begin to display properties normally associated with the condensed
phase (Castleman, 1982; Castleman et al., 1982; Kay et al., 1981).
Further impetus for studying ion clusters derives from implications of the results in furthering an understanding of many
applied processes and phenomenon. One application is in better
understanding the earth's environment, which consists of a weakly
ionizing plasma surrounding the globe (Smith and Adams, 1980).
Cluster ions are known to exist throughout most of the earth's
atmosphere (Ferguson et al., 1979) and are implicated as being
important in maintaining the charge balance as well as potentially
in effecting certain nucleation processes leading to aerosol
particle formation (Castleman and Keesee, 1981). Ion clusters
are also known to form during combustion processes and they have
been suggested as playing a role in soot formation. Further
areas of application concern interpreting certain reactions in
radiation and nuclear reactor chemistry, electrochemistry, and
biochemistry (Castleman et al., 1982).

The panel discussion on cluster ions was initiated by the
presentation of eight speakers whose remarks are summarized
herein. The general organization evolved around four topics
concerned with reactions and dynamics, bonding, elucidation of
transitions from the gas to the condensed phase, and application
to better understanding atmospheric processes.

DYNAMICS OF FORMATION AND REACTIVITY

Nigel G. Adams and David Smith: KINETICS OF ION MOLECULE
          ASSOCIATION REACTIONS

Cluster ions are generated by the association of positive
or negative ions with neutral molecules in third-body stabilized

collisions. The forward (association) rate coefficients, k_3, for many such reactions have been determined using a variety of techniques (see, for example, the reviews by Good, 1975, Meot-Ner, 1979 and Adams and Smith, 1983). Although many of these data were obtained only at room temperature, a significant fraction of measurements have been made over limited temperature ranges. It is usually assumed (and the available data corroborate the assumption) that k_3 varies inversely with temperature according to a power law of the form $k_3 \propto T^{-n}$. In cases where n has been determined for a given reaction in different experiments, an unacceptably large range of n values has often been obtained. A major cause of the discrepancies is probably the limited temperature of the experiments from which the values of n have been generally determined (in fact, it seems that the smaller the temperature range, the larger the derived value of n!).

Recent studies of the temperature dependence of k_3 using the variable temperature SIFT at Birmingham have allowed k_3 to be determined for several reactions over the wide temperature range 80 to 550 K (Adams and Smith, 1981). One important reason for this work was the need to provide accurate data with which to make a meaningful comparison with recent theories describing association reactions (Bates, 1979; Herbst, 1982). To this end k_3 has been determined for the helium stabilized reactions of CH_3^+ and CD_3^+ ions with H_2, D_2, O_2, CO and CO_2 and of C^+ with H_2 and D_2, e.g.,

$$CH_3^+ + O_2 + He \longrightarrow CH_3^+ \cdot O_2 + He \qquad (1)$$

The magnitudes of k_3 for these various reactions at a given temperature differ by more than two orders of magnitude, being smallest for the weakly-bound clusters and largest for the strongly-bound clusters. Thus, for example, k_3 (287 K) for reaction (1) is 7.3×10^{-30} $cm^6 s^{-1}$ whereas that for the reaction

$$CH_3^+ + CO + He \longrightarrow CH_3^+ \cdot CO + He \qquad (2)$$

is 2.4×10^{-27} $cm^6 s^{-1}$ at the same temperature. For each of the reactions studied, k_3 did indeed vary closely as T^{-n} and this permitted the values of n to be compared with the theories of Bates and Herbst. These theories predict that k_3 is proportional to $T^{-(\ell/2 + \delta)}$ where ℓ is the number of rotational degrees of freedom in the separate reactants and δ is a parameter which describes the temperature dependence of the collisional stabilization efficiency. Thus, for example, for the polyatomic CH_3^+ ions reacting with diatomic molecules, k_3 should vary as $T^{-(2.5 + \delta)}$. The experimental data indicate this to be so for the several reactions studied with δ being small in each case, varying from +0.3 to -0.2 (Adams and Smith, 1981). For the association of the polyatomic ion CH_5^+ with the polyatomic molecule CH_4, k_3 varied as $T^{-3.4}$ again indicating a relatively small value of δ of 0.4. Thus, the body of the data is in close agreement with theoretical predictions for the temperature dependence of k_3 (a more sophisticated development of these statistical theories has recently been given by Bass et al., 1979). Thus, for association reactions involving simple polyatomic ions and molecules, k_3 is not expected to vary more rapidly than T^{-3} to T^{-4}. However, for association reactions of weakly-bonded cluster ions, e.g.,

$$H_3O^+(H_2O)_n + H_2O + M \longrightarrow H_3O^+(H_2O)_n + M \qquad (3)$$

more substantial temperature variations might be expected and have indeed been observed (e.g., Meot-Ner, 1979; Kebarle, 1983). This difference might be due to floppy vibrational modes and internal rotations in the loosely-bonded intermediate complexes.

It is interesting to note that in the CD_3^+ association reactions with the several molecules listed above, the n values are identical to those for the corresponding CH_3^+ reactions. However, in each case the magnitude of k_3 at a given temperature is 50 to 100% larger for the CD_3^+ reactions, presumably due to the larger density of quantum states in the CD_3^+ intermediate complexes. We have also made a detailed study of association in the reactions of partially and totally deuterated CH_3^+ ions with H_2, HD and D_2 (Smith et al., 1982). These studies have revealed that while k_3 for the $CD_3^+ + D_2$ association reaction is about four times larger than that for the $CH_3^+ + H_2$ reaction at each temperature (again presumably due to the greater density of states in the CD_5^+ intermediate complex), k_3 for the reaction

$$CD_3^+ + H_2 + He \longrightarrow CH_2D_3^+ + He \qquad (4)$$

at 80 K is almost a factor of two larger than that for the $CD_3^+ + D_2$ reaction at this temperature. Thus an effect in addition to the density of quantum states is enhancing k_3 for reaction (4). We have tentatively attributed this phenomenon to an "isotope refrigerator" effect by which the endoergic rearrangement of H and D atoms within the excited intermediate $(CH_2D_3^+)^*$ complex increases its lifetime against unimolecular decomposition and hence increases k_3 above that for the $CD_3^+ + D_2$ reaction.

Some information relating to the structure of the $CH_3^+ \cdot O_2$ complex has been obtained. It has been found that this ion apparently retains the cluster ion structure indicated rather than isomerizing to a more stable structure. Evidence is provided by the observation that the ion rapidly undergoes ligand switching with N_2, i.e.,

$$CH_3^+ \cdot O_2 + N_2 \longrightarrow CH_3^+ \cdot N_2 + O_2 \qquad (5)$$

generating the cluster ion $CH_3^+ \cdot N_2$ [with k (80 K) equal to 3×10^{-10} cm^3s^{-1}]. Interestingly, however, the reaction

$$CH_3^+ \cdot O_2 + H_2 \longrightarrow CH_5^+ + O_2 \tag{6}$$

is relatively slow, k (80 K) $\sim 3 \times 10^{-12}$ cm^3s^{-1}, indicating that there is an activation energy barrier to the production of the more strongly bonded but non-cluster ion CH_5^+.

In some very recent experiments, a comparative study has been made of the association reactions

$$N_2^+ + N_2 + He \longrightarrow N_4^+ + He \tag{7}$$

and

$$N_2^+ + N_2 + He \longrightarrow N_4^+ + N_2 \tag{8}$$

using the variable temperature SIFT and variable temperature drift tube techniques (Böhringer et al., 1983). The results revealed, somewhat surprisingly, that k_3 varied as $T^{-2.3}$ for reaction (7) whereas it varied as $T^{-1.7}$ for reaction (8) indicating δ values of similar magnitudes but opposite signs. This may be a manifestation of a stabilization of the $(N_4^+)^*$ intermediate complex resulting from a hard sphere collision with helium atoms in reaction (7) and via a symmetrical switching of N_2 out of the $(N_4^+)^*$ complex in reaction (8). It should also be noted that the magnitude of k_3 for reaction (8) is about three times that for reaction (7). This difference between the stabilizing efficiency of helium atoms and of molecules has previously been observed in other ion-molecule association reactions (see the review by Good, 1975).

John F. Paulson and Michael J. Henchman: ON THE FORMATION OF
    H_3O^- IN AN ION-MOLECULE REACTION[1]

We have used a longitudinal double mass spectrometer system (DMS) (Paulson et al., 1970; Paulson and Gale, 1978) to study the formation of H_3O^- in reaction (9).

$$OH^- \cdot H_2O + H_2 \rightarrow H_3O^- + H_2O \tag{9}$$

In order to confirm the identity of the ion product, we have also studied the reactions of $OH^- \cdot H_2O$ with D_2 and of $OD^- \cdot D_2O$ with D_2. The latter reactant mixture was the most extensively studied because of minimal complications from the ^{18}O isotope. As an example of these studies, Fig. 1. shows cross sections for the formation of D_3O^- in reaction of $OD^- \cdot D_2O$ with D_2. In the case of both $OH^- \cdot H_2O + H_2$ and $OD^- \cdot D_2O + D_2$, the cross section increases rapidly to a value of 0.05×10^{-16} cm^2 at a relative energy of

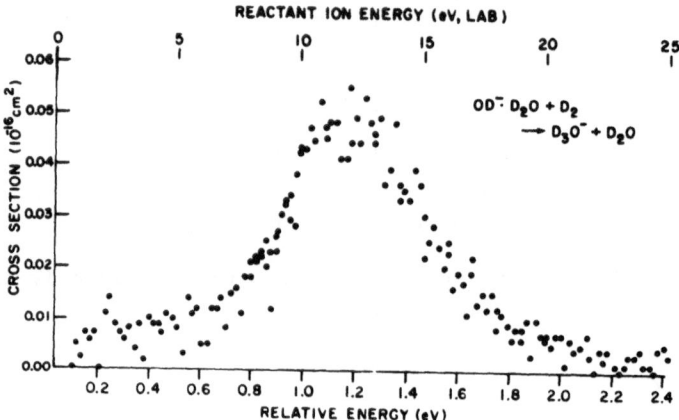

Figure 1.   Cross sections for reaction (9) as a function
of energy.

about 1.1 eV and then decreases rapidly at higher energies.   The
width of this peak in the cross section is about 0.5 eV (FWHM)
in the barycentric frame for all of the isotopic reactant
mixtures studied.

In the reaction of $OH^- \cdot H_2O$ with D_2, where both H_2DO^- and
HD_2O^- were observed as product ions, the sum of the cross sections for these reaction channels at a given barycentric energy
is approximately equal to the cross section for production of
H_3O^- from $OH^- \cdot H_2O + H_2$ or for production of D_3O^- from $OD^- \cdot D_2O +$
D_2 at the same energy.   If complete isotopic scrambling occurred
in the reaction of $OH^- \cdot H_2O$ with D_2, the abundance of the product
ions would be $n_{20} : n_{21} : n_{19} = 6 : 3 : 1$, where n_{20}, n_{21}, and n_{19} are the
relative abundances of H_2DO^-, and H_3O^-, respectively.   In contrast,
the observed abundances are $n_{20} : n_{21} = 1 : 2$, and there is thus some
preference for the reactant OH^- to appear in the product ion.
We did not observe the production of H_3O^- in this case and
believe it to be significantly smaller than that of either
H_2DO^- or HD_2O^-.

Reactions (10) and (11) were also observed in this work.

$$OH^- \cdot H_2O + H_2 \rightarrow OH^- + H_2O + H_2 - 1.08 \text{ eV} \qquad (10)$$

$$\rightarrow H^- + 2H_2O - 1.50 \text{ eV} \qquad (11)$$

In order to compare the interaction energies at which reactions
(9), (10),   and (11) become energetically possible, we show in
Fig. 2 the cross sections (in arbitrary units) for these reactions
in the $OD^- \cdot D_2O + D_2$ reactant system at low relative energies.

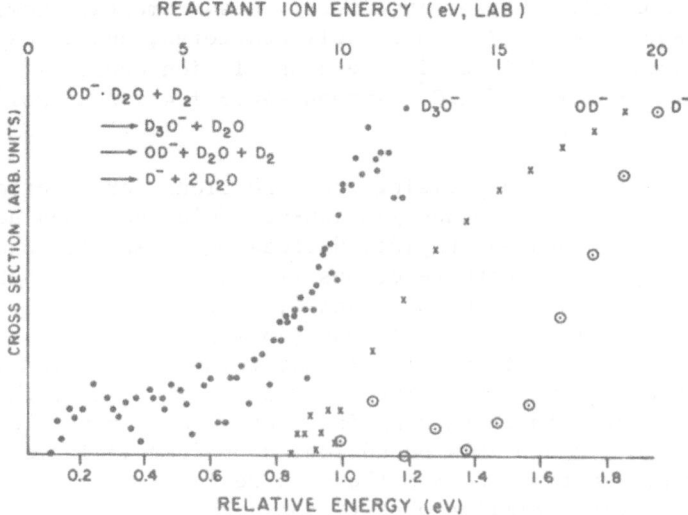

Figure 2.   Cross sections (arbitrary units) for reactions
(9), (10), and (11) near threshold.

In order for the species H_3O^- to be observed in the double
mass spectrometer used here, the H_3O^- ion must exist for at
least the flight time through the apparatus, i.e., about
15×10^{-6} s.   Reaction (12) is known (Betowski et al., 1975) to be
fast.

$$H^- + H_2O \rightarrow OH^- + H_2 + 0.44 \text{ eV} \tag{12}$$

The potential energy of a long lived state of H_3O^- must lie below
that of $OH^- + H_2$.   If the rapidly rising cross sections, for
which examples are shown in Figs. 1 and 2, are interpreted as
energy thresholds, then the threshold for H_3O^- formation lies
about 0.30 eV below that for $OH^- + H_2 + H_2O$, i.e., reaction (10),
and about 0.75 eV below that for $H^- + 2H_2O$, i.e., reaction (11).
Therefore, H_3O^- is bound by 0.30 eV with respect to dissociation
to $OH^- + H_2$ and by 0.75 eV with respect to dissociation to
$H^- + H_2O$.   An indication of the error in these measurements is
that the threshold observed here for reaction (10) is 0.96 eV,
to be compared with a more accurate equilibrium measurement of
1.08 eV (Payzant et al., 1971), while the threshold for reaction
(11) is 1.53 eV, compared with a predicted value of 1.50 eV
(Benson, 1976; Rosenstock et al., 1977; Payzant et al., 1971).
Errors of about ± 0.05 eV are expected in the thermochemical
parameters used in these calculations.   Relative energy measurements, such as those about comparing the threshold of reaction
(9) with the thresholds of reactions (10) and (11), are of

better accuracy than the absolute energy measurements. However, the cross sections for H_3O^- are badly scattered, and we estimate an uncertainty of ± 0.15 eV in the dissociation energies of H_3O^-. The heat of formation of H_3O^- is then -40.3 ± 4.1 kcal mol^{-1} or -169 ± 17 kJ mol^{-1}.

Although the sharply rising cross sections for reaction (9) (Fig. 1) are typical of endoergic ion-molecule reactions near threshold, the subsequent rapidly decreasing cross sections are unusual. Competition with reactions (10) and (11), which become energetically allowed at 0.30 eV and 0.75 eV, respectively, about the onset of reaction (9), may explain in part the decreasing probability of reaction (9) at relative energies above 1.1 eV. In addition, increasing internal excitation of the H_3O^- product ion with increasing interaction energy may lead to dissociation of this weakly bound ion, a process indistinguishable in these experiments from the occurrence of reactions (10) and (11) directly. Similar observations have recently been made for HNO_3^- and $NO_2^- \cdot H_2O$ formed in energetic collisions of $OH^- \cdot H_2O$ with NO_2 (Paulson and Dale, 1982).

The observation of $H^- \cdot nH_2O$, with n=5 and n=11, was recently reported by Armbruster, Haberland, and Schindler (1981). The formation of H_3O^- in an exoergic reaction between OH^- and H_2CO was reported at this same NATO Conference by Nibbering, Ingemann, and Kleingeld.

Douglas P. Ridge: CLUSTERS OF Fe^+ WITH HYDROCARBONS

Clusters of atomic metal ions with hydrocarbons are formed in ion molecule reactions which occur in mixtures of metal carbonyls with alkanes (Allison et al., 1979; Freas and Ridge, 1980). The properties of these clusters are of interest in connection with mechanisms of metal catalysis and organometallic chemistry. The cluster of an atomic metal ion with nearly any hydrocarbon RH can be formed in reaction (13) (Freas and Ridge,

$$MCO^+ + RH \longrightarrow MRH^+ + CO \tag{13}$$

1980). MCO^+ is formed by electron impact on a metal carbonyl. Even methane and ethane react in this way. In addition, atomic metal ions formed by electron impact on metal carbonyls react with alkanes to form metal ion hydrocarbon clusters. The butanes undergo reactions (14) to (19) with Fe^+ formed by electron impact

$$Fe^+ + i\text{-}C_4H_{10} \quad
\begin{cases}
\longrightarrow FeC_3H_6^+ + CH_4 & \tag{14} \\
\longrightarrow FeC_4H_8^+ + H_2 & \tag{15}
\end{cases}$$

$$Fe^+ + n-C_4H_{10} \longrightarrow FeC_2H_4 + C_2H_6 \tag{16}$$

$$\longrightarrow FeC_3H_6^+ + CH_4 \tag{17}$$

$$\longrightarrow FeC_4H_8^+ + H_2 \tag{18}$$

$$\longrightarrow FeC_4H_6^+ + 2H_2 \tag{19}$$

on $Fe(CO)_5$. Similar reactions are observed for Fe^+ and Co^+ formed in other ways (Armentrout et al., 1981; Byrd et al., 1982). These reactions have been observed in our ion cyclotron resonance studies and in our high pressure mass spectrometric studies.

The latter studies were done on the triple analyzer mass spectrometer at the Midwest Center for Mass Spectrometry (Gross et al., 1982). The high resolution capability of the triple analyzer made it possible to obtain collision induced decomposition spectra of the products of reactions (14) to (19) and of the products of a number of variants of reaction (13) with M=Fe.

Some of the conclusions from these CID studies are:

(1) $FeC_4H_{10}^+$ formed from (9) with RH = $i-C_4H_{10}$ differs from $FeC_4H_{10}^+$ formed from (13) with RH = $n-C_4H_{10}$. In both cases the hydrocarbon seems to fall apart on the metal. Structures such as $CH_3-Fe^+-i-C_3H_7$ and $CH_4Fe^+(C_3H_6)$ are suggested. The appearance of $FeCH_3^+$ in the CID spectra and the absence of $C_3H_7^+$ suggests that I.P.($FeCH_3$) < I.P.($i-C_3H_7$) = 7.36 eV (Houle and Beauchamp, 1979). This may be checked by examining the CID spectrum of $FeC_3H_6O^+$ formed in (13) with RH = acetone. $FeCH_3^+$ and CH_3CO^+ are both observed suggesting that I.P.($FeCH_3$) $\cong$ I.P.(CH_3CO) = 6.79 eV (Rosenstock et al, 1977).

(2) Four $FeC_4H_8^+$ structures appear to be obtainable. Reaction (13) with RH = 1 butene or 2 butene give a structure designated $(H_2)Fe^+(C_4H_6)$ since it loses H_2 on CID. Reaction (13) with RH = isobutene and Reaction (15) give a structure designated Fe^+(isobutene) which loses C_4H_8 on CID. A third structure is formed when $Fe(CO)_2$ reacts in two steps with C_2H_4 to form $Fe(C_2H_4)_2^+$ which loses C_2H_4 on CID. A fourth structure is formed in reaction of Fe^+ with cyclopentanone. This structure is probably a metalocycle and loses both H_2 and C_2H_4 on CID. The product of Reaction (18) seems to be about two-thirds $(H_2)Fe^+(C_4H_6)$ and one-third $Fe^+(C_2H_4)_2$. The predominant structure probably results from a metal insertion into one of the secondary C-H bonds in n-butane followed by shift of a β H atom to the metal and loss of H_2. The resulting $FeC_4H_8^+$ then probably rearranges via an allylic structure to $(H_2)Fe^+(C_4H_6)$. The $Fe(C_2H_4)_2^+$ probably results from Fe^+ insertion into the C2-C3 bond in n-butane

followed by transfer of H atoms from Cl and C_4 onto the metal followed by H_2 loss.

(3) C_3H_8 retains its integrity to a large extent when associated with Fe^+. The CID of the product of (13) with RH = C_3H_8 suggests that structures $(CH_4)Fe^+(C_2H_4)$ and $(H_2)Fe^+(C_3H_6)$ are present to some extent. CID spectra of FeC_2H_6 and $FeCH_4$ formed in reaction (13) with RH = C_2H_6 and RH = CH_4, respectively, show even less evidence that the metal attacks the hydrocarbon.

(4) The $FeC_3H_6^+$ species formed in Reactions (14), (17), and (13) with RH = propene seem to be the same structure.

(5) $FeC_2H_4^+$ formed in Reactions (16) and (13) with RH = C_2H_4 both seem to be the same structure.

A. W. Castleman, Jr. and D. E. Hunton: STUDIES OF PHOTO-
          DISSOCIATION

The subject of cluster ion photodissociation is a very new one for which there are results from only a limited number of research groups including those of Moseley, Vestal, Vanderhoff and ourselves. Most of the results presented in the literature have addressed the subject of spectroscopy and little attention has been given to the dynamics of photodissociation. During the course of a reinvestigation of the photodissociation of CO_3^-, which has been the subject of extensive controversy in the literature, we have undertaken measurements of the comparative collisional and photodissociation of the first two of its hydrates.

Experiments were conducted using a mass selected ion beam crossed with light from a tunable dye laser. As a result of discrepancies among previous studies published in the literature, questions have arisen in the interpretation of the photodissociation spectrum of CO_3^- due to the very low energy photon required to yield O^-, the apparent values being below the accepted bond energies for O^- with CO_2 (Dotan et al., 1977). The results of the CO_3^- spectra show features in excellent agreement with those reported by Moseley et al. (1975, 1976) and Smith et al. (1979). Measurements made at very high field parameters in the source have provided evidence for two states of CO_3^- in agreement with the collisional and dissociation measurements of Wu and coworkers (1981). The results of the present studies are consistent with the presence of at least one bound electronic excited state of CO_3^- which leads to a clarification of the findings of Vestal (1977) and Heller and Vestal (1980) suggesting that a two-photon process is involved in the photodissociation.

Experiments were carried out for the photodissociation of $CO_3^-(H_2O)$ and $CO_3^-(H_2O)_2$. Over the energy range 1.95 to 2.2 eV, the product ion was found to be largely CO_3^- in both cases. The spectra for the hydrates show features reminescent of the peaks in the bare ion spectrum, although the hydrate spectra are broader and red shifted. The cross sections are at least one order of magnitude greater for the hydrates. Collisional dissociation measurements were also performed in our apparatus which showed that only one water molecule was removed in contrast to two in the case of photons interacting with the double hydrated species.

The results are consistent with the bound electronic excited state of the parent ion being involved in both the first (intermediate) process of O^- loss, with a second photon necessary for its removal, and the same intermediate state involved in water loss from CO_3^- through either internal conversion or fluorescence back to the ground state into a vibrational level sufficiently high to enable hydrate loss. These results demonstrate the importance of energy transfer within the parent cluster ion that is responsible for the dynamics of cluster dissociation following photon absorption.

BONDING AND STRUCTURE

Robert R. Squires: AB INITIO STUDIES OF THE STRUCTURES AND
        ENERGIES OF SOME ANION-MOLECULE COMPLEXES

Ab initio molecular orbital (MO) calculations have become an increasingly reliable source of structural and energetic information for both neutral and ionic species (Schaefer, 1977; Del Bene et al., 1982). Computations of this type are especially relevant to gas phase ion chemistry since they directly refer to the isolated (albeit motionless) molecule or ion. Also, in the absence of general spectroscopic means for determining detailed ion structures, MO calculations have played a key role in our present understanding of structure-energy and structure-reactivity relationships for gas phase ions (Raghavachari et al., 1981; Wolfe et al., 1981). The generally recognized importance of long-lived ion-molecule complexes and ion clusters in gas phase ion chemistry makes these species attractive subjects for theoretical study. The following is a brief summary of results from ab initio self-consistent-field calculations of the structures and energies of some anion-molecule complexes and the corresponding ion solvation energies (eq. 20)

$$SH + A^- \xrightarrow{-\Delta E_{solv}} [SH \cdot A^-] \qquad\qquad (20)$$

$$SH = NH_3, \ H_2, \ H_2O, \ H_2C_2$$

$$A^- = NH_2^-, \ H^-, \ HO^-, \ HC_2^-$$

Computational Procedure: The ab initio calculations were performed with a modified version of the Gaussian/76 program (Binkley
et al., 1978) on a CDC 6600 at Purdue. Full geometry optimiza-
tions for each of the ions, neutrals and complexes were carried
out using analytically evaluated gradients (Schlegel, 1975). In
all cases except H_3O^- and NH_4^-, optimized geometries were determined using a 4-31G basis set. Total energies for each species
with the 4-31G geometry were then computed using the diffuse
function augmented 4-31+G basis set recently described by
Chandrasekhar, Andrade and Schleyer (1981) (procedure designated
4-31+G//4-31G). In the case of the H_3O^- and NH_4^- systems, optimum
geometries and total energies for H_2O, HO^-, NH_3, NH_2^-, H_3O^-, NH_4^-
and H_2 were determined using an augmented 4-31+G basis which
included diffuse S-type functions for each hydrogen (designated
4-31+G//4-31+G).

    Correction for correlation effects using second order Møller-
Plesset perturbation theory (MP2) (Møller and Plesset, 1934;
Binkley and Pople, 1977) produced only small changes in the
calculated anion solvation energies (i.e., -0.4 kcal·mol^{-1} for
$[HO^- \cdot H_2O]$, -1.2 kcal·mol^{-1} for $[HO^- \cdot NH_3]$ and -2.7 kcal·mol^{-1} for
$[NH_2^- \cdot NH_3]$). Zero point energy and temperature corrections to the
solvation energies likewise have been found to be generally small
($\sim$3-5 kcal·mol^{-1}) (Jorgensen and Ibrahim, 1981) and opposite in
sign to the correlation energy corrections. Thus, the anion
solvation energies determined in this study within the Hartee-
Fock approximation are expected to be reasonably good estimations
of the experimentally measurable solvation enthalpies.

Results: Table I gives a summary of the calculated solvation
energies along with the H-bond distances determined for each
complex. The calculated hydroxide ion hydration energy, -27.5
kcal·mol^{-1}, is in excellent agreement with the experimental
enthalpy value of -25 kcal·mol^{-1} determined by Payzant and co-
workers (1971). In addition, the optimized $H_3O_2^-$ geometry
obtained with the 4-31G basis set matches very closely with that
determined with larger basis sets which include polarization
functions (i.e., 6-31G*) (Jorgensen and Ibrahim, 1981). The calculated binding energy of H_3O in its hydrated hydride ion form,
-15.6 kcal·mol^{-1}, also agrees closely with the experimental value
of $\sim$ -17 kcal·mol^{-1} reported by J. F. Paulson at this panel
session.

Table I. Calculated Solvation Energies and Bond Distances for Negative Ion Clusters[a]

$$SH + A^- \xrightarrow{\Delta E_{solv}} [S \xrightarrow{r_1} H \cdots^{r_2} A^-]$$

SH	$H_{acid}(SH)$[b]	A^-	$PA(A^-)$[b]	$-\Delta E_{solv}$[c]	r_1	r_2
NH_3	404	HO^-	391	12.7	1.06	1.64
H_2	400	HO^-	391	4.5	0.75	1.90
H_2O	391	HO^-	391	27.5	1.10	1.36
H_2C_2	386	HO^-	391	16.7	$(1.05)^d$	1.68
H_2O	391	NH_2^-	404	29.9	$(1.10)^d$	1.43
H_2O	391	H^-	400	15.6	0.98	1.79
H_2O	391	HC_2^-	386	18.0	0.99	1.87
NH_3	404	NH_2^-	404	11.9	1.07	1.76
NH_3	404	H^-	400	5.3	1.01	2.25
H_2	400	NH_2^-	404	4.2	0.76	2.15

[a]Acidities (ΔH_{acid}), proton affinities (PA) and solvation energies (ΔE_{solv}) in kcal·mol^{-1}; bond distances in Å.

[b]Taken from J.E. Bartmess, R.T. McIver, Jr., in "Gas Phase Ion Chemistry," M.T. Bowers, ed., Academic Press, 1979, CHA. 11.

[c]4-31+G//4-31G; for H_3O^- and NH_4^- 4-31+G//4-31+G procedure included diffuse S-functions on all hydrogens.

[d]This parameter held fixed during geometry optimization; see text.

The ordering of the calculated anion solvation energies is in accord with the general relationship between hydrogen-bond donor acidity; acceptor basicity and binding energy as discussed by Yamdagni (1971) and others (Caldwell and Bartmess, 1981). Thus, replacement of water in [HO$^-$·H$_2$O] with the weaker acid ammonia to give [HO$^-$·NH$_3$] results in a decrease in the binding energy by roughly a factor of two. The weaker H-bond in the latter complex

is also manifested in the longer non-bonded A•••H distance r_2.
The theoretical results also suggest that donor acidity is more
important than acceptor basicity in determining ΔE_{solv} since the
binding energy of ammonia to amide ion is about the same as that
to hydroxide ion.   However, in comparing these clusters the
acceptor heteroatom is changing so a conclusion based upon acid/
base strengths alone may not be valid. Stable structures corresponding to hydrated amide ion, $[NH_2^- \cdots H_2O]$ and hydroxide ion
H-bonded to acetylene, $[H-C\equiv H \cdots OH^-]$, could not be found. Instead,
during geometry optimization of these structures proton transfer
from water and acetylene occurred to give $[NH_3 \cdots OH^-]$ and
$[H-C\equiv C^- \cdots H_2O]$, respectively. This indicates a small or non-
existent barrier to proton transfer as would be expected from
the high overall exothermicities. In order to obtain an estimate
of the anion solvation energies for $[NH_2^- \cdot H_2O]$ and $[H-C\equiv C-H\cdot OH^-]$
in the absence of the proton transfers, the donor-hydrogen bond
distances, r_1, were constrained during geometry optimization to
their values in the uncomplexed neutral molecule. The resulting
solvation energies recorded in Table I are probably maximum
values but they appear to be consistent with the previously
discussed correlation between acid/base strengths and solvation
energy.

Two stable forms of H_3O^- have been determined in the present
study. The hydride ion hydrate structure, $[H^- \cdots H_2O]$ is calculated to be more stable than the hydrogen-hydroxide ion form,
$[H_2 \cdots OH^-]$ by ~ 4 kcal•mol^{-1}. This theoretical finding is nicely
supported by Nibbering's observations that the hydrogens in H_3O^-
formed from HO$^-$ and CH_2O in an FT-ICR do not appear to scramble
(Kleingeld and Nibbering, 1982), and by Grabowski, Bierbaum and
DePuy's (1982) finding that DO$^-$ undergoes H/D exchange with H_2.
That the $[H^- \cdots H_2O]$ structure is more stable than $[H_2 \cdots OH^-]$
in spite of the 9 kcal•mol^{-1} greater basicity of hydride ion
relative to hydroxide ion is a dramatic manifestation of the
greater ion solvating ability of H_2O over H_2. In contrast, the
theoretical results for the NH_4^- system indicate that the net
binding energy of hydride ion to ammonia in $[H^- \cdots NH_3]$ is not
significantly greater than that of amide ion to hydrogen in
$[H_2 \cdots NH_2^-]$.

Robert G. Keesee and A. W. Castleman, Jr.:   SOME CONSIDERATIONS
     OF BONDING

An examination of the bonding of neutral molecules onto ions
in the gas phase has particular value in elucidating the properties which determine the ligand's ability to cluster onto ions.
For this purpose, the association of ammonia, water, sulfur
dioxide, and carbon dioxide with both Na$^+$ and Cl$^-$ has been
investigated. The ion-dipole interaction is the most important
electrostatically attractive force between an ion and a neutral

molecule. Both sodium and chloride ions have closed electronic configurations and spherical symmetry, although the ionic radius of Na^+ is considerably smaller. Consequently with other factors equal, a neutral molecule would be expected to bind more strongly to Na^+. However, factors such as quadrupole moment, polarizability, hydrogen bonding, and charge transfer may also be important. Table II lists some of the relevant electrostatic properties of the neutral molecules considered here.

Table II. Electrostatic properties

Molecule	Dipole Moment D_z (Debye)	Quadrupole Moment Q_{zz} Q_{yy} Q_{xx} (10^{-26} esu-cm^2)			Polarizability α_z α_y α_x (Å^3)			Axes $z \uparrow \rightarrow x$
H_2O	1.85	-0.26	-5.0	5.26	1.45	1.65	1.23	
SO_2	1.63	2.6	8.0	-10.6	2.65	4.27	4.17	
NH_3	1.47	-4.64	2.32	2.32	2.39	2.1	2.1	
CO_2	0	-8.64	4.32	4.32	4.05	1.95	1.95	

The experimentally determined stepwise heats of association are given in Table III. Some interesting comparisons are immediately apparent: (1) Only sulfur dioxide binds more strongly to Cl^- than Na^+ for the first association step, (2) ammonia exhibits the largest difference between Na^+ and Cl^-, (3) the relative stability of the ion-molecule complexes for NH_3, H_2O, and SO_2 are in reverse order for the two ions, and (4) the absolute magnitude of the enthalpy changes for the additional clustering of SO_2 onto Cl^- are smaller than for Na^+.

Many of these trends are in agreement with expectations from electrostatic properties. For instance water, sulfur dioxide, and ammonia have similar dipole moments; however, when the dipoles are aligned in the electrostatic field of the ion, the small quadrupole moments (Q_{zz}) of water and the considerably larger one of ammonia are repulsive to a negative charge and attractive to a positive one. In the case of sulfur dioxide, the situation is reversed. Thus, the ion-quadrupole interaction is consistent with the different ordering of $-\Delta H_{0-1}$ for water, ammonia, and

Table III.   Stepwise heats of association, $-\Delta H^{\circ}_{n,n+1}$, in
kcal/mole for $I \cdot B_n + B \rightleftharpoons I \cdot B_{n+1}$.

B	I	(0,1)	(1,2)	(2,3)	(3,4)	(4,5)	(5,6)
				(n, n+1)			
H_2O	Cl^- [a]	14.9	12.6	11.5	10.9	--	--
	Na^+ [b]	24.0	19.8	15.8	13.8	12.3	10.7
SO_2	Cl^- [a]	21.8	12.3	10.0	8.6	--	--
	$Na+$	18.9[c]	16.6[d]	14.3[d]	--	--	--
NH_3	Cl^- [e]	8.2	--	--	--	--	--
	Na^+ [f]	29.1	22.9	17.1	14.7	10.7	9.7
CO_2	Cl^- [a]	8.0	--	--	--	--	--
	Na^+ [d]	15.9	11.0	9.7	--	--	--

[a] Keesee et al., 1980   [d] Peterson, 1982

[b] Dzidic and Kebarle, 1970   [e] This work

[c] Perry et al., 1980   [f] Castleman et al., 1978b

sulfur dioxide between the positive ion Na^+ and the negative ion
Cl^-. From electrostatic considerations, $Na^+ \cdot CO_2$ is expected to
be a linear complex (Q_{zz} attractive) and $Cl^- \cdot CO_2$ should be a
T-complex (Q_{yy} and Q_{xx} attractive). As has been shown earlier
(Keesee et al., 1980), charge transfer or covalent bonding is
an additional consideration in the binding of CO_2 and particularly
SO_2 with Cl^-. This covalency increases the stability of the
$Cl^- \cdot SO_2$ complex but decreases subsequent clustering due to the
dispersal of charge over the SO_2Cl^- ion.

CONNECTING THE GAS AND CONDENSED PHASES

A. W. Castleman, Jr. and R. G. Keesee: SOME RECENT FINDINGS AND
    NEW TECHNIQUES

    Studies of ion clusters have contributed greatly to the
fields of interphase chemistry and physics which are concerned
with elucidating the molecular details of the collective effects
responsible for phase transitions (nucleation phenomena), and
ultimately solvation phenomena. The high pressure mass spectro-

metric technique enables accurate enthalpy and entropy changes to
be obtained for a sequence of successive clustering of a given
ligand to an ion.   These results can thereby be utilized in order
to investigate solvation complexes having analogies to those
known to exist in the condensed phase.   Furthermore, a step-
wise investigation of the energy barrier to nucleation can be
measured which is useful for comparison with various theoretical
models in assessing their range of validity.

  Nucleation: Studies of clustering about ions are important
in understanding nucleation induced by electrically charged sites,
and also contributes to a clarification of the general phenomena
as well.   Utilizing data obtained for a number of systems, it
has been possible to evaluate the limitations of the classical
liquid drop formulation of ion induced nucleation.   Castleman et
al. (1978a) compared the available experimental data on hydration
enthalpies with the appropriate derivatives of the free energy
relations (as expressed by the classical Thomson Equation),
obtaining excellent agreement for clusters containing as few as
four to six water molecules.   A similar result was also obtained
for the clustering of ammonia about a number of ionic species.
However, large discrepancies were found upon comparing the
entropies of cluster formation.   More negative values are found
experimentally than are predicted by theories which are based on
bulk liquid properties.   This finding reflects the more highly
ordered structure of small systems relative to the disordered
nature of liquids as implied in the classical nucleation expression.   The importance of ordered structures qualitatively explains
the finding that, while the Thomson Equation agrees with experiment for some systems, large differences are found for others.
The recent work has shown that a modified effective radius for
the clustering neutral molecule can be utilized to provide
a correlation suitable for accounting for both the enthalpy and
entropy contributions to the energy barrier to nucleation for a
variety of systems (Holland and Castleman, 1982a).

  Solvation: Examples of gas phase ion-molecule complexes
which have analogies in the condensed phase are becoming legion.
Several recent ones have been found in our measurements of the
clustering of ammonia to alkali metal and transition metal ions.
In the present work, $Ag^+(NH_3)_2$, $Cu^+(NH_3)_2$, and $Ag^+(pyridine)_2$
ions are examples of such complexes in the gas phase.   These
cluster ions further demonstrate very sharp decreases in binding
after the addition of the second ligand which correlates well
with their chemistry in the aqueous phase.

  Studies of the successive clustering of ammonia about Li^+
and Na^+ reveal a preferential coordination complex having four
ligands attached to the central ion, while those with larger
ions such as K^+ and Rb^+ do not display well defined structures

(Castleman et al., 1978b). These results are in very good agreement with Raman spectra published by Gans and Gill (1976) based
on liquid phase measurements. Similar findings of gas phase
complexes with solution phase analogies were obtained in the
transition metal studies (Holland and Castleman, 1982b). In
particular, the case of Cu^+ and Ag^+ ions it is known that bonding
strongly predisposes complexes to favor a linear structure. This
is thought to be due to the relatively small difference in
energy between the filled d and unfilled s orbitals. This allows
the d_{z^2} and s orbitals to form a hybrid orbital which has two
regions on opposite sides of the ion which are attractive to
electron donating ligand molecules. Further mixing with the p_z
orbital can then result in hybrid orbitals capable of forming
linear covalent bonds to the two ligands. Gas phase measurements
show that ammonia binds much stronger than water (11.5 kilo-
calories per mole for Ag^+ and over 20 for Cu^+). The differences
in binding between ammonia and water cluster ions of Ag^+ and
Cu^+ is a function of cluster size and show that beyond the second
ligand water bonds more strongly than ammonia, whereas ammonia
bonds more strongly to smaller sizes. This crossover revealed
by the gas phase data is consistent with the observed preferences
of the Ag^+ and Cu^+ ion for forming di-ligand complexes in the
aqueous phase.

Another interesting result from recent studies is the finding
that enthalpy measurements for gas phase complexes can be used
to derive quantitative values for both anion and cation solvation
in water and cation solvation in ammonia. A general correlation
was developed (Lee et al., 1980a) which shows that in each of these
systems, involving both complex ions, simple ions with open
electronic structure, and ions with closed electronic structure,
there is a specific ratio of ion solvation energy in the condensed
phase to the summed gas phase enthalpy values for the first four
to six clusters. These results, if found to be general, will find
wide application in deriving solution phase solvation values
directly from gas phase measurements.

Neutral Clusters; The Changing Properties of Electrolyte
Species Upon Clustering and Products of Ion-Ion Recombination:
The correlation developed in our laboratory which relates gas
phase bond energies of ion clusters to solvation energies in
the liquid phase offers promise that data on clustering about
individual ions can be suitably combined for pairs of anions
and cations such that the solvation of electrolytes can be
interpreted on a molecular basis. A related problem of interest
in the field of gas phase ion chemistry is ascertaining the
products of ion-ion recombination. Utilizing data from measurements of gas phase enthalpies including coulombic effects, it
is possible to demonstrate that for certain systems ion cluster-
ion cluster recombination may lead to neutral products having

large dipole moments. This would occur if the degree of
clustering prohibits electron transfer due to the stabilization
of the charge centers of unlike sign, thereby preventing electron
transfer.

In order to provide experimental techniques for investigating
the dipole moments of neutral complexes containing ion pairs, a
molecular beam system employing electrostatic focusing fields has
been constructed in our laboratory. Recent studies in our laboratory of acetic acid clusters have established the applicability
of the technique for measuring the changing dipole moments of
aggregated systems. The work has been extended to an investigation of electrolyte molecules such as HNO_3-H_2O complexes, H_2SO_4
and SO_3-water complexes, and the H_2SO_4 dimer. The results
demonstrate the usefulness of this method in relating data on
gas phase ion chemistry to the condensed phase.

Photoionization of Metal Atom and Metal Atom-Ligand Clusters:
Currently there is extensive interest in the photoionization of
small clusters with application to an understanding of the changing electronic states as a function of the degree of aggregation,
and ultimately providing a foundation for understanding the
physical basis for catalysis. Recent work has been completed
in our laboratory on studies of alkali metal clusters, alkali
metal oxide systems, and alkali metal dimer halide systems
(Peterson, 1982). An apparatus has been developed which enables
the determination of the appearance potentials of small clusters
by crossing them with either a laser or a wavelength selected
UV light source. An interesting finding was made in comparing
the electron impact ionization of sodium clusters with ones
determined by photoionization. In the former case, for a given
distribution, only monomer, dimer, and trimer clusters could be
detected; but, in the case of photoionization higher order
polymers were observed for the identical cluster distribution.
The electron impact ionization leads to cluster fragmentation
and photoionization is a much more useful technique in determining ionization potentials of easily fragmented cluster systems.

Recent single photon ionization spectra were taken for Na,
and Na_2 through Na_7. The respective ionization potentials for
the first three aggregates were found to be in very good agreement with recent work of Hermann et al (1982) and results for
higher order polymers were obtained at a greater degree of
resolution than attainable in the earlier work. Photoionization
curves for Na_2Cl were also obtained, where the molecule is formed
by the reaction of Na_2 with HCl and Cl_2, respectively. The
threshold for the photoionization was found to be 3.9 ± 0.1 eV.
More recent data were acquired for the system Na_xO with x =
1, 2, 3 and 4.

John Bartmess and G. Caldwell: SINGLE SOLVATION OF ALKOXIDES

Using ion cyclotron resonance spectrometry, we have determined the relative hydrogen bond strengths of a number of alkoxides
and other anions bonded to simple alcohols, by measuring equilibria such as (21).

$$RO^- \cdots HOR + R'OH \; \Longleftrightarrow \; R'O^- \cdots HOR + ROH \qquad (21)$$

$$RO^- + HCO_2R \; \longrightarrow \; CO + RO^- \cdots HOR \qquad (22)$$

The monosolvated anions are formed under the low pressure conditions in the ICR by the Riveros reaction (22). By constructing
a ladder of such equilibria, we are building a monosolvated gas
phase acidity scale. From a thermochemical cycle involving the
$\Delta G°$ (21) values and the gas phase acidities of the species involved, we can obtain the relative hydrogen bond strengths given
in Table IV and Figure 3. As other workers dealing with hydrogen

Table IV.   Relative Free Energies for Hydorgen Bonds
            $A^- \cdots HB$ in the Gas Phase

	$\Delta\Delta G°_{0,1}{}^a$		$\Delta\Delta G°_{0,1}{}^a$
$HO^- \cdots HOH$	$\geq +5.0$	$tBuO^- \cdots HOtBu$	-1.6
$MeO^- \cdots HOH$	≥ -2.6	$tBuCH_2O^- \cdots HOtBu$	-1.1
$MeO^- \cdots HOMe$	(0.0)		
$EtO^- \cdots HOMe$	-1.0	$tBuCH_2O^- \cdots HOCH_2tBu$	$+0.1$
$nPrO^- \cdots HOMe$	-2.0	$tBuCH(Me)O^- \cdots HOCH_2tBu$	-1.1
$tBuO^- \cdots HOMe$	-2.9	$tBuCH(Me)O^- \cdots HOCH(Me)tBu$	-0.1
$tBuCH_2O^- \cdots HOMe$	-3.0	$PhCH_2O^- \cdots HOCH(Me)tBu$	$+0.2$
		$PhCH_2O^- \cdots HOCH_2Ph$	$+0.8$
$EtO^- \cdots HOEt$	$-.12$		
$nPrO^- \cdots HOEt$	-1.5	$PhC\equiv C^- \cdots HOMe$	-6.5
$tBuO^- \cdots HOEt$	-2.2	$PhC\equiv C^- \cdots HOnPr$	-5.4
$tBuCH_2O^- \cdots HOEt$	-2.6	$PhC\equiv C^- \cdots HOCH_2tBu$	-3.4
		$PhC\equiv C^- \cdots HOCH_2Ph$	-1.0
$nPrO^- \cdots HOnPr$	-0.9	$Me_2C(CH_2S)_2CH^- \cdots HOMe$	-6.4^b
$tBuO^- \cdots HOnPr$	-1.6		
$tBuCH_2O^- \cdots HOnPr$	-1.7		

$^a \pm 0.3$ kcal/mole ^{b} 5,5-dimethyl-1,3 dithiane $(M-1)^-$

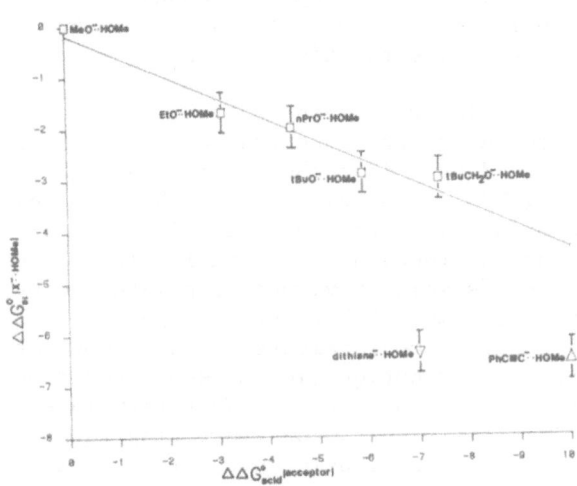

Figure 3. Correlation of acceptor ability with alkoxide basicity.

bond strengths for cationic species $ROH_2^+ \cdot \cdot HOR^-$ have noted, the symmetrical species in Table IV have nearly constant bond strengths. Figure 3 reveals a linear correlation of acceptor ability with alkoxide basicity; other donor alcohols form lines parallel to the $CH_3OH \cdot \cdot {}^-OR$ line. For the general case, we find $\Delta\Delta G^\circ_{0,1} = 0.34\Delta\Delta G^\circ_{acid}$ (acceptor) $- 0.29\Delta\Delta G^\circ_{acid}$ (donor) $- 0.5$ kcal/mole, with $r = 0.937$. The localized carbanions from $PhC\equiv CH$ and 1,3-dithiane are less effective as acceptors than alkoxides of equal basicity, consistent with the less electronegative nature of sp and sp^3 carbon compared to oxygen. Enolizable ketones, aldehydes, thiols, and oximes readily exchange into the alcohol-alkoxide cluster ions, while propene, cyclopentadinee, toluene, and DMSO do not. In general, it appears that heteroanions and localized carbanions are good acceptors, while delocalized carbanions are poor acceptors. If the C-H bond acts as a donor, it is much weaker than an O-H bond of comparable acidity. We are in the process of quantifying these latter species' hydrogen bonding ability.

APPLICATIONS

A. A. Viggiano and F. Arnold: THE STRATOSPHERE AS A LABORATORY

In situ stratospheric ion composition measurements have proven useful in learning about many properties of the atmosphere. The measurements are made with a balloon-borne cryopumped quadrupole mass spectrometer. The original measurements of the ambient ion composition of the stratosphere have been supplemented

by measurements of ions created by a filament placed directly
in the stratospheric air. Thus, information can be learned about
ions formed earlier in the reaction sequence.

One of the most important applications is the use of the
ion composition to derive trace neutral concentrations. Examples
of the gases detected by this technique include H_2SO_4 (Viggiano
and Arnold, 1982), HSO_3 (Arnold, et al., 1982), and HNO_3
(Henschen and Arnold, 1981) from negative ions and CH_3CN (Arnold
et al., 1980), and H_2O (Henschen and Arnold (1981) from positive
ions. Many of these species have not been detected by any other
means at the present time. Other information that can be learned
from the ion composition measurements include temperature from
the proton hydrate distribution, ion-molecule reaction rate
constants, properties of the aerosol layer (Arnold et al., 1982)
and thermodynamic properties of ion clustering (Arnold et al.,
1982). Here, attention is focused on the last topic.

The equilibrium constant for the reaction

$$A^{\pm} + B + M \rightleftharpoons A^{\pm} \cdot B + M \qquad\qquad (23)$$

is given by

$$K = \frac{[A^{\pm} \cdot B]}{[A^{\pm}][B]} \qquad\qquad (24)$$

where brackets denote concentration. The ion ratio is measured
directly. When the neutral concentration [B] can be determined
from the ion concentration measurements or some independent
means, K then can be calculated and ΔG derived.

This technique has the following advantages among others:
The reaction time, which is determined by the ion-ion recombination lifetime, is long ($\sim$ 1000 s) allowing equilibrium to be
reached for extremely low neutral concentrations. Also, ligands
such as H_2SO_4 and HSO_3 which are difficult to work with in the
laboratory can be studied. The primary disadvantages are
extreme selectivity and fixed experimental conditions (small
temperature range).

An example of the type of results that can be obtained
is shown in Table V. The data were obtained from a spectrum
taken at 32.9 km (Arnold et al., 1982). The first result to
notice is the good agreement obtained for $NO_3^- (HNO_3)_n$ cluster
bond strengths with the laboratory measurements (Davidson et al.,
1977; Lee et al., 1980b), thereby validating the technique. In
this case laboratory ΔS's were used to calculate ΔH. For all
other examples a typical $-\Delta S$ of 25 cal $mole^{-1}$ K^{-1} was assumed.

Table V. Results Obtained from a Spectrum Taken at 32.9 km.

Ion	kcal mole^{-1} $-\Delta G^o$	kcal mole^{-1} $-\Delta H^o$	kcal mole^{-1} $-\Delta H^o_{lab}$
$NO_3^-(HNO_3)-HNO_3$	12.6	18.4	18.3
$NO_3^-(HNO_3)_2-HNO_3$	10.2	16.0	16.1
$HSO_4^-(H_2SO_4)-H_2SO_4$	14	19.8	
$HSO_4^-(H_2SO_4)_3-H_2SO_4$	13	18.5	
$HSO_4^-(H_2SO_4)_4-H_2SO_4$	13	18.9	
$HSO_4^-(HNO_3)-HNO_3$	11.2	17.0	
$HSO^-(H_2SO_4)_2-HNO_3$	10.1	15.9	
$HSO_4^-(H_2SO_4)_3-HNO_3$	11.1	16.9	
$HSO_4^-(HNO_3)-H_2O$	7.2	13.0	
$HSO_4^-(H_2SO_4)_2-H_2O$	6.3	12.1	
$HSO_4^-(H_2SO_4)_3-H_2O$	8.0	13.8	
$HSO_4^-(H_2SO_4)_4-H_2O$	8.1	13.9	

H_2SO_4 is the most strongly bonded species where even for the fifth H_2SO_4 ligand, the bond strength is still 19 kcal mole^{-1}, comparable to the heat of vaporization of H_2SO_4. HNO_3 bonds less strongly than does H_2SO_4 with $-\Delta H$ typically being 16-17 kcal mole^{-1}. An interesting effect is seen in that the bond strength for HNO_3 bonding to $HSO_4^-(H_2SO_4)_n$ ions increases as n goes from 2 to 3. This may indicate the onset of cooperative bonding (Arnold et al., 1982).

This trend is also seen in the weakly bonded water clusters. Qualitatively, one may expect this increase due to the large heat of mixing between H_2O and H_2SO_4 although it is somewhat surprising to see this trend at the cluster sizes involved here. Preliminary laboratory data appear to verify this trend in the case of the H_2O bonding to $HSO_4^-(H_2SO4)_n$ (Glebe and Arnold, private communication).

In summary, *in situ* ion composition measurements have proven useful in studying a number of properties. Included among these

are the thermodynamics of ion clusters. The most interesting
result is that cooperative bonding effects may have been seen for
clusters involving as little as four ligands.

NOTE

[1]Some of these results were reported at the 34th Annual Gaseous
Electronics Conference, Boston, MA, 21 Oct. 1981. See J. F.
Paulson and M. J. Henchman, Bull. Am. Phys. Soc. $\underline{27}$, 108 (1982).

REFERENCES

Adams, N.G., and Smith, D.: 1981, Chem. Phys. Letts., 79, p. 563.

Adams, N.G., and Smith, D.: 1983, in "Reactions of Small Transient
        Species: Kinetics and Energetics" (M.A. Clyne and A.
        Fontijn. eds.) Academic, New York, in press.

Allison, J., Freas, R.B., and Ridge, D.P.: 1979, J. Am. Chem.
        Soc., 101, p. 1332.

Armbruster, M., Haberland, H., and Schindler, H. G.: 1981,
        Phys. Rev. Letts., 47, p. 323.

Armentrout, P.B., Halle, L.F., and Beauchamp, J.L.:1981, J.Am.
        Chem. Soc., 103, p. 6501.

Arnold, F., Fabian, R., Henschen, G., and Joos, W.: 1980, Planet.
        Space Sci., 28, p.681.

Arnold, F., Viggiano, A.A., and Schlager, H.: 1982, Nature, 297,
        p. 371.

Bass, L., Chesnavich, W.J., and Bowers, M.T.: 1979, J.Amer.Chem.
        Soc., 101, p. 5493.

Bates, D.R.: 1979, J.Phys.B., 12, p. 4135.

Benson, S.W.: 1976, "Thermochemical Kinetics", second edition,
        John Wiley and Sons, New York.

Betowski, D., Payzant, J.D., Mackay, G.I. and D.K. Bohme,:1975,
        Chem.Phys.Lett., 31, p. 321.

Binkley, J.S. and Pople, J.A.: 1977, Int.J.Quantum Chem.Symp.,
        99, p. 4899.

Binkley, J.S., Whitehead, R.A., Harihaven, P.C., Seeger, R., Pople, J.A., Hehre, W.J., and Newton, M.D.: 1978, Quantum Chemistry Program Exchange, 11, p. 368.

Böhringer, H., Arnold, F., Smith, D., and Adams, N.G.: 1983, Int.J.Mass Spectrom. Ion Phys., submitted.

Byrd, G.D., Bornier, R.C., and Freiser, B.S.: 1982, J.Am.Chem. Soc., 104, p. 3565.

Caldwell, G.W., and Bartmess, J.E., Proceedings of the 29th American Conference on Mass Spectrometry and Allied Topics; Minneapolis, MN, May 24-9, 1981, paper MPA10.

Castleman, A.W., Jr.,"SASP '82 Contributions", W. Lindinger, F. Howarka, T.D. Märk and F. Egger (eds.), Maria Alm/Salzburg, 1982, p.1.

Castleman, A.W., Jr., Holland, P.M., and Keesee, R.G.: 1978a, J.Chem.Phys., 68, p.1760.

Castleman, A.W., Jr., Holland, P.M., and Keesee, R.G.: 1982, Radiat.Phys.Chem., 20, p. 57.

Castleman, A.W., Jr., Holland, P.M., Lindsay, D.M., and Peterson, I.I., 1978b, J.Am.Chem.Soc., 100, p. 6039.

Castleman, A.W., Jr., and Keesee, R.G., "Annual Review of Earth and Planetary Sciences", Annual Reviews, Inc., Palo Alto, Vol. 9, 1981, p. 227.

Chandrasekhar, J., Andrade, J.G., and Schleyer, P.v.R.: 1981, J.Amer.Chem.Soc., 103, p. 5609.

Davidson, J.A., Fehsenfeld, F.C., and Howard, C.J.: 1977, Int. J.Mass Spectrom.Ion Phys., 9, p. 17.

Del Bene, J.E., Frisch, M.J., Raghavachari, K., and Pople, J.A.: 1982, Jour. Phys.Chem., 86, p. 1529.

Dotan, I., Davidson, J.A. Streit, G.E., Albritton, D.L., and Fehsenfeld, F.C.: 1977, J.Chem.Phys. 67, p. 2874.

Dzidic, I., and Kebarle, P.: 1970, J.Phys.Chem. 74, p. 1466.

Ferguson, E.E., Fehsenfeld, F.C., and Albritton, D.L., in"Gas Phase Ion Chemistry" (M.T. Bowers,ed.) Academic Press, New York, Vol. 1, 1979, p. 45.

Freas, R.B., and Ridge, D.P.: 1980, J.Am.Chem.Soc., 102, p. 7129.

Gans, P., and Gill, J.B.: 1976, J.Chem.Soc., Dalton Trans., p. 779.

Gatos. H.C., Ed., Proceedings of the Second International Meeting on Small Particles and Inorganic Clusters, "Surface Science", North Holland Publishing Co., Amsterdam, Vol. 106, 1981.

Glebe, W., and Arnold, F., private communication.

Good, A.: 1975, Chem. Revs., 75, p. 561.

Grabowski, J.J., DePuy, C.H., and Bierbaum, V.M.: 1982, J.Amer. Chem.Soc., in press.

Gross, M.L., Chess, E.K., Lyon, P.A., Crow, F.W., Evans, S., and Tudge, H.: 1982, Int.J.Mass Spectrom.Ion Phys., 42, p. 243.

Henschen, G., and Arnold, F.: 1981, Geophys. Res. Lett., 8, p. 999.

Herbst, E.: 1982, Chem. Phys., 68, p. 323.

Hermann, V., Kay, B.D., and Castleman, A.W., Jr.: 1982, Chem. Phys., submitted.

Hiller, J.F., and Vestal, M.L.: 1980, J. Chem. Phys., 72, p. 4713.

Holland, P.M., and Castleman, A.W., Jr.: 1982a, J.Phys. Chem., in press.

Holland, P.M. and Castleman, A.W., Jr.: 1982b, J. Chem. Phys., 76, p. 4195.

Houle, F.A., and Beauchamp, J.L.: 1979, J. Am. Chem. Soc., 101, p. 4067.

Jorgensen, W.L., and Ibrahim, M.: 1981, Jour. Computational Chem., 2, p. 7.

Kay, B.D., Hermann, V., and Castleman, A.W., Jr.: 1981, Chem. Phys. Lett., 80, p. 469.

Kebarle, P.: 1983, in "Chemistry of Ions in the Gas Phase", Vimeiro, Portugal.

Keese, R.G., Lee, N., and Castleman, A.W., Jr.: 1980, J. Chem. Phys., 73, p. 2195.

Kleingeld, J.C., and Nibbering, N.M.M.: 1982, Int. J. Mass Spectrom. Ion Phys., in press.

Lee, N., Keesee, R.G., and Castleman, A.W., Jr.: 1980a, K. Colloid Interface Sci., 75, p. 555.

Lee, N., Keesee, R.G., and Castleman, A.W., Jr.: 1980b, Chem. Phys., 72, p. 1089.

Meot-Ner, M.: 1979, in "Gas Phase Ion Chemistry (M.T. Bowers, ed.), Academic, New York, Vol. 1, p. 197.

Møller, C., and Plesset, M.S.: 1934, Phys. Rev., 46, p. 618.

Moseley, J.T., Cosby, P.C., Bennet, R.A., and Peterson, J.R.: 1975, J. Chem. Phys., 62, p. 19.

Moseley, J.T., Cosby, P.C. and Peterson, J.R.: 1976, J.Chem.Phys., 65, p. 2512.

Paulson, J.F., and Dale, F., and Studniarz, S.A.: 1970, Int. J. Mass Spectrom. Ion Phys. 5, p. 113.

Paulson, J.F., and Dale, F.: 1982, J. Chem. Phys. 77, p. 4006.

Paulson, J.F., and Gale, P.J.,: 1978, Adv. Mass Spectrom., 7a, p. 263.

Payzant, J.D., Yamdagni, R., and Kebarle, P.: 1971, Can. Jour. Chem., 49, p. 3309

Perry, R.A., Rowe, B.R., Viggiano, A.A., Albritton, D.L., Ferguson, E.E., and Fehsenfeld, F.C.: 1980, Geophys. Res. Lett., 7, p. 693.

Peterson, K.I.: 1982, Ph.D. Thesis, University of Colorado, Boulder.

Raghavachari, K., Whitesides, R.A., Pople, J.A. and Schleyer, P.v.R.: 1981, J. Amer. Chem. Soc., 103, p. 5649.

Rosenstock, H.M., Kraxl, K., Steiner, B.W., and Herron, J.T.: 1977, J. Phys. and Chem. Ref. Data, 6 Supplement No. 1.

Schaefer, H.F., III ed., "Applications of Electronic Structure Theory", Plenum Press, NY, 1977.

Schlegel, H.B., Wolfe, S., and Bernardi, F.: 1975, J.Phys. Chem., 63, 3632,

Smith, D., Adams, N.G., and Alge, E.: 1982, J. Chem. Phys., 77 p. 1261.

Smith, D., and Adams, N.G.: 1980, in "Topics in Current Chemistry", Springer-Verlag, Berlin, Vol. 89.

Smith, G.P., Lee, L.C., and Moseley, J.T.: 1979, J. Chem. Phys.,
        71, p. 4034.

Vestal, M.L., and Mauclaire, G.H.: 1977, J.Chem.Phys., 67, p. 3758.

Viggiano, A.A., and Arnold, F.: 1982, Geophys. Res. Lett., in press.

Wolfe, S., Mitchell, D.J, and Schlegel, H.B.: 1981, J. Amer. Chem.
        Soc., 103, p. 7692.

Wu, R.L.C., and Tiernan, T.O.: 1981, Planet. Space Sci., 29, p. 735.

Yamdagni, R., and Kebarle, P.: 1971, J. Amer. Chem. Soc., 93, 7139.

ION THERMOCHEMISTRY:

SUMMARY OF THE PANEL DISCUSSION

Sharon G. Lias

Center for Chemical Physics
National Bureau of Standards
Washington, D. C., 20234, U. S. A.

The Panel on Ion Thermochemistry included a discussion by Dr.
Tomas Baer of the problems inherent in detecting an ionization onset, results on ionization potentials of radicals from the laboratories of Dr. J. L. Beauchamp and of Dr. J. M. Dyke (presented in
his absence by S. G. Lias), new experimental data on the proton
affinity scale in the region below water from Dr. T. B. McMahon,
and a presentation of information about the dissociation of protonated dimers by Dr. R. E. March. In addition, Drs. S. G. Lias, J.
L. Holmes, and J. E. Bartmess gave details of a comprehensive evaluation of heats of formation of ions in progress at the time of
this writing. Synopses of these presentations are given below.

HOW ARE THE BEST APPEARANCE ENERGIES AND HEATS OF FORMATION
OBTAINED? (Tomas Baer, University of North Carolina, Chapel Hill,
North Carolina, 27514, U. S. A.)

The problem in determining fragment ion appearance energies,
even in photoionization experiments, is a very difficult one.
Choosing an onset in a slowly rising fragment ion signal is often
a matter of personal taste. This arbitrariness is, however, removed in a breakdown diagram when the fragment ion of interest is
the first one formed, and there are no complications due to kinetic
shift. This situation is illustrated in Figure 1, given on the
next page. The cross-over point, C_{298} in Figure 1, is the energy
at which precisely 50% of the parent ions have sufficient energy
to dissociate. When this energy has been identified, the 0 K onset,
C_0, can easily be derived by adding to the cross-over energy the
median thermal energy. The median is generally very close to the
mean for large molecules. Once we have the 0 K onset, the heat of
formation of the product ion can be obtained using the expression:

355

M. A. Almoster Ferreira (ed.), Ionic Processes in the Gas Phase, 355–360.
© *1984 by D. Reidel Publishing Company.*

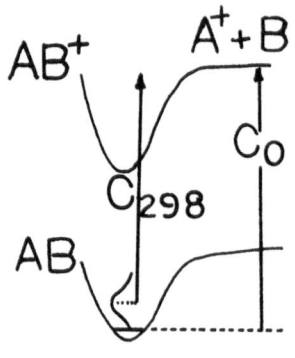 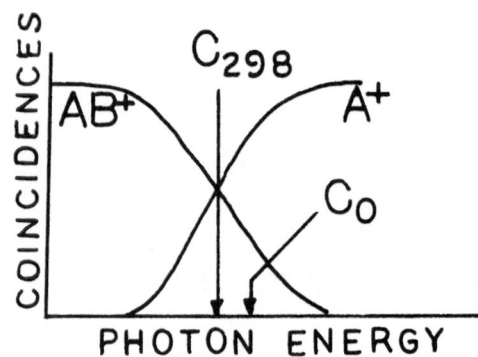

Figure 1.   Illustration of cross-over energy diagram.

$$\Delta H^\circ_{f0}(A^+) = \Delta H^\circ_{f0}(AB) + C_0 - \Delta H^\circ_{f0}(B)$$

where all the heats of formation refer to the species at 0 K.

To bring the 0 K heats of formation of some ion, $C_1H_mO_n^+$ for example, to 298 K, we use the thermochemical cycle as outlined in several recent publications (1,2,3):

$$\Delta H^\circ_{f298}(C_1H_mO_n^+) = \Delta H^\circ_{f0}(AB) + \int_0^{298} C_p(C_1H_mO_n^+)\,dT$$
$$- \int_0^{298} \left[1\ C_p(C) + m/2 C_p(H_2) + n/2 C_p(O_2)\right]dT$$

In this equation, we treat all ions and gases as perfect gases from 0 K to 298 K. In practice, the integral of the heat capacity at constant pressure is just the thermal energy content plus RT. For one mole of carbon atoms, this integral is 1.05 kJ/mol.

IONIZATION POTENTIALS OF RADICALS (J. L. Beauchamp, California Institute of Technology, Pasadena, California, 91125, U.S.A.) [Abstract of presentation prepared by S. G. Lias from the oral proceedings.]

Adiabatic ionization potentials of hydrocarbon radicals determined by photoelectron spectroscopy at the California Institute of Technology were briefly reviewed. The radicals were generated from organic nitrites:

$$RCH_2ONO \rightarrow RCH_2O + NO \rightarrow R\cdot + CH_2O + NO$$

A compilation of the ionization potentials of a number of radicals determined in this laboratory is available from Professor Beauchamp.

Some results of interest are given in Table 1.

Table 1.   Ionization Potentials of Some Hydrocarbon Radicals.

Radical	IP, eV	Radical	IP, eV
C_2H_5	8.39	$CH_2=CHCH_2$	8.13
$c-C_4H_7$	7.53	$CH_2=CHCH_2CH_2$	8.0
$t-C_4H_9$	6.70	$CH_2=CHCHCH_3$	7.39

It was pointed out that the heat of formation of the t-butyl ion
has been determined relative to that of the benzyl ion by measurement of the chloride transfer equilibrium constant. Accepting a
value of 213.8 kcal/mol for $\Delta H_f(C_6H_5CH_2^+)$, these results lead to a
value for $\Delta H_f(t-C_4H_9^+)$ of 165.6 kcal/mol. On this basis, the measured ionization potential of the $t-C_4H_9$ radical implies a heat of
formation for the radical of 11.1 kcal/mol, in good agreement with
recent determinations.

SOME RECENT RESULTS ON IONIZATION POTENTIALS OF RADICALS FROM THE
SOUTHAMPTON PHOTOELECTRON SPECTROSCOPY GROUP (J. M. Dyke, The University, Southampton, U. K.)

Determination of the first adiabatic ionization potential of
a small molecule usually leads to a determination of the heat of
formation of the ground state of the positive ion, provided that
the heat of formation of the neutral molecule is well established.
For example, in the photoelectron spectrum of HO_2 produced from the
reaction of F atoms with H_2O_2, the first adiabatic ionization energy
has been measured as 11.35 ± 0.01 eV (4). Probably the most reliable
determination of $\Delta H_{f298}(HO_2)$ has been made by Howard (5), who obtained a value of 10.5 ± 2.5 kJ/mol. Use of this value with the first
adiabatic ionization potential of HO_2 allows an improved estimate
for $\Delta H_{f298}(HO_2^+)$ of 1104.5 ± 4.0 kJ/mol to be made, which is approximately 30 kJ/mol lower than the previously accepted value. The
heat of formation of HO_2^+ obtained in this way leads to a value for
$D(H^+ - O_2)$ of 424.7 ± 4.2 kJ/mol, in good agreement with the proton
affinity of O_2 reported earlier (6).

In a similar study, the first band in the photoelectron spectrum of the HCO radical has been observed (7). A long vibrational
series was observed in this band, attributable to excitation of the
deformation mode in the ion. As the equilibrium geometries of HCO
and HCO^+ in their ground electronic states have both been determined independently using microwave spectroscopy, it is possible
to compute the vibrational band envelope in the first photoelectron
band of HCO fairly reliably. This calculation indicates that the
formyl radical provides an example of a rare case in photoelectron
spectroscopy where the adiabatic ionization potential is not directly observed. However, the Franck-Condon calculations allow the
first adiabatic ionization potential of HCO to be estimated as
8.27 ± 0.01 eV. Use of this value with the known heat of formation

of HCO$^+$ allows the heat of formation of HCO, ΔH_{f298}(HCO), and the C-H bond dissociation energy in formaldehyde to be calculated as 28.0±5.4 kJ/mol and 354.0±6.3 kJ/mol, respectively. The FCO radical has also been studied with photoelectron spectroscopy (8). In this case, use of the measured first adiabatic ionization potential with the established heat of formation of FCO allows an improved determination of the heat of formation of FCO$^+$.

The F + HN$_3$ reaction has also been studied recently with UV photoelectron spectroscopy. At short reaction times, three bands were observed corresponding to ionization of N$_3$(X^{2}Π) to the N$_3^+$ X^3 Σ$^-$, ^{1}Δ and ^{1}Σ$^+$ states. Use of the well-established value for the heat of formation of N$_3$ with the first adiabatic ionization potential of N$_3$ of 11.06±0.01 eV leads to an improved estimate of the heat of formation of N$_3^+$ (9).

THE PROTON AFFINITY OF WATER (T. B. McMahon, University of New Brunswick, Fredericton, New Brunswick, Canada E3B 6E2)

Ion cyclotron resonance proton transfer equilibrium constant measurements have been carried out for compounds less basic than water. Included in the compounds for which equilibrium measurements were performed is ethylene, whose absolute proton affinity has recently been accurately established as 162.6±0.5 kcal/mol, from photoionization appearance potential measurements and photoionization-photoelectron coincidence experiments (10). Using the scale of ΔG° values for proton transfer reactions determined in our study from observation of equilibria, and values of ΔS° estimated from rotational symmetry considerations, a scale of accurate proton affinities has been established relative to C$_2$H$_4$ as a primary standard. These experiments yield a value for the proton affinity of water of 166.7 kcal/mol, which is considerably less than the value of 170-171 kcal/mol previously accepted. This value was based on a proton affinity of NH$_3$ assigned relative to that of isobutene from proton transfer equilibrium constant measurements, and a proton affinity difference between H$_2$O and NH$_3$ of 34.1 kcal/mol. Our value is, however, in excellent agreement with other results: (i) PA(H$_2$O) has been referenced to PA(CH$_2$O) and an assignment of the proton affinity of CH$_2$O of 170.3 kcal/mol has been made from the appearance potential of CH$_2$OH$^+$ from alcohols. These experiments yield a PA(H$_2$O) of 166.4 kcal/mol (11). (ii) In similar fashion, PA(C$_3$H$_6$) has been assigned as 179.4 kcal/mol from appearance potential measurements of s-C$_3$H$_7^+$ from s-propyl halides (12). From data referencing PA(H$_2$O) to PA(C$_3$H$_6$) (13), the proton affinity of water may be assigned as 165.9 kcal/mol. (iii) Lee and co-workers have carried out photoionization appearance potential measurements of H$_3$O$^+$ from water dimers to obtain a proton affinity value of 167.3 kcal/mol. (iv) Bohme has directly measured forward and reverse rate constants for proton transfer between C$_2$H$_4$

and H_2O to give a proton affinity of H_2O of 165.1 kcal/mol (14).

Taking the above results together, we suggest that a value of 166.7±1.5 kcal/mol be adopted for the proton affinity of H_2O. These data imply that the proton affinity difference between NH_3 and H_2O should thus be taken as 36.5 kcal/mol, with a proton affinity of NH_3 of 203.3 kcal/mol.

MULTIPHOTON INDUCED DISSOCIATION OF PROTONATED DIMERS OF 2-PROPAN-
OL (Raymond E. March, Trent University, Peterborough, Ontario, Canada K9J 7B8)

It has been demonstrated previously by Beauchamp et al (15) that photodissociation probabilities of isolated gaseous ions de-pend only on energy fluence; thus, a relatively low power laser can effect dissociation processes in gaseous ions when irradiation is sufficiently prolonged under nearly collision-free conditions. Pro-tonated dimers of 2-propanol (and deuterated analogues), produced in a well-established sequence of ion-molecule reactions, were accumulated in a quadrupole ion store (QUESTOR) at a pressure of 5.3 mPa^2. Multiphoton dissociation of protonated dimers of each of the 2-propanol analogues was effected with irradiation for up to 110 ms with CW CO_2 laser intensities of approximately 20 Wcm^{-2} at 944 cm^{-1}. The relative order of degrees of photodissociation was found to be in inverse order of the single photon infrared ab-sorptivities of the corresponding neutral parent molecules at 944 cm^{-1}; thus, while gaseous ion absorptivities are similar to those of neutral precursors, they are not necessarily identical (16).

Of the three ionic photoproducts of multiphoton dissociation of proton-bound 2-propanol dimers, only the product formed by loss of H_2O was found to be photodissociative under the prevailing cond-itions. Protonated di-isopropyl ether which is isomeric with the photodissociative photoproduct could not be photodissociated under similar conditions, but the protonated mixed dimer of 2-propanol and D_2O was photodissociated. Thus, while the photoproducts from protonated 2-propanol corresponding to loss of H_2O and C_3H_6 are isomeric with $(s-C_3H_7^+)_2OH^+$ and $s-C_3H_7OH \cdot H^+ \cdot H_2O$, respectively, their behavior under the prevailing conditions for multiphoton dissociation indicates that they are of different structures.

Newly formed (nascent) protonated dimers of 2-propanol may possess up to 31 kcal/mol of vibrational energy. Multiphoton diss-ociation of completely stabilized protonated dimers would require 11 photons for dissociation to initial reactants, >5 photons for dissociation with loss of H_2O, and >8 photons for dissociation with loss of C_3H_6; the endothermicities for loss of H_2O and C_3H_6 are estimated at 14 and 22 kcal/mol, respectively. The predominant pathway observed in the multiphoton dissociation of fully relaxed protonated dimers is that which involves loss of H_2O; thus, in

this case, the least endothermic process must have the lowest activation energy. Irradiation of nascent protonated dimers for 10
ms produced not only a measurable degree of photodissociation, but
a predominance of dissociation to form the initial reactant ion
species, protonated 2-propanol. Thus, dissociation via the pathway
of greatest endothermicity can be induced readily by the absorption
of photons when the nascent ion possesses a degree of internal energy due to partial stabilization. As the process of collisional
and radiative stabilization proceeds, the dominance of the reaction
pathway of lowest activation energy increases.

REFERENCES

(1)   Rosenstock, H. M.: *"Kinetics of Ion-Molecule Reactions"*,
        p. 246 (Plenum Press, 1979, ed. P. Ausloos).

(2)   Fraser-Monteiro, M. L., Fraser-Monteiro, L., Butler, J. J.,
        Baer, T., and Hass, J. R.: 1982, J. Phys. Chem. 86, p. 739.

(3)   Traeger, J. C., and McLoughlin, R. G.: 1981, J. Am. Chem.
        Soc. 103, p. 3647.

(4)   Dyke, J. M., Jonathan, N. B. H., Morris, A., and Winter, M.
        J.: 1981, Molecular Physics 44, p. 1059.

(5)   Howard, C. J.: 1980, J. Am. Chem. Soc. 102, p. 6937.

(6)   Bohme, D. K.:*"Interactions between Ions and Molecules"*, p.
        489 (Plenum Press, 1975, ed. P. Ausloos).

(7)   Dyke, J. M., Jonathan, N., Morris, A., and Winter, M. J.:1980,
        Molecular Physics 39, p. 629.

(8)   Dyke, J. M., Jonathan, N., Morris, A., and Winter, M. J.:
        1981, J. C. S. Faraday II 77, p. 667.

(9)   Dyke, J. M., Jonathan, N., Lewis, A. E., and Morris, A.: 1982,
        Molecular Physics, accepted for publication.

(10)   Baer, T.: 1980, J. Am. Chem. Soc. 102, p. 2482.

(11)   Tanaka, K., MacKay, G. I., and Bohme, D. K.: 1978, Can. J.
        Chem. 56, p. 193.

(12)   Rosenstock, H. M., Buff, R., Ferreira, M. A. A., Lias, S. G.,
        Parr, A. C., Stockbauer, R. L., and Holmes, J. L.: 1982, J.
        Am. Chem. Soc. 104, p. 2337.

(13)   Lau, Y. K.: Ph. D. Thesis, 1979, University of Alberta.

(14)   Bohme, D. K., MacKay, G. I.: 1981, J. Am. Chem. Soc.103,p.2173

INSTRUMENTATION:

SUMMARY OF THE PANEL DISCUSSION

Melvin B. Comisarow

Department of Chemistry
University of British Columbia
Vancouver, B.C., Canada V6T 1Y6

Seven papers were presented on various recent developments in instrumentation for studying ion chemistry and related topics such as chemical ionization mass spectrometry.

A new instrument for conducting MS-MS experiments was described by M.L. Gross (University of Nebraska). The instrument (1) consists of a conventional Kratos MS-50 double focusing mass spectrometer to which has been added a collision chamber and electrostatic analyzer. The first two analyzers of the instrument, an electrostatic sector followed by a magnetic sector, provide ultrahigh mass resolution of up to 1 part in 100,00. Thus, ions, such as $C_9H_{18}S^+$ (m/z = 158.11292) from cyclohexyl n-propyl sulfide and $C_{12}H_{14}^+$ (m/z = 158.10955) from phenylcyclohexene which differ in mass by one part in 47,000 can be readily separated from each other. The CID spectra or the ion-molecule chemistry of the each of two m/z = 158 isobaric ions can thus be determined in the presence of each other (1). D.P. Ridge (University of Delaware) commented from the audience that he had used the Nebraska triple analyzer instrument to investigate the ion chemistry of $FeCO^+$ and $FeC_2H_4^+$ in the presence of each other and that the ultrahigh resolution capability of the instrument allowed the investigation to be performed with ease.

Fourier transform mass spectrometers (FT-MS) have been used for ultrahigh resolution mass spectrometry and also for collision induced decomposition (CID) experiments. High mass resolution in FT-MS will result if the excited cyclotron motion persists for a long time. In practice, this persistence will apply if the excited cyclotron motion is not interrupted by

361

M. A. Almoster Ferreira (ed.), Ionic Processes in the Gas Phase, 361–366.
© *1984 by D. Reidel Publishing Company.*

ion-molecule collisions. In FT-MS-CID experiments, the cyclotron motion of a selected ion is excited and collided with a neutral background gas. The ion fragments from such collisions are then mass analyzed in the usual FT-MS manner. Until now FT-MS-CID experiments were limited in mass resolution to about one part in 1000 by the high pressures of collision gas required for CID. B.S. Freiser (Purdue University) described a new approach (2) to the FT-MS-CID experiment which achieves the high background pressure needed for CID and the low background pressure needed for high mass resolution. The procedure is to ionize the sample, introduce the collision gas through a pulsed valve, excite the cyclotron motion of the ion to be decomposed, wait for the collision gas to be pumped away and then carry out FTMS excitation and detection in the normal manner (3). CID of a mixture of the isobaric ions $C_9H_{12}^+$ and $C_6H_5COCH_3^+$ (from mesitylene and acetophenone, respectively) can be collisionally decomposed to the isobaric ions $C_8H_9^+$ and $C_6H_5CO^+$ which differ in mass by one part in 3000. A mass resolution of 47,000 was observed in the mass spectrum of the CID product ions from this FTMS-CID pulsed pressure experiment.

The capabilities of the triple analyzer instrument described by M.L. Gross and the pulsed pressure FTMS instrument described by B.S. Freiser are complementary. For CID or ion-molecule reaction studies where a reactant ion, m_1, is reacted or collided with a neutral N to form a product ion, m_2,

$$m_1 + N \rightarrow m_2$$

the triple analyzer instrument has ultrahigh mass resolution for the reactant ion m_1 whereas the pulsed pressure FTMS instrument has ultrahigh mass resolution for the product ion m_2.

The use of lasers to effect ionization of involatile samples is well established in mass spectrometry and the first use of laser desorption/ionization in a FTMS spectrometer was reported by M.L. Gross (University of Nebraska) (3). An interresting feature of laser ionization FT-MS is that the signals are very large indicating that laser desorption FT-MS will be a very sensitive analytical technique.

Imre Szabo (University of Lund) discussed ion-molecule reaction studies on state selected ions as studied with a triple mass spectrometer. This study showed that continuous beams of ions in selected electronic states, with various degrees of vibrational excitation, can be produced by ionizing neutral molecules (M) in electron (or proton) transfer reactions with slow positive ions (A^+) in tandem and triple mass spectrometers. The ion chemistry of the state selected ions, M^+, formed in the first two stages of the triple mass spectrometer, can be examined with the third stage.

The first mass spectrometer of a tndem instrument produces a beam of slow incident ions (A^+) which is used to ionize molecules (M) in the ion source of the second mass spectrometer. The tandem instrument is said to have transverse geometry if the product ions are analyzed in a direction being perpendicular to the beam of incident ions. This geometry discriminates strongly such ions which are formed by the transfer of translational energy into internal energy or vice versa.

When the velocity of A^+ does not exceed a few electron volts and there is energy resonance for ionization by electron transfer, then the internal energy of the molecular ions giving rise to the formation of the first generation of the observed product ions is $E_{exc}(M^+) \approx E_{re}(A^+)-E_{ip}(M)$, where $E_{re}(A^+)$ is the recombination energy of A^+ and $E_{ip}(M)$ is the ionization potential of M. Thus, the internal energy of those molecular ions which originate the first generation of ions is known and can be altered by ionizing with different incident ions of known recombination energies in sucession. Consecutive ion-molecule rections occurring in the gas phase can be studied in detail as function of this energy (4).

The beam of state selected molecular or fragment ions obtained from the second mass spectrometer can be used for various types of investigations such as: (1) Coulomb explosion spectroscopy, (2) beam foil spectroscopy, (3) laser spectroscopy of ion beams, and (4) for comparative study of the reactivity of one particular ion originating from the same or different molecules at different values of internal energy of the parent molecular ion. The reactivity of ions can be used as a tool for distinguishing between isomers and to obtain information on ion structure and energetics.

Comparative study of the reactivity of state selected ions can be carried out in a triple mass spectrometer consisting of three mass spectrometers in series with a mass and energy selected ion beam in each. In such an instrument a molecular or fragment ion produced in the second mass spectrometer is used to ionize a "test gas" in the ion source of the third mass spectrometer. There are four basic types of fixed angle triple mass spectrometers assuming that the geometric angle between the beams of incident and product ions may be zero or 90 degrees in the ion source of the second and third mass spectrometer, respectively.

Recently, a triple mass spectrometer of in-line-transverse geometry has been used for comparing the reactivity of CH_2^+ from CH_4. The logarithmic ion-intensity pressure diagrams of CH_4 for incident CH_2^+ ions in different electronic

states are shown in Fig. 1. The following experimental
conditions apply for Fig. 1a. Ion beam 1: Ar$^+$ (E$_{re}$ ≈15.8 eV),
Ion beam 2: CH$^+$ from Ar$^+$ + CH$_4$, Ion beam 3: Product ions CH$_2^+$
+ CH$_4$. The results in Fig. 1b were obtained by using He$^+$
(E$_{re}$≈24.5 eV) instead of Ar$^+$ in the first mass spectrometer.
The experimental conditions were identical in all other
respects. As seen from the comparison of Fig. 1a and 1b, the
intensity distribution for the product ions differ considerably
due to the change in the population of internal energy states
of the CH$_2^+$ ions. He$^+$ produces mainly CH$_2^+$ ions in excited
energy states which can ionize CH$_4$ by charge exchange (5,6).

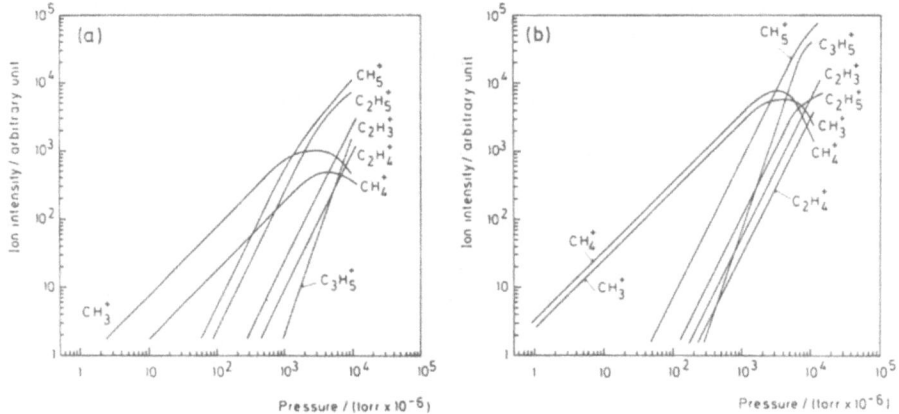

Fig. 1. Logarithmic ion-intensity pressure diagrams of CH$_4$
         for incident CH$_2^+$ ions in different electronic
         states.

    J. Bartmess (Indiana University) described an advanced
capacitance bridge ICR spectrometer developed in his laboratory.
This circuit has many advantages over the marginal oscillator
type of ICR detector such as faster scan rates, wider dynamic
range, fixed magnetic field operation and potentially greater
sensitivity.

    K. Jennings (University of Warwick) described his
Experiments on the negative ion chemical ionization mass
spectrometry of crude oil samples. By using O$^-$ from N$_2$O, OH$^-$
is formed from the hexane which is used to dissolve the samples.

The OH⁻ then abstracts a proton from the side chains of alkyl aromatics in the crude oil. A reference file of negative ion mass spectra of crude oils obtained in this manner provides a data base against which samples of unknown origin can be compared. It is possible to correctly identify crude oil samples from various producing regions (North Sea, Middle East, Nigeria, Venezuela) in this manner. Jennings also described some further work from his laboratory on the negative ion mass spectrometry of additives in lubricating oils. These additives are zinc dialkylthiophosphates where the alkyl groups may vary from sample to sample. Normal electron impact mass spectrometry of the lubricating oils does not show any molecular ion from the additive. In contrast, CI⁻ negative ion chemical ionization, where the Cl⁻ is attached to form a M+Cl⁻ ion, is very successful. No chloride attachment to the hydrocarbons of the oil was observed. The method is thus very convenient for identifying the additive.

In any spectrometer which uses Fourier transform techniques the spectrum, obtained by Fourier transformation of a time domain transient, is not a continuous spectrum but rather exists at only a finite number of points; i.e. the spectrum is discrete. If a peak maximum falls between two points of the discrete spectrum, the observed spectral intensity will be too small. This is a serious problem in Fourier transform ion cyclotron resonance mass spectrometers (FT-ICR or FTMS) where isotopic abundances will be inaccurately recorded. One solution to this problem is to zero-fill the transient signal prior to Fourier transformation (7). This procedure can give the true spectral intensities to any desired degree of accracy. Zero-filling is however very consumptive of machine time and a superior data reduction procedure which also gives the correct spectral intensities is desirable. Jan Kleingeld (University of Amsterdam) reported a new procedure, developed at the University of Amsterdam (8), which gives correct FT-ICR isotopic intensities but requires less machine time than zero-filling. In this procedure, the transient is multiplied by a sum of cosines to give a weighted transient. Fourier transformation of this weighted transient gives a discrete spectrum with isotopic intensities correct to 1 percent. This procedure however changes the lineshape and the spectral resolution is reduced somewhat. M. Comisarow (University of British Columbia) commented from the audience that another procedure is to interpolate the peak maxima with an interpolating function and that if the magnitude Lorentzian (9) function is used on FT-ICR spectra, the exact peak maximum and peak position can be obtained. Interpolating procedures do not cause any reduction in the spectral resolution but do require that there is no peak overlap of adjacent peaks.

K. Jennings (University of Warwick) described his work
which examined the dependence of CID product ratios as a
function of scarrering angle in the CID experiment. For small
scattering angles near zero the internal energy of the
scattered ion is proportional to the square of the scattering
angle. Thus it is reasonable to expect that the CID product
ion intensities will be a function of the scattering angle.
This functional dependence of the CID products on scattering
angle provides another dimension of data for identification
of ions. Various techniques for experimentally varying the
scattering angle were features of various methods discussed.
Examples of systems where the intensity radio of two CID
product ions was a function of scattering angle were given.

References

(1)   M.L. Gross, E.K. Chess, P.A. Lyon, F.W. Chow, S. Evans and
      H. Tudge: Int. J. Mass Spec. Ion Phys., 42, (1982) 243.

(2)   T.J. Carlin and B.S. Freiser: Anal. Chem., in press.

(3)   D.A. McCrery, E.B. Ledford, Jr. and M.L. Gross: Anal.
      Chem. 54 (1982) 1435.

(4)   I. Szabo: in "Advances in Mass Spectrometry", Vol. 7, (ed.
      N.R. Daly), Heyden, London, 1978, p. 258.

(5)   I. Szabo: Arkiv Fysik, 35, (1967) 339.

(6)   C. Galloy and J.C. Lorquet: Chem. Phys., 30 (1978) 169

(7)   M.B. Comisarow and J. Melka: Anal. Chem., 51 (1979) 2198.

(8)   A.J. Noest and C.W.F. Kort: Computers and Chemistry, 6
      (1982) 115.

(9)   C. Giancaspro and M.B. Comisarow: "Applied Spectroscopy",
      in press.